AMERICAN ORNITHOLOGY.

PRINTED BY BALLANTYNE AND COMPANY
EDINBURGH AND LONDON

1. Carolina Cuckoo. 2. Black-billed C. 3. Blue Yellow-backed Warbler. 4. Yellow Red-poll W.

2 B

AMERICAN ORNITHOLOGY;

OR,

THE NATURAL HISTORY

OF THE

BIRDS OF THE UNITED STATES.

BY

ALEXANDER WILSON

AND

PRINCE CHARLES LUCIEN BONAPARTE.

The Illustrative Notes and Life of Wilson

BY SIR WILLIAM JARDINE, Bart., F.R.S.E., F.L.S.

IN THREE VOLUMES.—VOL. II.

CASSELL PETTER & GALPIN:

LONDON, PARIS & NEW YORK.

CONTENTS OF THE SECOND VOLUME.

The names printed in italics are species not contained in the original, which have been introduced into the notes.

WILSON'S

AMERICAN ORNITHOLOGY.

YELLOW-BILLED CUCKOO.   (*Cuculus Carolinensis.*)

PLATE XXVIII.—Fig. 1.

Cuculus Americanus, *Linn. Syst.* 170.—*Catesb.* i. 9.—*Lath.* i. 537.—Le Coucou de la Caroline, *Briss.* iv. 112.—*Arct. Zool.* 265, No. 155.—*Peale's Museum,* No. 1778.

COCCYZUS AMERICANUS.—Bonaparte.*

Coccyzus Americanus, *Bonap. Synop.* p. 42.—The Yellow-billed Cuckoo, *Aud.* pl. 2. *Orn. Biog.* i. p. 18.

A STRANGER who visits the United States, for the purpose of examining their natural productions, and passes through our woods in the month of May or June, will sometimes hear, as

---

* Bonaparte has preferred restoring the specific name of Linnæus to that given by Catesby and Brisson, and by this it should stand in our systems.

This form will represent in America the true cuckoos, which otherwise range over the world ; it was first separated by Vaillant under the French name Conec, and the same division was adopted by Vieillot, under the name of *Coccyzus,* which is now retained. They differ from the cuckoos chiefly in habit,—building a regular nest, and rearing their young. North America possesses only two species, our present and the following, which are both migratory. Some beautiful species are met with in different parts of the southern continent.

Mr Audubon has added little to their history farther than confirming the accounts of Wilson. In their migrations northward, they move

he traverses the borders of deep, retired, high timbered hollows, an uncouth, guttural sound, or note, resembling the syllables, *kowe, kowe, kowe kowe kowe,* beginning slowly, but ending so rapidly, that the notes seem to run into each other; and *vice versa:* he will hear this frequently, without being able to discover the bird or animal from which it proceeds, as it is both shy and solitary, seeking always the thickest foliage for concealment. This is the yellow-billed cuckoo, the subject of the present account. From the imitative sound of its note, it is known in many parts by the name of the *cow-bird;* it is also called in Virginia the *rain crow,* being observed to be most clamorous immediately before rain.

This species arrives in Pennsylvania, from the south, about the 22nd of April, and spreads over the country, as far at least as Lake Ontario; is numerous in the Chickasaw and Chactaw nations; and also breeds in the upper parts of

singly; but when removing again to a warmer latitude, they appear to be gregarious, flying high in the air, and in loose flocks.

They appear to delight more in deep woody solitudes than the true cuckoos, or those which approach nearest to the form of the European species. They, again, though often found near woods, and in richly clothed countries, are fond of open and extensive heaths or commons, studded or fringed with brush and forest: here they may expect an abundant supply of the foster parent to their young. The gliding and turning motion when flying in a thicket, however, is similar to that of the American *Coccyzus.* Like them, also, they are seldom on the ground; but, when obliged to be near it, alight on some hillock or twig, where they will continue for a considerable time, swinging round their body in a rather ludicrous manner, with lowered wings and expanded tail, and uttering a rather low, monotonous sound, resembling the *kowe* of our American bird,—

Turning round and round with *cutty-coo.*

When suddenly surprised or disturbed from their roost at night, they utter a short, tremulous whistle, three or four times repeated; it is only on their first arrival, during the early part of incubation, when in search of a mate, that their well known and welcome note is heard; by the first of July all is silent. The idea that the common cuckoo destroys eggs and young birds, like the American *Coccyzus,* is also entertained; I have never seen them do so, but the fact is affirmed by most country persons, and many gamekeepers destroy them on this account.—ED.

Georgia : preferring, in all these places, the borders of solitary swamps and apple orchards. It leaves us, on its return southward, about the middle of September.

The singular—I will not say unnatural—conduct of the European cuckoo (*Cuculus canorus*), which never constructs a nest for itself, but drops its eggs in those of other birds, and abandons them to their mercy and management, is so universally known, and so proverbial, that the whole tribe of cuckoos have, by some inconsiderate people, been stigmatised as destitute of all parental care and affection. Without attempting to account for this remarkable habit of the European species, far less to consider as an error what the wisdom of Heaven has imposed as a duty on the species, I will only remark, that the bird now before us builds its own nest, hatches its own eggs, and rears its own young ; and, in conjugal and parental affection, seems nowise behind any of its neighbours of the grove.

Early in May, they begin to pair, when obstinate battles take place among the males. About the tenth of that month, they commence building. The nest is usually fixed among the horizontal branches of an apple tree ; sometimes in a solitary thorn, crab, or cedar, in some retired part of the woods. It is constructed, with little art, and scarcely any concavity, of small sticks and twigs, intermixed with green weeds, and blossoms of the common maple. On this almost flat bed, the eggs, usually three or four in number, are placed ; these are of a uniform greenish blue colour, and of a size proportionable to that of the bird. While the female is sitting, the male is generally not far distant, and gives the alarm, by his notes, when any person is approaching. The female sits so close, that you may almost reach her with your hand, and then precipitates herself to the ground, feigning lameness, to draw you away from the spot, fluttering, trailing her wings, and tumbling over, in the manner of the partridge, woodcock, and many other species. Both parents unite in providing food for the young. This consists, for the most part, of caterpillars, particularly such as infest apple trees. The same insects constitute the chief part of their own sustenance. They are accused, and with some justice, of sucking the eggs of other birds, like the crow, the blue jay, and other pillagers. They also occasionally eat various kinds of berries. But, from the circumstance of destroying such numbers of very noxious larvæ, they prove themselves the friends of the farmer, and are highly deserving of his protection.

The yellow-billed cuckoo is thirteen inches long, and sixteen inches in extent; the whole upper parts are of a dark, glossy drab, or what is usually called a quaker colour, with greenish silky reflections; from this must, however, be excepted the inner vanes of the wings, which are bright reddish cinnamon; the tail is long, composed of ten feathers, the two middle ones being of the same colour as the back, the others, which gradually shorten to the exterior ones, are black, largely tipt with white; the two outer ones are scarcely half the length of the middle ones. The whole lower parts are pure white; the feathers covering the thighs being large, like those of the hawk tribe; the legs and feet are light blue, the toes placed two before and two behind, as in the rest of the genus. The bill is long, a little bent, very broad at the base, dusky black above, and yellow below; the eye hazel, feathered close to the eyelid, which is yellow. The female differs little from the male; the four middle tail-feathers in her are of the same uniform drab; and the white, with which the others are tipt, not so pure as in the male.

In examining this bird by dissection, the inner membrane of the gizzard, which in many other species is so hard and muscular, in this is extremely lax and soft, capable of great distension; and, what is remarkable, is covered with a growth of fine down, or hair, of a light fawn colour. It is difficult to ascertain the particular purpose which Nature intends by this excrescence; perhaps it may serve to shield the tender parts from the irritating effects produced by the hairs of certain caterpillars, some of which are said to be almost equal to the sting of a nettle.

BLACK-BILLED CUCKOO. (*Cuculus erythropthalma.*)

PLATE XXVIII.—Fig. 2.

Peale's Museum, No. 1854.

COCCYZUS ERYTHROPTHALMUS.—Bonaparte.*

Coccyzus erythropthalmus, *Bonap. Synop.* p. 42.—The Black-billed Cuckoo,
Aud. pl. 32, male and female; *Orn. Biog.* i. p. 170.

THIS cuckoo is nearly as numerous as the former, but has
hitherto escaped the notice of European naturalists; or, from
its general resemblance, has been confounded with the preceding. Its particular markings, however, and some of its
habits, sufficiently characterise it as a distinct species. Its
general colour above is nearly that of the former, inclining
more to a pale ash on the cheeks and front; it is about an
inch less in length; the tail is of a uniform dark silky drab,
except at the tip, where each feather is marked with a spot
of white, bordered above with a slight touch of dull black;
the bill is wholly black, and much smaller than that of the
preceding; and it wants the bright cinnamon on the wings.
But what constitutes its most distinguishing trait is, a bare
wrinkled skin, of a deep red colour, that surrounds the eye.
The female differs little in external appearance from the
male.

The black-billed cuckoo is particularly fond of the sides of
creeks, feeding on small shell fish, snails, &c. I have also
often found broken pieces of oyster shells in its gizzard, which,
like that of the other, is covered with fine downy hair.

The nest of this bird is most commonly built in a cedar,
much in the same manner, and of nearly the same materials, as

* Wilson, I believe, deserves the credit of distinguishing this species.
It is closely allied to, but differs widely, both in its habits and feeding,
from its congeners and the true cuckoos. In addition to shells and water
insects, Audubon mentions having found in their stomachs a small black
frog, which appears after a summer shower.—ED.

that of the other ; but the eggs are smaller, usually four or five in number, and of a rather deeper greenish blue.

This bird is likewise found in the state of Georgia, and has not escaped the notice of Mr Abbot, who is satisfied of its being a distinct species from the preceding.

BLUE YELLOW-BACK WARBLER. *(Sylvia pusilla.)*

PLATE XXVIII.—Fig. 3.

Parus Americanus, *Linn. Syst.* 341.—Finch Creeper, *Catesb.* i. 64.—*Lath.* ii. 558.—Creeping Titmouse, *Arct. Zool.* 423, No. 326.—Parus varius, Various coloured little Finch Creeper, *Bart.* p. 292.—*Peale's Museum*, No. 6910.

SYLVICOLA AMERICANA.—Swainson.*

Sylvia Americana, *Lath. Ind. Orn.* ii. p. 520.—*Bonap. Synop.* p. 83.—Sylvicola pusilla, *Sw. Synop. Birds of Mex. Ann. of Phil.* p. 433.—*Zool. Journ.* No. 10, p. 169.—The Blue Yellow-backed Warbler, *Aud.* pl. 15, male and female ; *Orn. Biog.* i. p. 78.

Notwithstanding the respectability of the above authorities, I must continue to consider this bird as a species of warbler. Its habits, indeed, partake something of the titmouse ; but the form of its bill is decidedly that of the *Sylvia* genus. It is remarkable for frequenting the tops of the tallest trees, where it feeds on the small winged insects, and caterpillars that infest the young leaves and blossoms. It has a few feeble chirruping notes, scarcely loud enough to be heard at the foot of the tree. It visits Pennsylvania from the south, early in May ; is very abundant in the woods of Kentucky : and is also found in the northern parts of the state of New York. Its nest I have never yet met with.†

* There is nothing more annoying than the unravelling of names. That of *Americana*, without doubt, seems to have been the specific appellation first applied ; and if we are to adhere to any given rule in nomenclature, that should be now adopted. The present species has also been made typical of the group which is confined to the New World.—Ed.

† According to Audubon, the nest is small, formed of lichens, beautifully arranged on the outside, and lined with the cotton substances found

This little species is four inches and a half long, and six inches and a half in breadth; the front, and between the bill and eyes, is black; the upper part of the head and neck, a fine Prussian blue; upper part of the back, brownish yellow; lower, and rump, pale blue; wings and tail, black; the former crossed with two bars of white, and edged with blue; the latter marked on the inner webs of the three exterior feathers with white, a circumstance common to a great number of the genus; immediately above and below the eye, is a small touch of white: the upper mandible is black; the lower, as well as the whole throat and breast, rich yellow, deepening about its middle to orange red, and marked on the throat with a small crescent of black; on the edge of the breast is a slight touch of rufous; belly and vent, white; legs, dark brown; feet, dirty yellow. The female wants both the black and orange on the throat and breast; the blue, on the upper parts, is also of a duller tint.

YELLOW RED-POLL WARBLER. (*Sylvia petechia.*)

PLATE XXVIII.—Fig. 4.

Red-headed Warbler, *Turton,* i. 605.—*Peale's Museum,* No. 7124.

SYLVICOLA PETECHIA.—Swainson.

Lath. Ind. Orn. ii. p. 535.—Sylvia petechia, *Bonap. Synop.* p. 83.—Red-headed Warbler, *Penn. Arct. Zool.* ii. p. 401.—Sylvicola petechia, *North. Zool.* ii. p. 215.

This delicate little bird arrives in Pennsylvania early in April, while the maples are yet in blossom, among the branches of which it may generally be found at that season, feeding on the stamina of the flowers, and on small winged insects. Low swampy thickets are its favourite places of resort. It is not numerous, and its notes are undeserving the

on the edges of different mosses; it is placed in the fork of a small twig, near the extremity of the branch. The eggs are pure white, with a few reddish dots at the longer end. Mr Audubon thinks two broods are raised in the year.—Ed.

name of song. It remains with us all summer; but its nest has hitherto escaped me. It leaves us late in September. Some of them probably winter in Georgia, having myself shot several late in February, on the borders of the Savannah river.

Length of the yellow red-poll, five inches; extent, eight; line over the eye, and whole lower parts, rich yellow; breast, streaked with dull red; upper part of the head, reddish chestnut, which it loses in winter; back, yellow olive, streaked with dusky; rump, and tail-coverts, greenish yellow; wings, deep blackish brown, exteriorly edged with olive; tail, slightly forked, and of the same colour as the wings.

The female wants the red cap; and the yellow of the lower parts is less brilliant; the streaks of red on the breast are also fewer and less distinct.

IVORY-BILLED WOODPECKER. (*Picus principalis.*)

PLATE XXIX.—Fig. 1.

Picus principalis, *Linn. Syst.* i. p. 173. 2.—*Gmel. Syst.* i. p. 425.—Picus Niger Carolinensis, *Briss.* iv. p. 26. 9; *Id.* 8vo. ii. p. 49.—Pic noir à bec blanc, *Buff.* vii. p. 46. *Pl. enl.* 690.—King of the Woodpeckers, *Kalm*, ii. p. 85.—White-billed Woodpecker, *Catesb. Car.* i. 6. 16.—*Arct. Zool.* ii. No. 156.—*Lath. Syn.* ii. p. 553.—*Bartram*, p. 289.—*Peale's Museum*, No. 1884.

PICUS PRINCIPALIS.—Linnæus.*

Picus principalis, *Bonap. Synop.* p. 44.—*Wagl. Syst. Av. Picus*, No. 1.—The Ivory-billed Woodpecker, *Aud.* pl. 66, male and female; *Orn. Biog.* i. p. 341.

This majestic and formidable species, in strength and magnitude, stands at the head of the whole class of woodpeckers, hitherto discovered. He may be called the king or chief of

* The genus *Picus*, or woodpeckers, with the exception of the parrots, forms the most extensive group among the *Scansores*, and perhaps one of the most natural among the numerous divisions now assigned to the feathered race. In a former note we mentioned the difference of form, and corresponding modification of habit, that nevertheless existed among them. Most ornithologists have divided them into three groups only, taking the common form of woodpeckers for the type, making another

Head of the Pileated Woodpecker.

Head of the Ivory billed Woodpecker.

Engraved by W. H. Lizars.

1. Ivory-billed Woodpecker. 2. Pileated W. 3. Red-headed W.

his tribe; and Nature seems to have designed him a distinguished characteristic in the superb carmine crest and bill ot polished ivory with which she has ornamented him. His eye

of the golden-winged, and including in a third the very minute species which form Temminck's genus *Picumnus,* but which, I believe, will be found to rank in a family somewhat different. Mr Swainson, again, in following out the views which he holds regarding the affinities of living beings, has formed five groups,—taking our present form as typical, under the title *Picus;* that of the green woodpecker, under *Chrysoptilus;* that of the red-headed woodpecker, as *Melanerpes;* the golden-wings, as *Colaptes;* and *Malacolophus* as the soft-crested Brazilian and Indian species. Of these forms, the northern parts of America will contain only three: two we have had occasion already to remark upon; and the third forms the subject of our author's present description—the most powerful of the whole tribe, and showing all the forms and peculiarities of the true woodpecker developed to the utmost.

The *Pici* are very numerous, and are distributed over the whole world, New Holland excepted; America, however, including both continents, may be termed the land of woodpeckers. Her vast and solitary forests afford abundance to satisfy their various wants, and furnish a secluded retirement from the inroads of cultivation. Next in number, I believe, India and her islands are best stored; then Africa, and lastly, Europe. The numbers, however, are always greatest between the tropics, and generally diminish as we recede from and approach temperate or cold regions. They are mostly insectivorous; a few species only feed occasionally on different fruits and berries. The various *Coleoptera,* that form their abodes in dead and decaying timber, and beneath their bark and moss, with their eggs and large larvæ, form an essential part of their subsistence: for securing this prey, digging it out from their burrows in the wood, and the peculiar mode of life incident to such pursuits, they are most admirably adapted. The bill is strong and wedge-shaped; the neck possesses great muscularity. The tongue—fitted by the curious construction of its muscles and the *os hyoides,* and lubricated with a viscous saliva, either gently to secure and draw in the weaker prey, or with great force and rapidity to dart out, and, it is said, to transfix the larger and more nimble insects—joined to the short legs and hooked scansorial claws, with the stiff, bent tail, are all provisions beautifully arranged for their wants.

All the species are solitary, live in pairs only during the season of incubation, or are met with in small flocks, the amount of the years' brood, in the end of autumn, before they have separated. This solitary habit, and their haunts being generally gloomy and retired, has given rise to the opinion entertained by many, that the life of the woodpecker was

is brilliant and daring; and his whole frame so admirably adapted
for his mode of life, and method of procuring subsistence, as
to impress on the mind of the examiner the most reverential

hard and laborious, dragged on in the same unvaried tract for one purpose,—the supply of food. It has been painted in vivid and imaginary
colouring, and its existence has been described to be painful and bur-
densome in the extreme ; its cries have been converted into complaints,
and its search for food into exertions of no use. We cannot agree to
this. The cry of the woodpecker is wild, and no doubt the incessant
hewing of holes without an adequate object would be sufficiently miserable. These, however, are the pleasures of the bird. The knowledge
to search after food is implanted in it, and organs most admirably formed
to prevent exhaustion, and ensure success, have been granted to it. Its
cries, though melancholy to us, are so from association with the dark
forests, and the stillness which surrounds their haunts, but perhaps, at
the time when we judge, are expressive of the greatest enjoyment. An
answer of kindness in reply to a mate, the calling together of the newly
fledged brood, or exultation over the discovery of some favourite hoard
of food, are what are set down as painful and discontented.

Mr Audubon's remarks on this splendid species, " The king of the
woodpeckers," I have transcribed at some length, as indicating the particular manner of the typical family of this great group.

"The ivory-billed woodpecker confines its rambles to a comparatively
very small portion of the United States, it never having been observed
in the middle states within the memory of any person now living there.
In fact, in no portion of these districts does the nature of the woods appear suitable to its remarkable habits.

"Descending the Ohio, we meet with this splendid bird for the first
time near the confluence of that beautiful river and the Mississippi ; after
which, following the windings of the latter, either downwards toward
the sea, or upwards in the direction of the Missouri, we frequently observe it. On the Atlantic coast, North Carolina may be taken as the
limit of its distribution, although now and then an individual of the
species may be accidentally seen in Maryland. To the westward of the
Mississippi, it is found in all the dense forests bordering the streams which
empty their waters into that majestic river, from the very declivities of
the Rocky Mountains. The lower parts of the Carolinas, Georgia, Ala-
bama, Louisiana, and Mississippi, are, however, the most favourite resorts
of this bird, and in those states it constantly resides, breeds, and passes
a life of peaceful enjoyment, finding a profusion of food in all the deep,
dark, and gloomy swamps dispersed throughout them.

" The flight of this bird is graceful in the extreme, although seldom
prolonged to more than a few hundred yards at a time, unless when it

ideas of the Creator. His manners have also a dignity in them superior to the common herd of woodpeckers. Trees, shrubbery, orchards, rails, fence posts, and old prostrate logs, are alike interesting to those, in their humble and indefatigable search for prey ; but the royal hunter now before us, scorns the humility of such situations, and seeks the most towering trees of the forest ; seeming particularly attached to those prodigious cypress swamps, whose crowded giant sons stretch their bare and blasted or moss-hung arms midway to the skies. In these almost inaccessible recesses, amid ruinous piles of impending timber, his trumpet-like note and loud strokes resound through the solitary, savage wilds, of which he seems the sole lord and inhabitant. Wherever he frequents, he leaves numerous monuments of his industry behind him. We there see enormous pine trees with cartloads of bark lying around their roots, and chips of the trunk itself, in such quantities as to suggest the idea that half a dozen of axe-men

has to cross a large river, which it does in deep undulations, opening its wings at first to their full extent, and nearly closing them to renew the propelling impulse. The transit from one tree to another, even should the distance be as much as a hundred yards, is performed by a single sweep, and the bird appears as if merely swinging itself from the top of the one tree to that of the other, forming an elegantly curved line. At this moment all the beauty of the plumage is exhibited, and strikes the beholder with pleasure. It never utters any sound whilst on wing, unless during the love season ; but at all other times, no sooner has this bird alighted than its remarkable voice is heard, at almost every leap which it makes, whilst ascending against the upper parts of the trunk of a tree, or its highest branches. Its notes are clear, loud, and yet very plaintive. They are heard at a considerable distance, perhaps half a mile, and resemble the false high note of a clarionet. They are usually repeated three times in succession, and may be represented by the monosyllable *pait, pait, pait.* These are heard so frequently as to induce me to say that the bird spends few minutes of the day without uttering them, and this circumstance leads to its destruction, which is aimed at, not because (as is supposed by some) this species is a destroyer of trees, but more because it is a beautiful bird, and its rich scalp attached to the upper mandible forms an ornament for the war-dress of most of our Indians, or for the shot-pouch of our squatters and hunters, by all of whom the bird is shot merely for that purpose."—ED.

had been at work there for the whole morning. The body of
the tree is also disfigured with such numerous and so large
excavations, that one can hardly conceive it possible for the
whole to be the work of a woodpecker. With such strength,
and an apparatus so powerful, what havoc might he not
commit, if numerous, on the most useful of our forest trees!
and yet with all these appearances, and much of vulgar prejudice against him, it may fairly be questioned whether he is
at all injurious; or, at least, whether his exertions do not
contribute most powerfully to the protection of our timber.
Examine closely the tree where he has been at work, and you
will soon perceive, that it is neither from motives of mischief
nor amusement that he slices off the bark, or digs his way
into the trunk.—For the sound and healthy tree is the least
object of his attention. The diseased, infested with insects,
and hastening to putrefaction, are *his* favourites; there the
deadly crawling enemy have formed a lodgement between the
bark and tender wood, to drink up the very vital part of the
tree. It is the ravages of these vermin, which the intelligent
proprietor of the forest deplores as the sole perpetrators of the
destruction of his timber. Would it be believed that the
larvæ of an insect, or fly, no larger than a grain of rice, should
silently, and in one season, destroy some thousand acres of pine
trees, many of them from two to three feet in diameter, and
a hundred and fifty feet high! Yet whoever passes along the
high road from Georgetown to Charlestown, in South Carolina,
about twenty miles from the former place, can have striking
and melancholy proofs of this fact. In some places the whole
woods, as far as you can see around you, are dead, stripped of
the bark, their wintry-looking arms and bare trunks bleaching
in the sun, and tumbling in ruins before every blast, presenting
a frightful picture of desolation. And yet ignorance and prejudice stubbornly persist in directing their indignation against
the bird now before us, the constant and mortal enemy of these
very vermin; as if the hand that probed the wound to extract
its cause, should be equally detested with that which inflicted

it; or as if the thief-catcher should be confounded with the thief. Until some effectual preventive or more complete mode of destruction can be devised against these insects, and their larvæ, I would humbly suggest the propriety of protecting, and receiving with proper feelings of gratitude, the services of this and the whole tribe of woodpeckers, letting the odium of guilt fall upon its proper owners.

In looking over the accounts given of the ivory-billed woodpecker by the naturalists of Europe, I find it asserted, that it inhabits from New Jersey to Mexico. I believe, however, that few of them are ever seen to the north of Virginia, and very few of them even in that state. The first place I observed this bird at, when on my way to the south, was about twelve miles north of Wilmington in North Carolina. There I found the bird from which the drawing of the figure in the plate was taken. This bird was only wounded slightly in the wing, and, on being caught, uttered a loudly reiterated, and most piteous note, exactly resembling the violent crying of a young child; which terrified my horse so, as nearly to have cost me my life. It was distressing to hear it. I carried it with me in the chair, under cover, to Wilmington. In passing through the streets, its affecting cries surprised every one within hearing, particularly the females, who hurried to the doors and windows with looks of alarm and anxiety. I drove on, and on arriving at the piazza of the hotel, where I intended to put up, the landlord came forward, and a number of other persons who happened to be there, all equally alarmed at what they heard; this was greatly increased by my asking, whether he could furnish me with accommodations for myself and my baby. The man looked blank and foolish, while the others stared with still greater astonishment. After diverting myself for a minute or two at their expense, I drew my woodpecker from under the cover, and a general laugh took place. I took him up stairs and locked him up in my room, while I went to see my horse taken care of. In less than an hour I returned, and, on opening the door, he set up the same distressing shout, which now appeared to proceed from grief that he had been discovered in his attempts at escape. He had mounted along the side of the window, nearly as high as the ceiling, a little below which he had begun to break through. The bed was covered with large pieces of plaster; the lath was exposed for at least fifteen inches square, and a hole, large enough to admit the fist, opened to the weather-boards; so that, in less than another hour he would certainly have succeeded in making his way through. I now tied a string round his leg, and, fastening it to the table, again left him. I wished to preserve his life, and had gone off in search of suitable food for him. As I reascended the stairs, I heard him again hard at work, and on entering had the mortification to perceive that he had almost entirely ruined the mahogany table to which he was fastened, and on which he had wreaked his whole vengeance. While engaged in taking the drawing, he cut me severely in several places, and, on the whole, displayed such a noble and unconquerable spirit, that I was frequently tempted to restore him to his native woods. He lived with me nearly three days, but refused all sustenance, and I witnessed his death with regret.

The head and bill of this bird is in great esteem among the southern Indians, who wear them by way of amulet or charm, as well as ornament; and, it is said, dispose of them to the northern tribes at considerable prices. An Indian believes that the head, skin, or even feathers of certain birds, confer on the wearer all the virtues or excellencies of those birds. Thus I have seen a coat made of the skins, heads, and claws of the raven; caps stuck round with heads of butcher birds, hawks, and eagles; and as the disposition and courage of the ivory-billed woodpecker are well known to the savages, no wonder they should attach great value to it, having both beauty, and, in their estimation, distinguished merit to recommend it.

This bird is not migratory, but resident in the countries where it inhabits. In the low countries of the Carolinas it usually prefers the large timbered cypress swamps for breeding in. In the trunk of one of these trees, at a considerable height, the male and female alternately, and in conjunction, dig out a large and capacious cavity for their eggs and young. Trees thus dug out have frequently been cut down, with sometimes the eggs and young in them. This hole, according to information,—for I have never seen one myself,—is generally a little winding, the better to keep out the weather, and from two to five feet deep. The eggs are said to be generally four, sometimes five, as large as a pullet's, pure white, and equally thick at both ends—a description that, except in size, very nearly agrees with all the rest of our woodpeckers. The young begin to be seen abroad about the middle of June. Whether they breed more than once in the same season is uncertain.[*]

[*] The description of the nestling, &c., is thus also given by Audubon. Wilson observes, that he had no opportunity of ever seeing their holes, and the following will tend to render his account more complete :—

"The ivory-billed woodpecker nestles earlier in spring than any other species of its tribe. I have observed it boring a hole for that purpose in the beginning of March. The hole is, I believe, always made in the trunk of a live tree, generally an ash or a hagberry, and is at a great height. The birds pay great regard to the particular situation of the tree, and the inclination of its trunk ; first, because they prefer retirement, and again, because they are anxious to secure the aperture against the access of water during beating rains. To prevent such a calamity, the hole is generally dug immediately under the junction of a large branch with the trunk. It is first bored horizontally for a few inches, then directly downwards, and not in a spiral manner, as some people have imagined. According to circumstances, this cavity is more or less deep, being sometimes not more than ten inches, whilst at other times it reaches nearly three feet downwards into the core of the tree. I have been led to think that these differences result from the more or less immediate necessity under which the female may be of depositing her eggs, and again have thought that the older the woodpecker is, the deeper does it make its hole. The average diameter of the different nests which I have examined was about seven inches within, although the entrance, which is perfectly round, is only just large enough to admit the bird.

"Both birds work most assiduously at this excavation, one waiting outside to encourage the other, whilst it is engaged in digging, and when the latter is fatigued, taking its place. I have approached trees whilst these woodpeckers were thus busily employed in forming their nest, and by resting my head against the bark, could easily distinguish every blow given by the bird. I observed that in two instances, when the wood-

So little attention do the people of the countries where these birds inhabit pay to the minutiæ of natural history, that, generally speaking, they make no distinction between the

peckers saw me thus at the foot of the tree in which they were digging their nest, they abandoned it for ever. For the first brood there are generally six eggs. They are deposited on a few chips at the bottom of the hole, and are of a pure white colour. The young are seen creeping out of the hole about a fortnight before they venture to fly to any other tree. The second brood makes its appearance about the 15th of August.

" In Kentucky and Indiana, the ivory-bills seldom raise more than one brood in the season. The young are at first of the colour of the female, only that they want the crest, which, however, grows rapidly, and towards autumn—particularly in birds of the first breed—is nearly equal to that of the mother. The males have then a slight line of red on the head, and do not attain their richness of plumage until spring, or their full size until the second year. Indeed, even then, a difference is easily observed between them and individuals which are much older.

" The food of this species consists principally of beetles, larvæ, and large grubs. No sooner, however, are the grapes of our forests ripe than they are eaten by the ivory-billed woodpecker with great avidity. I have seen this bird hang by its claws to the vines, in the position so often assumed by a titmouse, and reaching downwards, help itself to a bunch of grapes with much apparent pleasure. Persimmons are also sought for by them, as soon as the fruit becomes quite mellow, as are hagberries.

" The ivory-bill is never seen attacking the corn, or the fruit of the orchards, although it is sometimes observed working upon and chipping off the bark from the belted trees of the newly cleared plantations. It seldom comes near the ground, but prefers at all times the tops of the tallest trees. Should it, however, discover the half-standing broken shaft of a large dead and rotten tree, it attacks it in such a manner as nearly to demolish it in the course of a few days. I have seen the remains of some of these ancient monarchs of our forests so excavated, and that so singularly, that the tottering fragments of the trunk appeared to be merely supported by the great pile of chips by which its base was surrounded. The strength of this woodpecker is such, that I have seen it detach pieces of bark seven or eight inches in length at a single blow of its powerful bill, and by beginning at the top branch of a dead tree, tear off the bark, to an extent of twenty or thirty feet, in the course of a few hours, leaping downwards, with its body in an upward position, tossing its head to the right and left, or leaning it against the bark to ascertain the precise spot where the grubs were concealed, and imme-

ivory-billed and pileated woodpecker, represented in the same plate; and it was not till I showed them the two birds together, that they knew of any difference. The more intelligent and observing part of the natives, however, distinguish them by the name of the large and lesser *logcocks*. They seldom examine them but at a distance, gunpowder being considered too precious to be thrown away on woodpeckers; nothing less than a turkey being thought worth the value of a load.

The food of this bird consists, I believe, entirely of insects and their larvæ.* The pileated woodpecker is suspected

diately after renewing its blows with fresh vigour, all the while sounding its loud notes, as if highly delighted.

"This species generally moves in pairs, after the young have left their parents. The female is always the most clamorous and the least shy. Their mutual attachment is, I believe, continued through life. Excepting when digging a hole for the reception of their eggs, these birds seldom, if ever, attack living trees, for any other purpose than that of procuring food, in doing which they destroy the insects that would otherwise prove injurious to the trees.

"I have frequently observed the male and female retire to rest for the night, into the same hole in which they had long before reared their young. This generally happens a short time after sunset.

"When wounded and brought to the ground, the ivory-bill immediately makes for the nearest tree, and ascends it with great rapidity and perseverance until it reaches the top branches, when it squats and hides, generally with great effect. Whilst ascending, it moves spirally round the tree, utters its loud *pait, pait, pait,* at almost every hop, but becomes silent the moment it reaches a place where it conceives itself secure. They sometimes cling to the bark with their claws so firmly as to remain cramped to the spot for several hours after death. When taken by the hand, which is rather a hazardous undertaking, they strike with great violence, and inflict very severe wounds with their bill as well as claws, which are extremely sharp and strong. On such occasions, this bird utters a mournful and very piteous cry."—ED.

* Mr Audubon says, that though the greater part of their food consists of insects and their larvæ, no sooner are the grapes of our forests ripe, than they are eaten with the greatest avidity. I have seen this bird hang by its claws to the vines, in the position so often assumed by the titmouse, and, reaching down, help itself to a bunch of grapes. Persimmons are also sought by them, as soon as the fruit becomes quite mellow, and hagberries.—ED.

of sometimes tasting the Indian corn : the ivory-billed never. His common note, repeated every three or four seconds, very much resembles the tone of a trumpet, or the high note of a clarionet, and can plainly be distinguished at the distance of more than half a mile ; seeming to be immediately at hand, though perhaps more than one hundred yards off. This it utters while mounting along the trunk or digging into it. At these times it has a stately and novel appearance ; and the note instantly attracts the notice of a stranger. Along the borders of the Savannah river, between Savannah and Augusta, I found them very frequently ; but my horse no sooner heard their trumpet-like note, than, remembering his former alarm, he became almost ungovernable.

The ivory-billed woodpecker is twenty inches long, and thirty inches in extent ; the general colour is black, with a considerable gloss of green when exposed to a good light ; iris of the eye, vivid yellow ; nostrils, covered with recumbent white hairs ; fore part of the head, black ; rest of the crest, of a most splendid red, spotted at the bottom with white, which is only seen when the crest is erected, as represented in the plate ; this long red plumage being ash-coloured at its base, above that white, and ending in brilliant red ; a stripe of white proceeds from a point, about half an inch below each eye, passes down each side of the neck, and along the back, where they are about an inch apart, nearly to the rump ; the first five primaries are wholly black ; on the next five the white spreads from the tip, higher and higher, to the secondaries, which are wholly white from their coverts downward. These markings, when the wings are shut, make the bird appear as if his back were white : hence he has been called by some of our naturalists the large white-backed woodpecker. The neck is long ; the beak an inch broad at the base, of the colour and consistence of ivory, prodigiously strong and elegantly fluted. The tail is black, tapering from the two exterior feathers, which are three inches shorter than the middle ones, and each feather has the singularity of being greatly concave below; the wing is lined with yellowish white; the legs are about an inch and a quarter long, the exterior toe about the same length, the claws exactly semicircular and remarkably powerful,—the whole of a light blue or lead colour. The female is about half an inch shorter, the bill rather less, and the whole plumage of the head black, glossed with green; in the other parts of the plumage, she exactly resembles the male. In the stomachs of three which I opened, I found large quantities of a species of worm called borers, two or three inches long, of a dirty cream colour, with a black head; the stomach was an oblong pouch, not muscular like the gizzards of some others. The tongue was worm-shaped, and for half an inch at the tip as hard as horn, flat, pointed, of the same white colour as the bill, and thickly barbed on each side.*

PILEATED WOODPECKER. (*Picus pileatus.*)

PLATE XXIX.—Fig. 2.

Picus niger, crista rubra, *Lath. Ind. Orn.* i. p. 225, 4.—Picus pileatus, *Linn. Syst.* i. p. 173, 3.—*Gmel. Syst.* i. p. 425.—Picus Virginianus pileatus, *Briss.* iv. p. 29, 10.—*Id.* 8vo, ii. p. 50.—Pic noir à huppé rouge, *Buff.* vii. p. 48.— Pic noir huppé de la Louisiana, *Pl. enl.* 718.—Larger Crested Woodpecker, *Catesb. Car.* i. 6, 17.—Pileated Woodpecker, *Arct. Zool.* ii. No. 157.—*Lath. Syn.* ii. p. 554, 3.—*Id. Supp.* p. 105.—*Bartram,* p. 289.—*Peale's Museum,* No. 1886.

PICUS PILEATUS.—Linnæus.†

Picus pileatus, *Bonap. Synop.* p. 44.—*Wagl. Syst. Av.* No. 2.—Picus (dryotomus) pileatus, *North. Zool.* ii. p. 304.

This American species is the second in size among his tribe, and may be styled the great northern chief of the woodpeckers,

---

* Wilson seems to have been in some uncertainty regarding the nidification of this species, and probably never saw the nest. The account of Mr Audubon will fill up what is here wanting.—Ed.

† As we remarked in our last note, Mr Swainson, according to the views he entertains, has divided the large family *Picianæ* into five great divisions, and the different forms in these again into groups of lesser

though, in fact, his range extends over the whole of the United States from the interior of Canada to the Gulf of Mexico. He is very numerous in the Gennesee country, and in all the tracts of high-timbered forests, particularly in the neighbourhood of our large rivers, where he is noted for making a loud and almost incessant cackling before wet weather; flying at such times in a restless uneasy manner from tree to tree, making the woods echo to his outcry. In Pennsylvania and the northern states, he is called the black woodcock; in the southern states, the logcock. Almost every old trunk in the forest where he resides bears the marks of his chisel. Wherever he perceives a tree beginning to decay, he examines it round and round with great skill and dexterity, strips off the bark in sheets of five or six feet in length, to get at the hidden cause of the disease, and labours with a gaiety and activity really surprising. I have seen him separate the greatest part of the bark from a large dead pine tree, for twenty or thirty feet, in less than a quarter of an hour. Whether engaged in flying from tree to tree, in digging, climbing, or barking, he seems perpetually in a hurry. He is extremely hard to kill, clinging close to the tree even after he has received his mortal wound; nor yielding up his hold but with his expiring breath. If slightly wounded in the wing, and dropt while flying, he instantly makes for the nearest tree, and strikes with great bitterness at the hand stretched out to seize him; and can rarely be reconciled to confinement. He is sometimes observed among the hills of Indian corn, and it is said by some that he frequently feeds on it. Complaints of this kind are, however, not general; many farmers doubting the fact, and conceiving that at these times he is in search of insects which lie concealed in the husk. I will not be positive that they never occasionally taste maize; yet I have opened and examined great numbers

value. For the type of one of them, he has chosen the *Picus pileatus*, under the title of *Dryotomus*, differing from *Picus*, in the *exterior* outer toe being shorter than the *anterior* external one, exactly the reverse of the proportions of *Picus.*—ED.

of these birds, killed in various parts of the United States, from Lake Ontario to the Alatamaha river, but never found a grain of Indian corn in their stomachs.

The pileated woodpecker is not migratory, but braves the extremes of both the arctic and torrid regions. Neither is he gregarious, for it is rare to see more than one or two, or at the most three, in company. Formerly they were numerous in the neighbourhood of Philadelphia; but gradually, as the old timber fell, and the country became better cleared, they retreated to the forest. At present few of those birds are to be found within ten or fifteen miles of the city.

Their nest is built, or rather the eggs are deposited, in the hole of a tree, dug out by themselves, no other materials being used but the soft chips of rotten wood. The female lays six large eggs of a snowy whiteness; and, it is said, they generally raise two broods in the same season.

This species is eighteen inches long, and twenty-eight in extent; the general colour is a dusky brownish black; the head is ornamented with a conical cap of bright scarlet; two scarlet mustaches proceed from the lower mandible; the chin is white; the nostrils are covered with brownish white hair-like feathers, and this stripe of white passes from thence down the side of the neck to the sides, spreading under the wings; the upper half of the wings are white, but concealed by the black coverts; the lower extremities of the wings are black, so that the white on the wing is not seen but when the bird is flying, at which time it is very prominent; the tail is taper-ing, the feathers being very convex above, and strong; the legs are of a leaden gray colour, very short, scarcely half an inch; the toes very long; claws, strong and semicircular, and of a pale blue; the bill is fluted, sharply ridged, very broad at the base, bluish black above, below and at the point bluish white; the eye is of a bright golden colour, the pupil black; the tongue, like those of its tribe, is worm-shaped, except near the tip, where for one-eighth of an inch it is horny, pointed, and beset with barbs.

The female has the forehead, and nearly to the crown, of a light brown colour, and the mustaches are dusky, instead of red. In both a fine line of white separates the red crest from the dusky line that passes over the eye.

RED-WINGED STARLING. (*Sturnus predatorius.*)

PLATE XXX.—Fig. 1. Male; Fig. 2. Female.

Bartram, 291.—Oriolus phœniceus, *Linn. Syst.* 161.—Red-winged Oriole, *Arct. Zool.* 255, No. 140.—Le Troupiale à aisles rouges, *Briss.* ii. 97.—Le commandeur, *Buff.* iii. 214, *Pl. enl.* 402.—*Lath.* i. 428.—Acolchichi, *Fernand. Nov. Hisp.* p. 14.—*Peale's Museum*, No. 1466, 1467.

AGLAIUS PHŒNICEUS.—Vieillot.*

Aglaius phœniceus, *Vieill. Gall. des Ois.*—*North. Zool.* ii. p. 280.—Icterus phœniceus, *Bonap. Synop.* p. 52.- The Red-Winged Starling, or Marsh Blackbird, *Aud.* pl. 67., male in different states, female and young ; *Orn. Biog.* i. p. 348.

This notorious and celebrated corn thief, the long reputed plunderer and pest of our honest and laborious farmers, now

* This bird, I believe, will rank under the *Icteri* of Brisson, but seems first mentioned by Daudin under that title. Like the others of this intricate family, it has been described under a multitude of names ; but the above seems the preferable one to be adopted. Wilson also changed the specific name to *Predatorius*, taken from its plundering habits, whereas, without doubt, he should have retained its original designation. North America possesses another beautiful species, figured in the continuation of the *Ornithology* by Bonaparte.

Wilson is somewhat puzzled in what genus to place this bird, and is only reconciled to join it with our common starling, which it much resembles in its congregated flights. In this country, we cannot expect to see a flight of such numbers as Wilson mentions ; still they are sometimes very numerous, and one might almost conceive the appearance of the one, from their recollections of the other. In the low meadows of Holland, again, some relative proportion may be found. I have seen an extent of flat surface, as far as the eye could reach around, covered with flocks of starlings, associated with lapwings and golden plovers ; and the flocks that rose on the approach of night, were sometimes immense. In the islands of Sardinia, and those adjacent, and where they may be augmented by the presence of another species, the *St unicolor* of Temminck, I am told that the assemblage of birds is

1.Red winged Starling. 2.Female. 3.Black-poll Warbler. 4.Lesser Red poll.

30.

presents himself before us, with his copartner in iniquity, to receive the character due for their very active and distinguished services. In investigating the nature of these, I shall endeavour to render strict historical justice to this noted pair ; adhering to the honest injunctions of the poet,

> " Nothing extenuate,
> Nor set down aught in malice. "

Let the reader divest himself equally of prejudice, and we shall be at no loss to ascertain accurately their true character.

The red-winged starlings, though generally migratory in the states north of Maryland, are found during winter in immense flocks, sometimes associated with the purple grakles, and often by themselves, along the whole lower parts of Virginia, both Carolinas, Georgia, and Louisiana, particularly near the sea coast, and in the vicinity of large rice and corn fields. In the months of January and February, while passing through the former of these countries, I was frequently entertained with the aërial evolutions of these great bodies of starlings. Sometimes they appeared driving about like an enormous black cloud carried before the wind, varying its shape every moment ; sometimes suddenly rising from the fields around me with a noise like thunder ; while the glittering of innumerable wings of the brightest vermilion amid the black cloud they formed, produced on these occasions a very striking and splendid effect. Then, descending like a torrent,

innumerable in the lower valleys, and among the lakes and reedy marshes which cover so much of the lower parts of these countries. In their evolutions before retiring to rest among reeds or bushes, the two birds also resemble each other. That of Europe is thus described by an observing naturalist :—" There is something singularly curious and mysterious in the conduct of these birds, previous to their nightly retirement, by the variety and intricacy of the evolutions they execute at that time. They will form themselves, perhaps, into a triangle, then shoot into a long, pear-shaped figure, expand like a sheet, wheel into a ball, as Pliny observes, each individual striving to get into the centre, &c., with a promptitude more like parade movements, than the actions of birds." I have known them watched for, when coming to roost, and shot in considerable numbers. Their wings afford favourite feather for fishers.—Ed.

and covering the branches of some detached grove, or clump of trees, the whole congregated multitude commenced one general concert or chorus, that I have plainly distinguished at the distance of more than two miles; and, when listened to at the intermediate space of about a quarter of a mile, with a slight breeze of wind to swell and soften the flow of its cadences, was to me grand, and even sublime. The whole season of winter, that, with most birds, is passed in struggling to sustain life in silent melancholy, is, with the red-wings, one continued carnival. The profuse gleanings of the old rice, corn, and buckwheat fields, supply them with abundant food, at once ready and nutritious; and the intermediate time is spent either in aërial manœuvres, or in grand vocal performances, as if solicitous to supply the absence of all the tuneful summer tribes, and to cheer the dejected face of nature with their whole combined powers of harmony.

About the 20th of March, or earlier, if the season be open, they begin to enter Pennsylvania in numerous, though small parties. These migrating flocks are usually observed from daybreak to eight or nine in the morning, passing to the north, chattering to each other as they fly along; and, in spite of all our antipathy, their well known notes and appearance, after the long and dreary solitude of winter, inspire cheerful and pleasing ideas of returning spring, warmth, and verdure. Selecting their old haunts, every meadow is soon enlivened by their presence. They continue in small parties to frequent the low borders of creeks, swamps, and ponds, till about the middle of April, when they separate in pairs to breed; and, about the last week in April, or first in May, begin to construct their nest. The place chosen for this is generally within the precincts of a marsh or swamp, meadow, or other like watery situation,—the spot, usually a thicket of alder bushes, at the height of six or seven feet from the ground; sometimes in a detached bush, in a meadow of high grass; often in a tussock of rushes, or coarse rank grass; and not unfrequently on the ground: in all of which situations I have repeatedly found them. When in a bush, they are generally composed outwardly of wet rushes, picked from the swamp, and long tough grass in large quantity, and well lined with very fine bent. The rushes, forming the exterior, are generally extended to several of the adjoining twigs, round which they are repeatedly and securely twisted; a precaution absolutely necessary for its preservation, on account of the flexible nature of the bushes in which it is placed. The same caution is observed when a tussock is chosen, by fastening the tops together, and intertwining the materials of which the nest is formed with the stalks of rushes around. When placed on the ground, less care and fewer materials being necessary, the nest is much simpler and slighter than before. The female lays five eggs, of a very pale light blue, marked with faint tinges of light purple, and long straggling lines and dashes of black. It is not uncommon to find several nests in the same thicket, within a few feet of each other.

During the time the female is sitting, and still more particularly after the young are hatched, the male, like most other birds that build in low situations, exhibits the most violent symptoms of apprehension and alarm on the approach of any person to its near neighbourhood. Like the lapwing of Europe, he flies to meet the intruder, hovers at a short height over-head, uttering loud notes of distress; and, while in this situation, displays to great advantage the rich glowing scarlet of his wings, heightened by the jetty black of his general plumage. As the danger increases, his cries become more shrill and incessant, and his motions rapid and restless; the whole meadow is alarmed, and a collected crowd of his fellows hover around, and mingle their notes of alarm and agitation with his. When the young are taken away, or destroyed, he continues for several days near the place, restless and dejected, and generally recommences building soon after, in the same meadow. Towards the beginning or middle of August, the young birds begin to fly in flocks, and at that age nearly resemble the female, with the exception of some reddish or orange, that marks the shoulders of the males, and which increases in space and brilliancy as winter approaches. It has been frequently remarked, that, at this time, the young birds chiefly associate by themselves, there being sometimes not more than two or three old males observed in a flock of many thousands. These, from the superior blackness and rich red of their plumage, are very conspicuous.

Before the beginning of September, these flocks have become numerous and formidable ; and the young ears of maize, or Indian corn, being then in their soft, succulent, milky state, present a temptation that cannot be resisted. Reinforced by numerous and daily flocks from all parts of the interior, they pour down on the low countries in prodigious multitudes. Here they are seen, like vast clouds, wheeling and driving over the meadows and devoted corn-fields, darkening the air with their numbers. Then commences the work of destruction on the corn, the husks of which, though composed of numerous envelopments of closely wrapt leaves, are soon completely or partially torn off ; while from all quarters myriads continue to pour down like a tempest, blackening half an acre at a time ; and, if not disturbed, repeat their depredations, till little remains but the cob and the shrivelled skins of the grain ; what little is left of the tender ear, being exposed to the rains and weather, is generally much injured. All the attacks and havoc made at this time among them with the gun, and by the hawks,—several species of which are their constant attendants,—has little effect on the remainder. When the hawks make a sweep among them, they suddenly open on all sides, but rarely in time to disappoint them of their victims ; and, though repeatedly fired at, with mortal effect, they only remove from one field to an adjoining one, or to another quarter of the same enclosure. From dawn to nearly sunset, this open and daring devastation is carried on, under the eye of the proprietor ; and a farmer, who has any considerable extent of corn, would require half-a-dozen men at least, with guns, to guard it ; and even then, all their vigilance and activity would not prevent a good tithe of it from becoming the prey of the blackbirds. The Indians, who usually plant their corn in one general field, keep the whole young boys of the village all day patrolling round and among it ; and each being furnished with bow and arrows, with which they are very expert, they generally contrive to destroy great numbers of them.

It must, however, be observed, that this scene of pillage is principally carried on in the low countries, not far from the sea coast, or near the extensive flats that border our large rivers ; and is also chiefly confined to the months of August and September. After this period, the corn having acquired its hard shelly coat, and the seeds of the reeds or wild oats, with a profusion of other plants, that abound along the river shores, being now ripe, and in great abundance, they present a new and more extensive field for these marauding multitudes. The reeds also supply them with convenient roosting places, being often in almost unapproachable morasses; and thither they repair every evening, from all quarters of the country. In some places, however, when the reeds become dry, advantage is taken of this circumstance to destroy these birds, by a party secretly approaching the place, under cover of a dark night, setting fire to the reeds in several places at once, which being soon enveloped in one general flame, the uproar among the blackbirds becomes universal; and, by the light of the conflagration, they are shot down in vast numbers, while hovering and screaming over the place. Sometimes straw is used for the same purpose, being previously strewed near the reeds and alder bushes, where they are known to roost, which being instantly set on fire, the consternation and havoc is prodigious ; and the party return by day to pick up the slaughtered game. About the first of November, they begin to move off towards the south ; though, near the sea-coast, in the states of New Jersey and Delaware, they continue long after that period.

Such are the general manners and character of the red-winged starling; but there remain some facts to be mentioned, no less authentic, and well deserving the consideration of its enemies, more especially of those whose detestation of this species would stop at nothing short of total extirpation.

It has been already stated, that they arrive in Pennsylvania late in March. Their general food at this season, as well as during the early part of summer (for the crows and purple grakles are the principal pests in planting time), consists of grub-worms, caterpillars, and various other larvæ, the silent, but deadly enemies of all vegetation, and whose secret and insidious attacks are more to be dreaded by the husbandman than the combined forces of the whole feathered tribes together. For these vermin, the starlings search with great diligence; in the ground, at the roots of plants, in orchards, and meadows, as well as among buds, leaves, and blossoms; and, from their known voracity, the multitudes of these insects which they destroy must be immense. Let me illustrate this by a short computation: If we suppose each bird, on an average, to devour fifty of these larvæ in a day (a very moderate allowance), a single pair, in four months, the usual time such food is sought after, will consume upwards of twelve thousand. It is believed, that not less than a million pair of these birds are distributed over the whole extent of the United States in summer; whose food, being nearly the same, would swell the amount of vermin destroyed to twelve thousand millions. But the number of young birds may be fairly estimated at double that of their parents; and, as these are constantly fed on larvæ for at least three weeks, making only the same allowance for them as for the old ones, their share would amount to four thousand two hundred millions; making a grand total of sixteen thousand two hundred millions of noxious insects destroyed in the space of four months by this single species! The combined ravages of such a hideous host of vermin would be sufficient to spread famine and desolation over a wide extent of the richest and best cultivated country on earth. All this, it may be said, is mere supposition. It is, however, supposition founded on known and acknowledged facts. I have never dissected any of these birds in spring without receiving the most striking and satisfactory proofs of those facts; and though, in a matter of this kind, it is impossible to ascertain precisely the amount of the benefits derived by agriculture from this, and many other species of our birds, yet, in the present case, I cannot resist the belief, that the services of this species, in spring, are far more important and beneficial than the value of all that portion of corn which a careful and active farmer permits himself to lose by it.

The great range of country frequented by this bird extends from Mexico, on the south, to Labrador. Our late enterprising travellers across the continent to the Pacific Ocean, observed it numerous in several of the valleys at a great distance up the Missouri. When taken alive, or reared from the nest, it soon becomes familiar, sings frequently, bristling out its feathers, something in the manner of the cow bunting. These notes, though not remarkably various, are very peculiar. The most common one resembles the syllables *conk-quer-rèe ;* others, the shrill sounds produced by filing a saw : some are more guttural ; and others remarkably clear. The usual note of both male and female is a single *chuck.* Instances have been produced where they have been taught to articulate several words distinctly ; and, contrary to what is observed of many birds, the male loses little of the brilliancy of his plumage by confinement.

A very remarkable trait of this bird is, the great difference of size between the male and female ; the former being nearly two inches longer than the latter, and of proportionate magnitude. They are known by various names in the different states of the Union ; such as the *swamp blackbird, marsh blackbird, red-winged blackbird, corn or maize thief, starling,* &c. Many of them have been carried from this to different parts of Europe ; and Edwards relates, that one of them, which had, no doubt, escaped from a cage, was shot in the neighbourhood of London ; and, on being opened, its stomach was found to be filled with grub-worms, caterpillars, and beetles ; which Buffon seems to wonder at, as, " in their own country," he observes, " they feed exclusively on grain and maize."

Hitherto this species has been generally classed by naturalists with the orioles. By a careful comparison, however, of its bill with those of that tribe, the similarity is by no means sufficient to justify this arrangement ; and its manners are altogether different. I can find no genus to which it makes so near an approach, both in the structure of the bill and in food, flight, and manners, as those of the stare ; with which, following my judicious friend Mr Bartram, I have accordingly placed it. To the European, the perusal of the foregoing pages will be sufficient to satisfy him of their similarity of manners. For the satisfaction of those who are unacquainted with the common starling of Europe, I shall select a few sketches of its character, from the latest and most accurate publication I have seen from that quarter.* Speaking of the stare, or starling, this writer observes, " In the winter season, these birds fly in vast flocks, and may be known at a great distance by their whirling mode of flight, which Buffon compares to a sort of vortex, in which the collective body performs an uniform circular revolution, and, at the same time, continues to make a progressive advance. The evening is the time when the stares assemble in the greatest numbers, and betake themselves to the fens and marshes, where they roost among the reeds : they chatter much in the evening and morning, both when they assemble and disperse. So attached are they to society, that they not only join those of their own species, but also birds of a different kind ; and are frequently seen in company with red-wings (a species of thrush), fieldfares, and even with crows, jackdaws, and pigeons. Their principal food consists of worms, snails, and caterpillars ; they likewise eat various kinds of grain, seeds, and berries."

* Bewick's " British Birds," part i. p. 119. Newcastle, 1809.

He adds, that, " in a confined state, they are very docile, and may easily be taught to repeat short phrases, or whistle tunes with great exactness."

The red-winged starling (fig. 1.) is nine inches long, and fourteen inches in extent ; the general colour is a glossy black, with the exception of the whole lesser wing-coverts, the first, or lower row of which is of a reddish cream colour, the rest a rich and splendid scarlet ; legs and bill, glossy brownish black ; irides, hazel ; bill, cylindrical above, compressed at the sides, straight, running considerably up the forehead, where it is prominent, rounding and flattish towards the tip, though sharp-pointed ; tongue, nearly as long as the bill, tapering and lacerated at the end ; tail, rounded, the two middle feathers also somewhat shorter than those immediately adjoining.

The female (fig. 2.) is seven inches and a quarter in length, and twelve inches in extent ; chin, a pale reddish cream ; from the nostril over the eye, and from the lower mandible, run two stripes of the same, speckled with black ; from the posterior angle of the eye backwards, a streak of brownish black covers the auriculars ; throat, and whole lower parts, thickly streaked with black and white, the latter inclining to cream on the breast ; whole plumage above, black, each feather bordered with pale brown, white, or bay, giving the bird a very mottled appearance ; lesser coverts, the same ; bill and legs as in the male.

The young birds at first greatly resemble the female ; but have the plumage more broadly skirted with brown. The red early shows itself on the lesser wing-coverts of the males, at first pale, inclining to orange, and partially disposed. The brown continues to skirt the black plumage for a year or two, so that it is rare to find an old male altogether destitute of some remains of it ; but the red is generally complete in breadth and brilliancy by the succeeding spring. The females are entirely destitute of that ornament.

The flesh of these birds is but little esteemed, being, in general, black, dry, and tough. Strings of them are, however, frequently seen exposed for sale in our markets.

BLACK-POLL WARBLER. (*Sylvia striata.*)

PLATE XXX.—Fig. 3.

Lath. ii. 460.—*Arct. Zool.* 401.—*Turton,* 600.—*Peale's Museum,* No. 7054.

SYLVICOLA STRIATA.*—Swainson.

Sylvia striata, *Bonap. Synop.* p. 81.—Sylvicola striata, *North. Zool.* ii. p. 218.

THIS species has considerable affinity to the flycatchers in its habits. It is chiefly confined to the woods, and even there, to the tops of the tallest trees, where it is descried skipping from branch to branch, in pursuit of winged insects. Its note is a single screep, scarcely audible from below. It arrives in Pennsylvania about the 20th of April, and is first seen on the tops of the highest maples, darting about among the blossoms. As the woods thicken with leaves, it may be found pretty generally, being none of the least numerous of our summer birds. It is, however, most partial to woods in the immediate neighbourhood of creeks, swamps, or morasses, probably from the greater number of its favourite insects frequenting such places. It is also pretty generally diffused over the United States, having myself met with it in most quarters of the Union; though its nest has hitherto defied all my researches.

This bird may be considered as occupying an intermediate station between the flycatchers and the warblers, having the manners of the former, and the bill, partially, of the latter. The nice gradations by which nature passes from one species to another, even in this department of the great chain of beings, will for ever baffle all the artificial rules and systems of man. And this truth every fresh discovery must impress more forcibly on the mind of the observing naturalist. These birds leave us early in September.

The black-poll warbler is five and a half inches long, and

* This is an aberrant *Sylvicola,* approaching *Setophaga* in the form and bristling of the bill, and also in the manners of the flycatchers.—ED.

eight and a half in extent; crown and hind head, black; cheeks, pure white; from each lower mandible runs a streak of small black spots, those on the side, larger; the rest of the lower parts, white; primaries, black, edged with yellow; rest of the wing, black, edged with ash; the first and second row of coverts, broadly tipt with white; back, ash, tinged with yellow ochre, and streaked laterally with black; tail, black, edged with ash, the three exterior feathers marked on the inner webs with white; bill, black above, whitish below, furnished with bristles at the base; iris, hazel; legs and feet, reddish yellow.

The female differs very little in plumage from the male.

LESSER REDPOLL. (*Fringilla linaria.*)

PLATE XXX.—FIG. 4.

Lath. ii. 305.—*Arct. Zool.* 379.—Le Sizeren, *Buff.* iv. 216. *Pl. enl.* 151, 2.— *Peale's Museum*, No. 6579.

LINARIA MINOR.—WILLOUGHBY.

Fringilla linaria, *Bonap. Synop.* p. 112.

THIS bird corresponds so exactly in size, figure, and colour of plumage, with that of Europe of the same name, as to place their identity beyond a doubt. They inhabit, during summer, the most northern parts of Canada, and still more remote northern countries, from whence they migrate at the commencement of winter. They appear in the Gennesee country with the first deep snow, and on that account are usually called by the title of snow birds. As the female is destitute of the crimson on the breast and forehead, and the young birds do not receive that ornament till the succeeding spring, such a small proportion of the individuals that form these flocks are marked with red, as to induce a general belief among the inhabitants of those parts that they are two different kinds associated together. Flocks of these birds have been occasionally seen in severe winters in the neighbourhood of

Philadelphia. They seem particularly fond of the seeds of the common alder, and hang, head downwards, while feeding, in the manner of the yellow bird. They seem extremely unsuspicious at such times, and will allow a very near approach without betraying any symptoms of alarm.

The specimen represented in the plate was shot, with several others of both sexes, in Seneca county, between the Seneca and Cayuga lakes. Some individuals were occasionally heard to chant a few interrupted notes, but no satisfactory account can be given of their powers of song.

This species extends throughout the whole northern parts of Europe, is likewise found in the remote wilds of Russia, was seen by Steller in Kamtschatka, and probably inhabits corresponding climates round the whole habitable parts of the northern hemisphere. In the Highlands of Scotland they are common, building often on the tops of the heath, sometimes in a low furze bush, like the common linnet, and sometimes on the ground. The nest is formed of light stalks of dried grass, intermixed with tufts of wool, and warmly lined with feathers. The eggs are usually four, white, sprinkled with specks of reddish.*

* I have not been able to procure American specimens of this bird, but comparing the description of Wilson and of Ord, there seems little doubt of their identity. Wilson is certainly confounding the mountain linnet (*L. montium*), when he says, " In the Highlands of Scotland they are common, building often on the tops of the heath, sometimes in a low furze bush, like the common linnet, and sometimes on the ground." This is exactly the habit of the mountain linnet, and Mr Ord is wrong in saying the young possess the crimson head ; I have many in my possession without it, and have shot them at all seasons ; they receive that mark at the commencement of the first breeding season, when the adult birds also receive an addition of plumage and lustre. They seem very fond of the beech, as well as of the birch and alder, and appear to find insects in the husks of the old mast, which they are constantly picking and looking into. I have found their nests also pretty frequently in a young fir plantation : it was in a low situation, but they were invariably lined with the wool of willow catkins. I shall here add Mr Selby's correct description of the manners of this species, which are in every way confirmed by my own observations. " It is only known in the

[Mr Ord has added to the description of Wilson as follows :
—" Contrary to the usual practice of Mr Wilson, he omitted

southern parts of Britain as a winter visitant, and is at that period
gregarious, and frequently taken in company with the other species by
the bird-catchers, by whom it is called the stone redpoll. In the
northern counties of England, and in Scotland, and its isles, it is
resident through the year. It retires, during the summer, to the underwood that covers the basis of many of our mountains and hills, and that
often fringes the banks of their precipitous streams, in which sequestered
situations it breeds. The nest is built in a bush or low tree (such as
willow, alder, or hazel), of moss and the stalks of dry grass, intermixed
with down from the catkin of the willow, which also forms the lining,
and renders it a particularly soft and warm receptacle for the eggs and
young. From this substance being a constant material of the nest, it
follows, that the young are produced late in the season, and are seldom
able to fly before the end of June, or the beginning of July. The eggs
are four or five in number ; their colour, pale bluish green, spotted with
orange brown, principally towards the larger end. In winter, the lesser
redpoll descends to the lower grounds, in considerable flocks, frequenting
woods and plantations, more especially such as abound in birch or alder
trees, the catkins of which yield it a plentiful supply of food. When
feeding, its motion affords both interest and amusement ; since, in order
to reach the catkins, which generally grow near the extremities of the
smaller branches, it is obliged, like the titmouse, to hang with its back
downwards, and assume a variety of constrained attitudes, and, when
thus engaged, it is so intent upon its work, as frequently to allow itself to
be taken by a long stick smeared with bird-lime, in which way I have
occasionally captured it when in want of specimens for examination.
It also eats the buds of trees, and (when in flocks) proves in this way
seriously injurious to young plantations. Its call note is very frequently
repeated when on wing, and by this it may be always distinguished
from the other species. The notes it produces during the pairing
season, although few, and not delivered in continuous song, are sweet
and pleasing."

"This bird is widely diffused through all the northern parts of
Europe ; inhabits Northern Asia as far as Siberia and Kamtschatka ;
and is also abundant in North America."

The authors of the " Northern Zoology " describe another bird allied
to the linnets, of which one individual only was obtained in the last
northern expedition. It is said to be new, and is described as *Linaria*
(*Leocosticte*) *Teprocotis*, Sw. grey-crowned linnet. It is an aberrant form
of *Linaria*, which Mr Swainson proposes to designate under the above
sub-generic title.— ED.

to furnish a *particular* description of this species. But this sup-
plementary notice would not have been considered necessary, if
our author had not fallen into a mistake respecting the markings of the female and the young male ; the former of which
he describes as ' destitute of the crimson on the forehead,' and
the latter, ' not receiving that ornament till the succeeding
spring.' When Mr Wilson procured his specimens, it was in
the autumn, previously to their receiving their perfect winter
dress ; and he was never afterwards aware of his error, owing
to the circumstance of these birds seldom appearing in the
neighbourhood of Philadelphia. Considerable flocks of them,
however, have visited us this winter (1813–14) ; and we have
been enabled to procure several fine specimens of both sexes,
from the most perfect of which we have taken the following
description. We will add, that having had the good fortune to
observe a flock, consisting of nearly an hundred, within a few
feet of them, as they were busily engaged in picking the seeds
of the wild orache,* we can, with confidence, assert, that they
all had the red patch on the crown ; but there were very few
which had the red rump and breast : the young males, it is
probable, are not thus marked until the spring, and the females
are destitute of that ornament altogether.

" The lesser redpoll is five inches and a quarter in length,
and eight inches and a half in breadth ; the bill is pale yellow,
ridged above and below with dark horn colour, the upper
mandible projecting somewhat over the lower at the tip ; irides,
dark hazel ; the nostrils are covered with recumbent, hair-like
feathers, of drab colour ; a line of brown extends from the
eyes, and encircles the base of the bill, forming, in some
specimens, a patch below the chin ; the crown is ornamented
with a pretty large spot of deep shining crimson ; the throat,
breast, and rump, stained with the same, but of a more delicate
red ; the belly is of a very pale ash, or dull white ; the sides
are streaked with dusky ; the whole upper parts are brown or
dusky ; the plumage, edged with yellowish white and pale ash,

* Atriplex hastata, Linn.

1. American Crossbill. 2. Female. 3. White-winged Crossbill. 4. White-crowned Bunting. 5. Bay-winged B.

the latter most predominant near the rump ; wings and tail, dusky ; the latter is forked, and consists of twelve feathers edged with white ; the primaries are very slightly tipt and edged with white, the secondaries more so ; the greater and lesser coverts are also tipt with white, forming the bars across the wings ; thighs, cinereous ; legs and feet, black ; hind claw, considerably hooked, and longer than the rest. The female is less bright in her plumage above ; and her under parts incline more to an ash colour ; the spot on her crown is of a golden crimson, or reddish saffron colour. One male specimen was considerably larger than the rest ; it measured five inches and three quarters in length, and nine inches and a quarter in extent ; the breast and rump were tawny ; its claws were uncommonly long, the hind one measured nearly three-eighths of an inch ; and the spot on the crown was of a darker hue than that of the rest.

" The call of this bird exactly resembles that of the *Fringilla iristis,* or common yellow bird of Pennsylvania. The redpolls linger in the neighbourhood of Philadelphia until about the middle of April; but whither they retire for the business of incubation, we cannot determine. In common with almost all our finches, the redpolls become very fat, and are then accounted delicious eating. During the last winter, many hundreds of them were exposed to sale in the Philadelphia market, and were readily purchased by those epicures, whose love of variety permits no delicacy to escape them."]

AMERICAN CROSSBILL. (*Curvirostra Americana.*)

PLATE XXXI.—Fig. 1. male ; Fig. 2. female.

Peale's Museum, No. 5640.

LOXIA CURVIROSTRA?—Bonaparte.*

Loxia curvirostra, *Bonap. Synop.* p. 117.

On first glancing at the bill of this extraordinary bird, one is apt to pronounce it deformed and monstrous ; but on atten-

* Brisson first limited the crossbills to a genus, and proposed for them the title *Loxia,* which has been adopted by most ornithologists. *Cruci-*

tively observing the use to which it is applied by the owner, and the dexterity with which he detaches the seeds of the pine tree from the cone, and from the husks that enclose them, we

rostra and *Curvirostra*, have also been formed for it from the shape of the bill; but ought to be rejected, from the priority of the former. They are a very limited group, being composed of at most four species, provided that of America be proved distinct, or one differing from those of Europe be found in the former continent. Their distribution appears to extend pretty generally over the north of Europe, decreasing in numbers to the south, and over North America. In form, all the members are similar. They are endowed with considerable power of flight ; are of a thick, stout make, and in addition to the curiously formed bill, possess scansorial habits, using their bills and feet to disengage the seeds from the fir cones, when in confinement, holding their food like a parrot in the latter member, and by the same means climbing about the wires of the cage.

Regarding the identity of our author's species with that of this country, I am uncertain, not having a specimen of the bird from America. Wilson thinks it distinct, and I have been told the same thing by Audubon. On the other hand, we have the authority of Bonaparte, who thus writes in his *Observations on Wilson's Nomenclature :*—" I think Wilson was in error when he considered this bird a new species, and stated that it differs considerably from the European. He probably compared it with the *L. pytiopsittacus,* and not with the *curvirostra,* with which latter it is identical. Wilson's new names must therefore be rejected, and the name of *Loxia curvirostra* must be restored to this bird." Our author was also incorrect in remarking, that " the young males, as is usual with most other birds, very much resemble the female." The fact is, that the young of all the crossbills, as well as that of *Pyrrhula enucleator,* contrary to the habit of the generality of birds, lose their red colour as they advance in age, instead of gaining an additional brilliancy of plumage. The figure which our author gives as that of an adult male, represents a young bird of about one year, and his supposed female is a remarkably fine adult male.

The species of this group, then, are,—*L. pytiopsittacus,* or parrot-billed crossbill of Europe, and which Bonaparte also hints the possibility of finding in America, a circumstance I should think very likely,—the *L. leucoptera,* and the *L. curvirostra ;* but I fear we must remain uncertain whether the last constitutes one or two, until the examination of numerous specimens from both countries decide the point. The haunts of our common species in Europe are the immense northern pine forests, where their chief food is the seeds of the fir cones ; from thence, after breeding, they appear to migrate to various parts southward, in comparatively

are obliged to confess, on this, as on many other occasions, where we have judged too hastily of the operations of Nature, that no other conformation could have been so excellently adapted to the purpose; and that its deviation from the common form, instead of being a defect or monstrosity, as the celebrated French naturalist insinuates, is a striking proof of the wisdom and kind superintending care of the great Creator.

small flocks, at uncertain intervals. This is the case with those which visit Britain. They must hatch very early, arriving in this country by the middle of June; the females at that time bear all the marks of incubation, but have never yet been authentically proved to breed in this country, as supposed by Mr Knap, from the bareness of the breast. They descend, at the same season, to the orchards, where they do considerable damage, by splitting the apples for the pips, thus leaving the fruit useless, and incapable of farther growth; and, at the same time, giving us a good instance of the power of their bills. Some old writers accuse them of visiting Worcester and Herefordshire, "in great flocks, for the sake of the seeds of the apple. Repeated persecution on this account perhaps lessened their numbers, and their depredations at the present day are unnoticed or unknown:" their visitations, at least, are less frequent; for a later writer in *Loudon's Magazine* observes, that, in 1821, and the commencement of 1822 (the same season of their great appearance mentioned by Mr Selby), a large flock of crossbills frequented some fir groves at Cothoridge, near Worcester, where they used to visit the same spot pretty regularly twice a-day, delighting chiefly on the Weymouth pines. When feeding, they seem in this country, as well as with our author, to be remarkably tame, or so much engrossed with their food, as to be unmindful of danger. Montague relates, that a birdcatcher at Bath had taken a hundred pairs in the month of June and July, 1791; and so intent were these birds when picking out the seeds of a cone, that they would suffer themselves to be caught with a hair noose at the end of a long fishing-rod. In 1821, this country was visited with large flocks; they appeared in June, and gradually moved northward, as they were observed by Mr Selby in September among the fir tracts of Scotland, after they had disappeared to the southward of the river Tweed. In 1828, a pretty large flock visited the vicinity of Ambleside, Westmoreland. Their favourite haunt was a plantation of young larches, where they might be seen disporting almost every day, particularly between the hours of eleven and one.

I have quoted no synonyms which belong to our British species. The American birds appear to me much smaller; that is, to judge from our author's plate, and the usually correct drawings of Mr Audubon.—ED.

This species is a regular inhabitant of almost all our pine forests situated north of 40°, from the beginning of September to the middle of April. It is not improbable that some of them remain during summer within the territory of the United States to breed. Their numbers must, however, be comparatively few, as I have never yet met with any of them in summer; though lately I took a journey to the Great Pine Swamp beyond Pocano mountain, in Northampton county, Pennsylvania, in the month of May, expressly for that purpose; and ransacked, for six or seven days, the gloomy recesses of that extensive and desolate morass, without being able to discover a single crossbill. In fall, however, as well as in winter and spring, this tract appears to be their favourite rendezvous; particularly about the head waters of the Lehigh, the banks of the Tobyhanna, Tunkhannock, and Bear Creek, where I have myself killed them at these seasons. They then appear in large flocks, feeding on the seeds of the hemlock and white pine, have a loud, sharp, and not unmusical note; chatter as they fly; alight, during the prevalence of deep snows, before the door of the hunter, and around the house, picking off the clay with which the logs are plastered, and searching in corners where urine, or any substance of a saline quality, had been thrown. At such times they are so tame as only to settle on the roof of the cabin when disturbed, and a moment after descend to feed as before. They are then easily caught in traps; and will frequently permit one to approach so near as to knock them down with a stick. Those killed and opened at such times are generally found to have the stomach filled with a soft greasy kind of earth or clay. When kept in a cage, they have many of the habits of the parrot; often climbing along the wires; and using their feet to grasp the cones in, while taking out the seeds.

This same species is found in Nova Scotia, and as far north as Hudson's Bay, arriving at Severn River about the latter end of May: and, according to accounts, proceeding farther north to breed. It is added by Pennant, that "they return at the first setting in of frost."

Hitherto this bird has, as usual, been considered a mere variety of the European species; though differing from it in several respects, and being nearly one-third less, and although the singular conformation of the bill of these birds, and their peculiarity of manners, are strikingly different from those of the grosbeaks, yet many, disregarding these plain and obvious discriminations, still continue to consider them as belonging to the genus *Loxia;* as if the particular structure of the bill should, in all cases but this, be the criterion by which to judge of a species; or perhaps, conceiving themselves the wiser of the two, they have thought proper to associate together what Nature has, in the most pointed manner, placed apart.

In separating these birds, therefore, from the grosbeaks, and classing them as a family by themselves, substituting the specific for the generic appellation, I have only followed the steps and dictates of that great Original, whose arrangements ought never to be disregarded by any who would faithfully copy her.

The crossbills are subject to considerable changes of colour; the young males of the present species being, during the first season, olive yellow, mixed with ash; then bright greenish yellow, intermixed with spots of dusky olive, all of which yellow plumage becomes, in the second year, of a light red, having the edges of the tail inclining to yellow. When confined in a cage, they usually lose the red colour at the first moulting, that tint changing to a brownish yellow, which remains permanent. The same circumstance happens to the purple finch and pine grosbeak, both of which, when in confinement, exchange their brilliant crimson for a motley garb of light brownish yellow; as I have had frequent opportunities of observing.

The male of this species, when in perfect plumage, is five inches and three quarters long, and nine inches in extent;

the bill is a brown horn colour, sharp, and single-edged
towards the extremity, where the mandibles cross each other;
the general colour of the plumage is a red-lead colour,
brightest on the rump, generally intermixed on the other
parts with touches of olive; wings and tail, brown black, the
latter forked, and edged with yellow; legs and feet, brown;
claws, large, much curved, and very sharp; vent, white,
streaked with dark ash; base of the bill, covered with recumbent down, of a pale brown colour; eye, hazel.

The female is rather less than the male; the bill of a paler
horn colour; rump, tail-coverts, and edges of the tail, golden
yellow; wings and tail, dull brownish black; the rest of the
plumage, olive yellow mixed with ash; legs and feet, as in
the male. The young males, during the first season, as is
usual with most other birds, very much resemble the female.
In moulting, the males exchange their red for brownish
yellow, which gradually brightens into red. Hence, at different seasons, they differ greatly in colour.

WHITE-WINGED CROSSBILL. (*Curvirostra leucoptera.*)

PLATE XXXI.—Fig. 3.

Turton, Syst. i. p. 515.

LOXIA LEUCOPTERA.—Gmelin.*

Loxia leucoptera, *Bonap. Synop.* p. 117.

THIS is a much rarer species than the preceding; though
found frequenting the same places, and at the same seasons;

* Bonaparte has fulfilled Wilson's promise, and figured the female of
this species, with some valuable remarks regarding its first discovery
and habits, which will be found in Vol. III. From these it appears to
be very like its congeners, performing its migrations at uncertain periods
and in various abundance, enjoying the pine forests, though not farther
known by any destructive propensities among orchards. It may be
looked upon yet as exclusively North American. The only record of

differing, however, from the former in the deep black wings and tail, the large bed of white on the wing, the dark crimson of the plumage ; and a less and more slender conformation of body. The bird represented in the plate was shot in the neighbourhood of the Great Pine Swamp, in the month of September, by my friend Mr Ainsley, a German naturalist, collector in this country for the emperor of Austria. The individual of this species, mentioned by Turton and Latham, had evidently been shot in moulting time. The present specimen was a male in full and perfect plumage.

The white-winged crossbill is five inches and a quarter long, and eight inches and a quarter in extent ; wings and tail, deep black, the former crossed with two broad bars of white ; general colour of the plumage dark crimson, partially spotted with dusky ; lores and frontlet, pale brown ; vent, white, streaked with black ; bill, a brown horn colour, the mandibles crossing each other as in the preceding species, the lower sometimes bending to the right, sometimes to the left, usually to the left in the male, and to the right in the female of the American crossbill. The female of the present species will be introduced as soon as a good specimen can be obtained, with such additional facts relative to their manners as may then be ascertained.

its being found in another country is in extracts from the minute book of the Linnæan Society for 1803. " Mr Templeton, A.L.S. of Orange-grove, near Belfast, in a letter to Mr Dawson Turner, F.L.S., mentions that the white-winged crossbill, *Loxia falcirostra* of Latham, was shot within two miles of Belfast, in the month of January 1802. It was a female, and perfectly resembled the figure in Dixon's *Voyage to the North-west Coast of America.*" Such is the only record we have of this bird as a British visitor. When Ireland becomes more settled, and her naturalists more devoted to actual observation, we may hear more of *L. leucoptera, Cypselus melba,* &c. Bonaparte, in his description of the female, has entered fully into the reasons for adopting the specific name of *leucoptera.*—ED.

WHITE-CROWNED BUNTING. (*Emberiza leucophrys.*)

PLATE XXXI.—Fig. 4.

Turton, Syst. p. 536.—*Peale's Museum,* No. 6587.

ZONOTRICHIA LEUCOPHRYS.—Swainson.

Fringilla leucophrys, *Bonap. Synop.* p. 107.—Fringilla (Zonotrichia) leucophrys, *North. Zool.* ii. p. 255.

This beautifully marked species is one of the rarest of its tribe in the United States, being chiefly confined to the northern districts, or higher interior parts of the country, except in severe winters, when some few wanderers appear in the lower parts of the state of Pennsylvania. Of three specimens of this bird, the only ones I have yet met with, the first was caught in a trap near the city of New York, and lived with me several months. It had no song, and, as I afterwards discovered, was a female. Another, a male, was presented to me by Mr Michael of Lancaster, Pennsylvania. The third, a male, and in complete plumage, was shot in the Great Pine Swamp, in the month of May, and is faithfully represented in the plate. It appeared to me to be unsuspicious, silent, and solitary ; flitting in short flights among the underwood and piles of prostrate trees, torn up by a tornado, that some years ago passed through the swamp. All my endeavours to discover the female or nest were unsuccessful.

From the great scarcity of this species, our acquaintance with its manners is but very limited. Those persons who have resided near Hudson's Bay, where it is common, inform us, that it makes its nest in June, at the bottom of willows, and lays four chocolate-coloured eggs. Its flight is said to be short and silent ; but, when it perches, it sings very melodiously.*

The white-crowned bunting is seven inches long, and ten inches in extent ; the bill, a cinnamon brown ; crown, from

---

* Arctic Zoology.

the front to the hind head, pure white, bounded on each side by a stripe of black proceeding from each nostril ; and these again are bordered by a stripe of pure white passing over each eye to the hind head, where they meet ; below this, another narrow stripe of black passes from the posterior angle of the eye, widening as it descends to the hind head ; chin, white ; breast, sides of the neck, and upper parts of the same, very pale ash ; back, streaked laterally with dark rusty brown and pale bluish white ; wings, dusky, edged broadly with brown ; the greater and lesser coverts tipt broadly with white, forming two handsome bands across the wing ; tertials, black, edged with brown and white ; rump and tail-coverts, drab, tipt with a lighter tint ; tail, long, rounded, dusky, and edged broadly with drab ; belly, white ; vent, pale yellow ochre ; legs and feet, reddish brown ; eye, reddish hazel ; lower eyelid, white.

The female may easily be distinguished from the male, by the white on the head being less pure, the black also less in extent, and the ash on the breast darker ; she is also smaller in size.

There is a considerable resemblance between this species and the white-throated sparrow, already described in this work. Yet they rarely associate together ; the latter remaining in the lower parts of Pennsylvania in great numbers, until the beginning of May, when they retire to the north and to the high inland regions to breed ; the former inhabiting much more northern countries, and though said to be common in Canada, rarely visiting this part of the United States.

BAY-WINGED BUNTING. (*Emberiza graminea.*)

PLATE XXXI.—Fig. 5.

Grass Finch, *Arct. Zool.* No. 253.—*Lath.* iii. 273.—*Turton, Syst.* i. p. 565.
ZONOTRICHIA GRAMINEA.—Swainson.
Fringilla graminea, *Bonap. Synop.* p. 108.—Fringilla (Zonotrichia) graminea,
North. Zool. ii. p. 254.

The manners of this bird bear great affinity to those of the common bunting of Britain. It delights in frequenting grass and clover fields, perches on the tops of the fences, singing, from the middle of April to the beginning of July, with a clear and pleasant note, in which particular it far excels its European relation. It is partially a bird of passage here, some leaving us, and others remaining with us, during the winter. In the month of March I observed them numerous in the lower parts of Georgia, where, according to Mr Abbot, they are only winter visitants. They frequent the middle of fields more than hedges or thickets; run along the ground like a lark, which they also resemble in the great breadth of their wings. They are timid birds, and rarely approach the farmhouse.

Their nest is built on the ground, in a grass or clover field, and formed of old withered leaves and dry grass, and lined with hair. The female lays four or five eggs, of a grayish white. On the first week in May, I found one of their nests with four young, from which circumstance I think it probable that they raise two or more broods in the same season.

This bird measures five inches and three quarters in length, and ten inches and a half in extent; the upper parts are cinereous brown, mottled with deep brown or black; lesser wing-coverts, bright bay; greater, black, edged with very pale brown; wings, dusky, edged with brown; the exterior primary, edged with white; tail, sub-cuneiform, the outer feather white on the exterior edge, and tipt with white; the next, tipt and edged for half an inch with the same; the rest, dusky, edged with pale brown; bill, dark brown above, paler below; round the eye is a narrow circle of white; upper part of the breast, yellowish white, thickly streaked with pointed spots of black that pass along the sides; belly and vent, white; legs and feet, flesh-coloured; third wing-feather from the body, nearly as long as the tip of the wing when shut.

I can perceive little or no difference between the colours and markings of the male and female.

1. *Snow Owl*. 2. *Male Sparrow Hawk*.

32.

Drawn from Nature by A. Wilson

Engraved by W.H.Lizars

SNOW OWL. (*Strix nyctea.*)

PLATE XXXII.—Fig. 1. Male.

Lath. i. 132. No. 17.—*Buff.* i. 387.—Great White Owl, *Edw.* 61.—Snowy Owl,
Arct. Zool. 233. No. 121.—*Peale's Museum*, No. 458.

SURNIA NYCTEA.—Dumeril.

Snowy Owl, *Mont. Orn. Dict. Supp.*—*Bewick's Brit. Birds, Supp.*—Snowy Owl,
Strix nyctea, *Selby's Brit. Orn.* p. 58, pl. 23.—Strix nyctea, *Temm. Man.* i.
p. 82.—*Flem. Br. Anim.* p. 58.—*Bonap. Synop.* p. 36.—*North. Zool.* ii. p. 88.

The snow owl represented in the plate is reduced to half its
natural size. To preserve the apparent magnitude, the other
accompanying figures are drawn by the same scale.

This great northern hunter inhabits the coldest and most
dreary regions of the northern hemisphere on both continents.
The forlorn mountains of Greenland, covered with eternal
ice and snows, where, for nearly half the year, the silence of
death and desolation might almost be expected to reign,
furnish food and shelter to this hardy adventurer; whence
he is only driven by the extreme severity of weather towards
the sea-shore. He is found in Lapland, Norway, and the
country near Hudson's Bay, during the whole year; is said to
be common in Siberia, and numerous in Kamtschatka. He is
often seen in Canada and the northern districts of the United
States; and sometimes extends his visits to the borders of
Florida. Nature, ever provident, has so effectually secured
this bird from the attacks of cold, that not even a point is left
exposed. The bill is almost completely hid among a mass of
feathers that cover the face; the legs are clothed with such
an exuberance of long, thick, hair-like plumage, as to appear
nearly as large as those of a middle-sized dog, nothing being
visible but the claws, which are large, black, much hooked,
and extremely sharp. The whole plumage below the surface
is of the most exquisitely soft, warm, and elastic kind, and so
closely matted together as to make it a difficult matter to
penetrate to the skin.

The usual food of this species is said to be hares, grouse, rabbits, ducks, mice, and even carrion. Unlike most of his tribe, he hunts by day as well as by twilight, and is particularly fond of frequenting the shores and banks of shallow rivers, over the surface of which he slowly sails, or sits on a rock a little raised above the water, watching for fish. These he seizes with a sudden and instantaneous stroke of the foot, seldom missing his aim. In the more southern and thickly settled parts, he is seldom seen ; and when he appears, his size, colour, and singular aspect, attract general notice.*

In the month of October, I met with this bird on Oswego River, New York State, a little below the Falls, vigilantly watching for fish. At Pittsburg, in the month of February, I saw another, which had been shot in the wing some time before. At a place on the Ohio, called Long Reach, I examined another, which was the first ever recollected to have been seen there. In the town of Cincinnati, State of Ohio, two of these birds alighted on the roof of the court house, and alarmed the whole town. A people more disposed to superstition would have deduced some dire or fortunate prognostication from their selecting such a place ; but the only solicitude was how to get possession of them, which, after several volleys, was at length effected. One of these, a female,

* The following observations by Mr Bree of Allesly, taken from *Loudon's Magazine of Natural History*, will show that other owls also fish for their prey :—" Probably it may not be generally known to naturalists, that the common brown owl *(Strix stridula)*, is in the habit—occasionally, at least—of feeding its young with live fish,—a fact which I have ascertained beyond doubt. Some years since several young owls were taken from the nest, and placed in a yew tree, in the rectory garden here. In this situation, the parent birds repeatedly brought them live fish, bull heads *(Cottus gobbius)*, and loach *(Cobitis barbatula)*, which had doubtless been procured from a neighbouring brook, in which these species abound. Since the above period, I have, upon more than one occasion, found the same fish, either whole or in fragments, lying under the trees on which I have observed the young owls to perch after they have left the nest, and where the old birds were accustomed to feed them."—ED.

I afterwards examined, when on my way through that place to New Orleans. Near Bairdstown, in Kentucky, I met with a large and very beautiful one, which appeared to be altogether unknown to the inhabitants of that quarter, and excited general surprise. A person living on the eastern shore of Maryland, shot one of these birds a few months ago, a female; and, having stuffed the skin, brought it to Philadelphia, to Mr Peale, in expectation, no doubt, of a great reward. I have examined eleven of these birds within these fifteen months last past, in different and very distant parts of the country, all of which were shot either during winter, late in the fall, or early in spring; so that it does not appear certain whether any remain during summer within the territory of the United States; though I think it highly probable that a few do, in some of the more northern inland parts, where they are most numerous during winter.

The colour of this bird is well suited for concealment, while roaming over the general waste of snows; and its flight strong and swift, very similar to that of some of our large hawks. Its hearing must be exquisite, if we judge from the largeness of these organs in it; and its voice is so dismal, that, as Pennant observes, it adds horror even to the regions of Greenland, by its hideous cries, resembling those of a man in deep distress.

The male of this species measures twenty-two inches and a half in length, and four feet six inches in breadth; head and neck, nearly white, with a few small dots of dull brown interspersed; eyes, deep sunk, under projecting eyebrows, the plumage at their internal angles, fluted or prest in, to admit direct vision; below this it bristles up, covering nearly the whole bill; the irides are of the most brilliant golden yellow, and the countenance, from the proportionate smallness of the head, projection of the eyebrow, and concavity of the plumage at the angle of the eye, very different from that of any other of the genus; general colour of the body, white, marked with lunated spots of pale brown above, and with semicircular dashes below; femoral feathers, long, and legs covered, even over the claws, with long shaggy hair-like down, of a dirty white; the claws, when exposed, appear large, much hooked, of a black colour, and extremely sharp pointed; back, white; tail, rounded at the end, white, slightly dotted with pale brown near the tips; wings, when closed, reach near the extremity of the tail; vent-feathers, large, strong shafted, and extending also to the point of the tail; upper part of the breast and belly, plain white; body, very broad and flat.

The female, which measures two feet in length, and five feet two inches in extent, is covered more thickly with spots of a much darker colour than those on the male; the chin, throat, face, belly, and vent, are white; femoral feathers, white, long, and shaggy, marked with a few heart-shaped spots of brown; legs, also covered to the claws with long white hairy down; rest of the plumage white, every feather spotted or barred with dark brown, largest on the wing-quills, where they are about two inches apart; fore part of the crown, thickly marked with roundish black spots; tail, crossed with bands of broad brownish spots; shafts of all the plumage, white; bill and claws, as in the male, black; third and fourth wing-quill the longest; span of the foot, four inches.

From the various individuals of these birds which I have examined, I have reason to believe that the male alone approaches nearly to white in his plumage, the female rarely or never. The bird from which the figure in the plate was drawn, was killed at Egg Harbour, New Jersey, in the month of December. The conformation of the eye of this bird forms a curious and interesting subject to the young anatomist. The globe of the eye is immoveably fixed in its socket, by a strong elastic hard cartilaginous case, in form of a truncated cone; this case being closely covered with a skin, appears at first to be of one continued piece; but, on removing the exterior membrane, it is found to be formed of fifteen pieces, placed like the staves of a cask, overlapping a little at the base, or narrow end, and seem as if capable of being enlarged or contracted, perhaps by the muscular membrane with which they are encased. In five other different species of owls, which I have since examined, I found nearly the same conformation of this organ, and exactly the same number of staves. The eye being thus fixed, these birds, as they view different objects, are always obliged to turn the head; and Nature has so excellently adapted their neck to this purpose, that they can, with ease, turn it round, without moving the body, in almost a complete circle.*

* In prefixing the generic appellations to this curious family, I must at once confess my inability to do it in a manner satisfactory to myself. They have been yet comparatively unstudied ; and the organs of greatest importance have been seemingly most neglected. Neither my own collection, nor those accessible in Britain, contain sufficient materials to decide upon : I will, therefore, consider any attempt now to divide them in the words of Mr Swainson, "as somewhat speculative, and certainly not warranted by any evidence that has yet been brought forward on the subject." The names are applied, then, on the authority of ornithologists of high standing.

This owl, and some others, will form the genus *Noctua* of Savigny and Cuvier, and are closely allied to the *Surnia* of Dumeril. In fact, the characters of the latter appear to me to agree better than those of *Noctua ;* and Lesson says, "Les chevèches ne se font pas reconnaître très nettement des chouettes." The snowy owl feeds by day as well as by night, and is much more active than the night feeding birds ; it approaches nearer to the hawk owls. The head is less ; the tail and wings, elongated, and the plumage is more compact and rigid. It appears to extend as far north in America as any inhabited country, and is found in the coldest districts of Europe. It is also mentioned by Pennant to reach beyond the Asiatic frontier to the hot latitude of Astracan (*a contrast, if it should turn out the same species*), and was discovered to breed in Orkney and Shetland by Mr Bullock, who procured several specimens. Its visits to the mainland of Britain are, again, more rare ; indeed, I believe one of the only instances on record is that of a male and female killed near Rothbury in Northumberland, in January 1823, —a winter remarkable for a severe snow storm. They were killed on an open moor, in a wild and rocky part of the country, and were generally seen perched upon the snow, or upon some large stone projecting from it. Both now form beautiful specimens in the collection of Mr Selby.

They become very familiar in winter, approaching close to the dwellings of the Indians. In Lapland they are shot with ball when hunting after moles and lemmings, and in that country, like many other owls,

AMERICAN SPARROW HAWK. (*Falco sparverius.*)

PLATE XXXII.—Fig. 2, Male.

Little Hawk, *Arct. Zool.* 211, No. 110.—Emerillon de Cayenne, *Buff.* i. 291.
Pl. enl. No. 444.—*Lath.* i. 110.—*Peale's Museum*, No. 340.

FALCO SPARVERIUS.—Linnæus.

Falco sparverius, *Bonap. Synop.* p. 27.—Falco sparverius, Little Rusty-crowned
Falcon, *North. Zool.* ii. p. 31.

The female of this species has been already figured and described in Vol. I. of this work. As they differ considerably in the markings of their plumage, the male is introduced here, drawn to one half its natural size, to conform with the rest of the figures on the plate.

The male sparrow hawk measures about ten inches in length, and twenty-one in extent; the whole upper parts of the head are of a fine slate blue, the shafts of the plumage being black, the crown excepted, which is marked with a spot of bright rufous; the slate tapers to a point on each side of the neck; seven black spots surround the head, as in the female, on a

they are looked upon with superstition. They utter a sound at night when perched, like the grunting of pigs, which, by the common and uninformed people, is thought to be some apparition or spectre. By Hearne the snow owl is said to be known to watch the grouse shooters a whole day, for the purpose of sharing in the spoil. On such occasions, it perches on a high tree, and when a bird is shot, skims down and carries it off before the sportsman can get near it. We have the following remarks by Dr Richardson in the "Northern Zoology":—
"Frequents most of the arctic lands that have been visited, but retires with the ptarmigan, on which it preys, to more sheltered districts in winter; hunts by day. When I have seen it on the barren grounds, it was generally squatting on the earth; and if put up, it alighted again after a short flight, but was always so wary as to be approached with difficulty. In woody districts it shows less caution. I have seen it pursue the American hare on the wing, making repeated strokes at the animal with its feet. In winter, when this owl is fat, the Indians and white residents in the Fur Countries esteem it to be good eating. Its flesh is delicately white." By the Cree Indians it is called Wapow-keethoo, or Wapahoo; by the Esquimaux, Oookpēēguak; by the Norwegians, Lemensgrüs and Gysfugl; by the Swedes, Harfang.—Ed.

reddish white ground, which also borders each sloping side of the blue; front, lores, line over and under the eye, chin, and throat, white; femoral and vent-feathers, yellowish white; the rest of the lower parts, of the same tint, each feather being streaked down the centre with a long black drop; those on the breast, slender, on the sides, larger; upper part of the back and scapulars, deep reddish bay, marked with ten or twelve transverse waves of black; whole wing-coverts and ends of the secondaries, bright slate, spotted with black; primaries, and upper half of the secondaries, black, tipt with white, and spotted on their inner vanes with the same; lower part of the back, the rump, and tail-coverts, plain bright bay; tail rounded, the two exterior feathers, white, their inner vanes beautifully spotted with black; the next, bright bay, with a broad band of black near its end, and tipt for half an inch with yellowish white; part of its lower exterior edge, white, spotted with black, and its opposite interior edge, touched with white; the whole of the others are very deep red bay, with a single broad band of black near the end, and tipt with yellowish white; cere and legs, yellow; orbits, the same; bill, light blue; iris of the eye, dark, almost black; claws, blue black.

The character of this corresponds with that of the female, given at large in Vol. I. p. 262. I have reason, however, to believe, that these birds vary considerably in the colour and markings of their plumage during the first and second years; having met with specimens every way corresponding with the above, except in the breast, which was a plain rufous white, without spots; the markings on the tail also differing a little in different specimens. These I uniformly found, on dissection, to be males; from the stomach of one of which I took a considerable part of the carcass of a robin (*Turdus migratorius*), including the unbroken feet and claws; though the robin actually measures within half an inch as long as the sparrow hawk.*

* Bonaparte has separated the small American falcons from the larger kinds, characterising the group as having the wings shorter than the tail,

ROUGH-LEGGED FALCON.   (*Falco lagopus.*)

PLATE XXXIII.—Fig. 1.

Arct. Zool. p. 200, No. 92.—*Lath.* i. 75.—*Peale's Museum*, No. 116.

BUTEO LAGOPUS.—Bechstein?

Rough-legged Falcon, *Mont. Ornith. ¦Dict. Supp.*—*Bew. Br. Birds, Supp.*—
Rough-legged Buzzard, *Selby's Illust. Br. Ornith.* i. p. 20. pl. 7.—Falco
lagopus, *Temm. Man.* i. p. 65.—*Bonap. Synop.* p. 32.—Buteo lagopus, *Flem.
Br. Anim.* p. 54.—*North. Zool.* ii. p. 52.

This handsome species, notwithstanding its formidable size
and appearance, spends the chief part of the winter among

tarsi scutellated ; and Mr Swainson says, that the group seems natural,
differing somewhat in their manners from the larger falcons, and having
analogies in their habits to the shrikes.

With both these we agree. It is long since we thought the general
form and habits of our common kestrel—analogous to Wilson's bird in
Europe—differed from those of the true falcons, as much, certainly, as
Astur does from *Accipiter*, and both should be only by subordinate divisions. The manner of suspending itself in the air is exactly similar to
that of our windhover ; and I am not aware that this peculiar manner
of hunting is made use of by any other of the *Falconidæ*, with the exception of the kestrels, that is, those of Europe or Africa, *F. rupicola, tinun-
culoides,* &c. The true falcons survey the ground by extensive sweeps,
or a rapid flight, and stoop at once on their prey with the velocity and
force of lightning; the others quietly watch their quarry when suspended
or perched on a bare eminence or tree in the manner described, and take
it by surprise. Insects, reptiles, and small animals form part of their
food ; and to the old falconists they were known by the name of
" Ignoble." The whole of the kestrels are very familiar, easily tamed,
and when in confinement become even playful. Their great breeding-
place is steep rocks, clothed with ivy, and fringed with the various wild
plants incident to the different climes ; in the chinks and hearts of these
they nestle, often in security from any clamberer that has not the assistance of a rope ; though the appearance of a stranger immediately calls
forth peculiarly shrill and timid notes of alarm. When the young are
hatched, and partly advanced, they may be seen stretching out from their
hole, and, on the appearance of their parent, mutual greetings are heard,
and in a tone at once different from those before mentioned. Our native
species, in addition to rocks, delights in ruined buildings as a breeding-
place ; and it is remarkable, that perhaps more kestrels build and bring

1 Rough-legged Falcon. 2. Barred Owl. 3. Short eared C.   Engraved by W.H.Lizars

33.

our low swamps and meadows, watching for mice, frogs, lame ducks, and other inglorious game. Twenty or thirty individuals of this family have regularly taken up their winter quarters, for several years past, and probably long anterior to that date, in the meadows below this city, between the rivers Delaware and Schuylkill, where they spend their time watching along the dry banks like cats ; or sailing low and slowly over the surface of the ditches. Though rendered shy from the many attempts made to shoot them, they seldom fly far, usually from one tree to another at no great distance, making a loud squeeling as they arise, something resembling the neighing of a young colt, though in a more shrill and savage tone.

The bird represented in the plate was one of this fraternity,

to maturity their young in *London*, than in any space of the same dimensions : the breeding-places there are the belfries of the different churches, where neither the bustle beneath, nor the *jingle* of the bells, seems to have any effect upon them.

We have the following characteristic observations on this species in the "Northern Zoology" :—

"In the vicinity of Carlton House, where the plains are beautifully ornamented by numerous small clumps of aspens, that give a rich picturesque effect to the landscape, which I have never seen equalled in an English park, this small falcon was frequently discovered, perched upon the most lofty tree in the clump, sitting with his eye apparently closed, but, nevertheless, sufficiently awake to what was going on, as it would occasionally evince, by suddenly pouncing upon any small bird that happened to come within its reach. It is the least shy of any of the American hawks ; and, when on its perch, will suffer the fowler to advance to the foot of the tree, provided he has the precaution to make a slow and devious approach. He is not, however, unnoticed ; for the bird shows, by the motion of its head, that he is carefully watching his manœuvres, though, unless he walks directly towards it, it is not readily alarmed. When at rest, the wings are closely applied to the sides, with their tips lying over the tail, about one-third from its end ; and the tail itself, being closely shut up, looks long and narrow. If its suspicion be excited, it raises and depresses its head quickly two or three times, and spreads its tail, but does not open its wings until the instant it takes its flight. The individuals shot at Carlton House, had mice and small birds in their stomachs. They were not observed by the expedition beyond the 54th degree of latitude."—ED.

and several others of the same association have been obtained and examined during the present winter. On comparing these with Pennant's description referred to above, they correspond so exactly, that no doubts remain of their being the same species. Towards the beginning of April, these birds abandon this part of the country, and retire to the north to breed.

They are common, during winter, in the lower parts of Maryland, and numerous in the extensive meadows below Newark, New Jersey; are frequent along the Connecticut River; and, according to Pennant, inhabit England, Norway, and Lapmark. Their flight is slow and heavy. They are often seen coursing over the surface of the meadows, long after sunset, many times in pairs. They generally roost on the tall detached trees that rise from these low grounds; and take their stations, at day-break, near a ditch, bank, or hay stack, for hours together, watching, with patient vigilance, for the first unlucky frog, mouse, or lizard, to make its appearance. The instant one of these is descried, the hawk, sliding into the air, and taking a circuitous course along the surface, sweeps over the spot, and in an instant has his prey grappled and sprawling in the air.

The rough-legged hawk measures twenty-two inches in length, and four feet two inches in extent; cere, sides of the mouth, and feet, rich yellow; legs, feathered to the toes, with brownish yellow plumage, streaked with brown; femorals, the same; toes, comparatively short; claws and bill, blue black; iris of the eye, bright amber; upper part of the head, pale ochre, streaked with brown; back and wings, chocolate, each feather edged with bright ferruginous; first four primaries, nearly black about the tips, edged externally with silvery in some lights; rest of the quills, dark chocolate; lower side, and interior vanes, white; tail-coverts, white; tail, rounded, white, with a broad band of dark brown near the end, and tipt with white; body below, and breast, light yellow ochre, blotched and streaked with chocolate. What constitutes a characteristic mark of this bird, is a belt, or girdle, of very dark brown, passing round the belly just below the breast, and reaching under the wings to the rump ; head, very broad, and bill uncommonly small, suited to the humility of its prey.

The female is much darker, both above and below, particularly in the belt, or girdle, which is nearly black ; the tail-coverts are also spotted with chocolate ; she is also something larger.*

* From their different form, *Buteo* has been now adopted for the buzzards. They will also rank in two divisions ; those with clothed, and those with bare tarsi. The American species belonging to the first, will be our present one, Wilson's *Falco niger*, and Audubon's *F. Harlanii* ; [1] to the second, Wilson's *B. borealis*, *hyemalis*, and the common European buzzard, which was met with in the last Overland Arctic Expedition. The buzzards are sluggish and inactive in their habits ; their bills, feet, and claws, comparatively weak ; the form heavy, and the plumage more soft and downy, as if a smooth flight was to supply in part their want of activity. Their general flight is in sweeping circles, after mounting from their resting-place. They watch their prey either from the air, or on some tree or eminence, and sometimes pounce upon it when sailing near the ground. When satiated, they again return to their perch, and if undisturbed, will remain in one situation until hunger again calls them forth. Our present species is one of the more active, and is common also to the European continent. In Britain, it is an occasional visitant. They seem to appear at uncertain intervals, in more abundance ; thus, in 1823, I received two beautiful specimens from East Lothian ; and, in the same year, two or three more were killed on that coast. Mr Selby mentions, that in the year 1815, Northumberland was visited by them, and several specimens were obtained. He remarks, "Two of these birds, from having attached themselves to a neighbouring marsh, passed under my frequent observation. Their flight was smooth but slow, and not unlike that of the common buzzard ; and they seldom continued for any length of time on the wing. They preyed upon wild ducks and other birds, frogs and mice, which they mostly pounced upon on the ground." They appear to prefer trees for their breeding-place, whereas rocks, and the sides of deep ravines, are more frequently selected by the common buzzard. No instance has occurred of them breeding in this country. In plumage, they vary as much as the common species, the colour of the upper parts being of lighter or darker shades ; the

[1] See description of *F. Niger*.

BARRED OWL. *(Strix nebulosa.)*

PLATE XXXIII.—Fɪɢ. 2.

Turton, Syst. 169.—*Arct. Zool.* p. 234, No. 122.—*Lath.* 133.—Strix acclamator,
The Whooting Owl, *Bartram,* 289.—*Peale's Museum,* No. 464.

STRIX NEBULOSA.—Fᴏʀsᴛᴇʀ.*

La chouette du Canada (Ulula), *Cuv. Regn. Anim.* i. p. 328.—Strix nebulosa,
(sub-gen. Ulula, *Cuv.) Bonap. Synop.* p. 38.—Chouette nébuleuse, *Temm.
Man.* i. p. 86.—Strix nebulosa, *North. Zool.* ii. p. 81.

Tʜɪs is one of our most common owls.   In winter particularly,
it is numerous in the lower parts of Pennsylvania, among the

breast sometimes largely patched with deep brown, and sometimes entirely of that colour ; and the white bar at the base of the tail, though
always present, is of various dimensions.   Dr Richardson says it arrives
in the Fur Countries in April and May ; and having reared its young,
retires southward early in October.   They were so shy, that only one
specimen could be got by the Expedition.—Eᴅ.

    * Cuvier places this bird in his genus *Ulula.*   It may be called
nocturnal, though it does show a greater facility of conducting itself
during the day than the really night-living species, and will approach
nearer to the tawny owl of this country than any other ; indeed, it
almost seems the American representative of that species.   The tawny
owl, though not so abundant, has the very same manners ; and when
raised from its dormitory in a spruce or silver fir, or holly, or oak that
still carries its leaves, it will flit before one for half a day, moving its
station whenever it thinks the aggressor too near.   It does not utter
any cry during flight.

    It is common to both continents, visiting, however, only the more
northern parts of the European, and does not extend so generally as
many of those which inhabit both.

    According to Mr Audubon, this owl was a most abundant visitor to
his various solitary encampments, often a most amusing one ; and, by
less accustomed travellers, might easily have been converted into some
supposed inhabitant of another world.

    " How often," says this distinguished ornithologist, " when snugly
settled under the boughs of my temporary encampment, and preparing
to roast a venison steak, or the body of a squirrel, on a wooden spit,
have I been saluted with the exulting bursts of this nightly disturber
of the peace, that, had it not been for him, would have prevailed around
me, as well as in my lonely retreat !   How often have I seen this ·

woods that border the extensive meadows of Schuylkill and
Delaware. It is very frequently observed flying during day,
and certainly sees more distinctly at that time than many of

nocturnal marauder alight within a few yards of me, exposing his whole
body to the glare of my fire, and eye me in such a curious manner, that,
had it been reasonable to do so, I would gladly have invited him to
walk in and join me in my repast, that I might have enjoyed the
pleasure of forming a better acquaintance with him. The liveliness of
his motions, joined to their oddness, have often made me think that his
society would be at least as agreeable as that of many of the buffoons
we meet with in the world. But as such opportunities of forming
acquaintance have not existed, be content, kind reader, with the imperfect information which I can give you of the habits of this Sancho
Pança of our woods.

"Such persons as conclude, when looking upon owls in the glare of
day, that they are, as they then appear, extremely dull, are greatly
mistaken. Were they to state, like Buffon, that woodpeckers are
miserable beings, they would be talking as incorrectly ; and, to one who
might have lived long in the woods, they would seem to have lived only
in their libraries.

"The barred owl is found in all those parts of the United States which
I have visited, and is a constant resident. In Louisiana, it seems to be
more abundant than in any other state. It is almost impossible to
travel eight or ten miles in any of the retired woods there, without
seeing several of them even in broad day ; and, at the approach of
night, their cries are heard proceeding from every part of the forest
around the plantations. Should the weather be lowering, and indicative of the approach of rain, their cries are so multiplied during the
day, and especially in the evening, and they respond to each other in
tones so strange, that one might imagine some extraordinary fête about
to take place among them. On approaching one of them, its gesticulations are seen to be of a very extraordinary nature. The position of
the bird, which is generally erect, is immediately changed. It lowers
its head and inclines its body, to watch the motions of the person
beneath ; throws forward the lateral feathers of its head, which thus
has the appearance of being surrounded by a broad ruff ; looks towards
him as if half blind, and moves its head to and fro in so extraordinary
a manner, as almost to induce a person to fancy that part dislocated
from the body. It follows all the motions of the intruder with its eyes ;
and should it suspect any treacherous intentions, flies off to a short distance, alighting with its back to the person, and immediately turning
about with a single jump, to recommence its scrutiny. In this manner,
the barred owl may be followed to a considerable distance, if not shot

its genus.   In one spring, at different times, I met with more
than forty of them, generally flying or sitting exposed.   I also
once met with one of their nests, containing three young, in

at, for to halloo after it does not seem to frighten it much.   But if shot
at and missed, it removes to a considerable distance, after which, its
whah-whah-whah is uttered with considerable pomposity.   This owl
will answer the imitation of its own sounds, and is frequently decoyed
by this means.

   " The flight of the barred owl is smooth, light, noiseless, and capable
of being greatly protracted.   I have seen them take their departure
from a detached grove in a prairie, and pursue a direct course towards
the skirts of the main forest, distant more than two miles, in broad daylight.   I have thus followed them with the eye until they were lost in
the distance, and have reason to suppose that they continued their flight
until they reached the woods.   Once, whilst descending the Ohio, not
far from the well known *Cave-in-rock*, about two hours before sunset,
in the month of November, I saw a barred owl teased by several crows,
and chased from the tree in which it was.   On leaving the tree, it
gradually rose in the air, in the manner of a hawk, and at length
attained so great a height, that our party lost sight of it.   It acted, I
thought, as if it had lost itself, now and then describing small
circles, and flapping its wings quickly, then flying in zigzag lines.   This
being so uncommon an occurrence, I noted it down at the time.   I felt
anxious to see the bird return towards the earth, but it did not make
its appearance again.   So very lightly do they fly, that I have frequently
discovered one passing over me, and only a few yards distant, by first
seeing its shadow on the ground, during clear moonlight nights, when
not the faintest rustling of its wings could be heard.

   " Their power of sight during the day seems to be rather of an equivocal character, as I once saw one alight on the back of a cow, which it
left so suddenly afterwards, when the cow moved, as to prove to me that
it had mistaken the object on which it had perched for something else.
At other times, I have observed that the approach of the gray squirrel
intimidated them, if one of these animals accidentally jumped on a
branch close to them, although the owl destroys a number of them during the twilight.''

   Audubon has heard it said, in addition to small animals and birds,
and a peculiar sort of frog, common in the woods of Louisiana, that the
barred owl catches fish.   He never saw this performed, though it may
be as natural for it as those species which have been ascertained to feed
on them.   It is often exposed for sale in the New Orleans market, and
the creoles make *gumbo* of it, and pronounce it palatable.

   In this place may be introduced another species, mentioned by

the crotch of a white oak, among thick foliage. The nest was rudely put together, composed outwardly of sticks, intermixed with some dry grass and leaves, and lined with smaller twigs. At another time, in passing through the woods, I perceived something white, on the high shaded branch of a tree, close to the trunk, that, as I thought, looked like a cat asleep. Unable to satisfy myself, I was induced to fire, when, to my surprise and regret, four young owls, of this same species, nearly full grown, came down headlong, and, fluttering for a few moments, died at my feet. Their nest was probably not far distant. I have also seen the eggs of this species, which are nearly as large as those of a young pullet, but much more globular, and perfectly white.

Bonaparte as inhabiting Arctic America, and met with by Dr Richardson during the last northern expedition. It is the largest of the American owls, exceeding even the size of the Virginian horned owl, and seems to have been first noticed and described by Dr Latham, from Hudson's Bay specimens. Dr Richardson has more lately given the following sketch of its manners :—" It is by no means a rare bird in the Fur Countries, being an inhabitant of all the woody districts lying between Lake Superior and latitudes 67° or 68°, and between Hudson's Bay and the Pacific. It is common on the borders of Great Bear Lake ; and there and in the higher parallels of latitude it must pursue its prey, during the summer months, by daylight. It keeps, however, within the woods, and does not frequent the barren grounds, like the snowy owl, nor is it so often met with in broad daylight as the hawk owl, but hunts principally when the sun is low ; indeed, it is only at such times, when the recesses of the woods are deeply shadowed, that the American hare and the marine animals, on which this owl chiefly preys, come forth to feed. On the 23d of May, I discovered a nest of this owl, built, on the top of a lofty balsam poplar, of sticks, and lined with feathers. It contained three young, which were covered with a whitish down. We could get at the nest only by felling the tree, which was remarkably thick ; and whilst this operation was going on, the two parent birds flew in circles round the objects of their care, keeping, however, so high in the air as to be out of gunshot : they did not appear to be dazzled by the light. The young ones were kept alive for two months, when they made their escape. They had the habit, common also to other owls, of throwing themselves back, and making a loud snapping noise with their bills, when any one entered the room in which they were kept." —Ed.

These birds sometimes seize on fowls, partridges, and young rabbits; mice and small game are, however, their most usual food. The difference of size between the male and female of this owl is extraordinary, amounting sometimes to nearly eight inches in the length. Both scream during day, like a hawk.

The male barred owl measures sixteen inches and a half in length, and thirty-eight inches in extent; upper parts a pale brown, marked with transverse spots of white; wings, barred with alternate bands of pale brown, and darker; head, smooth, very large, mottled with transverse touches of dark brown, pale brown, and white; eyes, large, deep blue, the pupil not perceivable; face, or radiated circle of the eyes, gray, surrounded by an outline of brown and white dots; bill, yellow, tinged with green; breast, barred transversely with rows of brown and white; belly, streaked longitudinally with long stripes of brown, on a yellowish ground; vent, plain yellowish white; thighs and feathered legs, the same, slightly pointed with brown; toes, nearly covered with plumage; claws, dark horn colour, very sharp; tail, rounded, and remarkably concave below, barred with six broad bars of brown, and as many narrow ones of white; the back and shoulders have a cast of chestnut; at each internal angle of the eye, is a broad spot of black; the plumage of the radiated circle round the eye ends in long black hairs; and the bill is encompassed by others of a longer and more bristly kind. These probably serve to guard the eye when any danger approaches it in sweeping hastily through the woods; and those usually found on flycatchers may have the same intention to fulfil; for, on the slightest touch of the point of any of these hairs, the nictitant membrane was instantly thrown over the eye.

The female is twenty-two inches long, and four feet in extent; the chief difference of colour consists in her wings being broadly spotted with white; the shoulder being a plain chocolate brown; the tail extends considerably beyond the tips of the wings ; the bill is much larger, and of a more golden yellow ; iris of the eye, the same as that of the male.

The different character of the feathers of this, and, I believe, of most owls, is really surprising. Those that surround the bill differ little from bristles ; those that surround the region of the eyes are exceedingly open, and unwebbed ; these are bounded by another set, generally proceeding from the external edge of the ear, of a most peculiar small, narrow, velvety kind, whose fibres are so exquisitely fine, as to be invisible to the naked eye ; above, the plumage has one general character at the surface, calculated to repel rain and moisture ; but, towards the roots, it is of the most soft, loose, and downy substance in nature—so much so, that it may be touched without being felt ; the webs of the wing-quills are also of a delicate softness, covered with an almost imperceptible hair, and edged with a loose silky down, so that the owner passes through the air without interrupting the most profound silence. Who cannot perceive the hand of God in all these things !

SHORT-EARED OWL. (*Strix Brachyotos.*)

PLATE XXXIII.—Fig. 3.

Turton, Syst. p. 167.—*Arct. Zool.* p. 229, No. 116.—*Lath.* i. 124.—La chouette, ou la grand chevêche, *Buff.* i. *Pl. enl.* 438.—*Peale's Museum*, No. 440.

OTUS BRACHYOTOS.—Cuvier.*

Short-eared Owl, *Bew. Br. Birds*, i. p. 48, 50.—*Selby, Illust. Br. Orn.* i. p. 54, pl. 21.—Hibou brachyote, *Temm. Man.* i. p. 99.—La Chouette, ou le moyen duc, à Huppes courtes, *Cuv. Regn. Anim.* i. p. 328.—Otus brachyotus, *Flem. Br. Anim.* p. 56.—Strix brachyotos, *Bonap. Synop.* p. 37.—Strix brachyota, *North. Zool.* p. 75.

THIS is another species common to both continents, being found in Britain as far north as the Orkney Isles, where it

---

* This owl, as Wilson observes, is also common to both continents, but the British history of it is comparatively unknown. The following observations may perhaps advance some parts of it :--

In England it bears the name of woodcock owl, from its appearance

also breeds, building its nest. upon the ground, amidst the heath ; arrives and disappears in the south parts of England with the woodcock, that is, in October and April ; consequently does not breed there. It is called at Hudson's Bay,

nearly about the same time with that bird, and its reappearance again in the spring. Very few, if any, remain during the whole season, and they are only met with in their migrations to and from the north, their breeding-places, similar to the appearance, for a few days, of the ringousels and dotterels ; in spring, singly or in pairs ; and in the fall, in small groups, the amount of their broods when again retiring. They do not appear to be otherwise gregarious ; and it is only in this way that we can account for the flock of twenty-eight in a turnip field, quoted by our author, and the instances of five or six of these birds frequently found roosting together, as mentioned by Mr Selby. They appear at the same seasons (according to Temminck), and are plentiful in Holland. It is only in the north of England, and over Scotland, that they will rank as summer visitants. Hoy, and the other Hebrides, where they were first discovered to breed, were considered the southern limit of their incubation. It extends, however, much farther ; and may be, perhaps, stated as the extensive muirland ranges of Cumberland, Westmoreland, and Northumberland. Over all the Scottish muirs, it occurs in considerable abundance ; there are few sportsmen who are unacquainted with it ; many are killed during the grouse season, and those individuals which Mr Selby mentions as found on upland moors, I have no doubt bred there. On the extensive moors at the Head of Dryfe (a small rivulet in Dumfriesshire), I have, for many years past, met with one or two pairs of these birds, and the accidental discovery of their young first turned my attention to the range of their breeding ; for, previous to this, I also held the opinion, that they had commenced their migration southward. The young was discovered by one of my dogs pointing it ; and, on the following year, by searching at the proper season, two nests were found with five eggs. They were formed upon the ground among the heath ; the bottom of the nest scraped until the fresh earth appeared, on which the eggs were placed, without any lining or other accessory covering. When approaching the nest or young, the old birds fly and hover round, uttering a shrill cry, and snapping with their bills. They will then alight at a short distance, survey the aggressor, and again resume their flight and cries. The young are barely able to fly by the 12th of August, and appear to leave the nest some time before they are able to rise from the ground. I have taken them, on that great day to sportsmen, squatted on the heath like young black game, at no great distance from each other, and always attended by the parent birds. Last year (1831) I found them in their old haunts, to which

the mouse hawk; and is described as not flying, like other owls, in search of prey, but sitting quiet, on a stump of a tree, watching for mice. It is said to be found in plenty in the woods near Chatteau Bay, on the coast of Labrador. In the United States, it is also a bird of passage, coming to us from the north in November, and departing in April. The bird represented in the plate was shot in New Jersey, a few miles below Philadelphia, in a thicket of pines. It has the stern aspect of a keen, vigorous, and active bird; and is reputed to be an excellent mouser. It flies frequently by day, and particularly in dark cloudy weather, takes short flights; and, when sitting and looking sharply around, erects the two slight feathers that constitute its horns, which are at such times very noticeable; but, otherwise, not perceivable. No person on slightly examining this bird after being shot, would suspect it to be furnished with horns; nor are they discovered but by careful search, or previous observation, on the living bird. Bewick, in his "History of British Birds," remarks that this species is sometimes seen in companies,—twenty-eight of them having been once counted in a turnip field in November.

Length, fifteen inches; extent, three feet four inches; general colour above, dark brown, the feathers broadly skirted with pale yellowish brown; bill, large, black; irides, rich

they appear to return very regularly; and the female, with a young bird, was procured; the young could only fly for sixty or seventy yards.[1]

In form, this species will bear the same analogy to those furnished with horns which the snowy owl bears to the earless birds. The name of *hawk owl* implies more activity and boldness, and a different make; and we find the head small, the body more slender, the wings and tail powerful. They hunt regularly by day, and will sometimes soar to a great height. They feed on small birds, and destroy young game, as well as mice and moles.

It seems to have a pretty extensive geographical range. Pennant mentions it as inhabiting the Falkland Isles. It extends to Siberia; and I have received it from the neighbourhood of Canton, in China.—ED.

[1] A specimen was shot in December (1831) on the same ground, and one was seen when drawing a whin covert for a fox, on 31st January 1832. I believe some reside during the whole year.—ED.

golden yellow, placed in a bed of deep black, which radiates outwards all around, except towards the bill, where the plumage is whitish ; ears, bordered with a semi-circular line of black, and tawny yellow dots ; tail, rounded, longer than usual with owls, crossed with five bands of dark brown, and as many of yellow ochre—some of the latter have central spots of dark brown, the whole tipt with white quills also banded with dark brown and yellow ochre ; breast and belly streaked with dark brown, on a ground of yellowish ; legs, thighs, and vent, plain dull yellow ; tips of the three first quill-feathers, black ; legs, clothed to the claws, which are black, curved to about the quarter of a circle, and exceedingly sharp.

The female I have never seen ; but she is said to be somewhat larger, and much darker, and the spots on the breast larger, and more numerous.*

LITTLE OWL.   (*Strix passerina.*)

PLATE XXXIV.—Fig. 1.

Arct. Zool. 236, No. 126.—*Turton, Syst.* 172.—*Peale's Museum*, No. 522.

STRIX ACADICA.—Gmelin.†

Chouette chevêchette, *Temm. Man.* i. p. 96.—Strix acadica, *Bonap. Synop.* p. 38. —Monog. sinot strigi inauric. osserv. sulla, 2d edit. *del. Reg. Anim. Cuv.* p. 52. —Strix acadica, American Sparrow Owl, *North. Zool.* p. 97.

THIS is one of the least of its whole genus ; but, like many other little folks, makes up, in neatness of general form

---

* The female is nearly of the same size with the male ; the colours are all of a browned tinge, the markings more clouded and indistinct ; the white of the lower parts, and under the wings, is less pure, and the belly and vent are more thickly dashed with black streaks ; the ears are nearly of the same length with the other feathers, but can be easily distinguished. She is always foremost to attack any intruder on her nest or young.—ED.

† There is so much alliance between many of the small owls, that it is a matter of surprise more species have not been confounded. Wilson appears to have been mistaken, or to have confounded the name at least of the little owl ; and on the authority of Temminck and Bonaparte, we

1. Little Owl. 2. Sea-side Finch. 3. Sharp tailed F. 4. Savannah F.

34.

and appearance, for deficiency of size, and is, perhaps, the most shapely of all our owls. Nor are the colours and markings of its plumage inferior in simplicity and effect to most others. It also possesses an eye fully equal in spirit and brilliancy to the best of them.

This species is a general and constant inhabitant of the middle and northern states ; but is found most numerous in the neighbourhood of the sea-shore, and among woods and swamps of pine trees. It rarely rambles much during day ; but, if disturbed, flies a short way, and again takes shelter from the light: at the approach of twilight it is all life and activity, being a noted and dexterous mouse catcher. It is found as far north as Nova Scotia, and even Hudson's Bay ; is frequent in Russia ; builds its nest generally in pines, half way up the tree, and lays two eggs, which, like those of the

have given it as above, that of *acadica.* It is a native of both continents, but does not yet appear to have reached the British shores. According to Temminck, it is found in the deep German forests, though rarely, but is plentiful in Livonia. Bonaparte hints at the probability of the *St passerina* being yet discovered in America, which seems very likely, considering the similarity of its European haunts. The last Overland Arctic Expedition met with this and another allied species, *St Tengmalmi,* which will rank as an addition to the ornithology of that continent. Dr Richardson has the following observations regarding the latter: "When it accidentally wanders abroad in the day, it is so much dazzled by the light of the sun as to become stupid, and it may be easily caught by the hand. Its cry in the night is a single melancholy note, repeated at intervals of a minute or two, and it is one of the superstitious practices of the natives to whistle when they hear it. If the bird is silent when thus challenged, the speedy death of the inquirer is thus augured ; hence its Cree appellation of *Death Bird.*

On the banks of the Sascatchewan it is so common, that its voice is heard almost every night by the traveller, wherever he selects his bivouac.

Both the latter species extend over the north of Europe, and are found occasionally in Britain. The specimens which I have seen in confinement seem to sleep or dose away the morning and forenoon, but are remarkably active when roused, and move about with great agility. Both are often exposed for sale, with other birds, in the Dutch and Belgian markets.—ED.

rest of its genus, are white. The melancholy and gloomy umbrage of those solitary evergreens forms its favourite haunts, where it sits dozing and slumbering all day lulled by the roar of the neighbouring ocean.

The little owl is seven inches and a half long, and eighteen inches in extent; the upper parts are a plain brown olive, the scapulars and some of the greater and lesser coverts being spotted with white; the first five primaries are crossed obliquely with five bars of white; tail, rounded, rather darker than the body, crossed with two rows of white spots, and tipt with white; whole interior vanes of the wings, spotted with the same; auriculars, yellowish brown; crown, upper part of the neck, and circle surrounding the ears, beautifully marked with numerous points of white on an olive brown ground; front, pure white, ending in long blackish hairs; at the internal angle of the eyes, a broad spot of black radiating outwards; irides, pale yellow; bill, a blackish horn colour; lower parts, streaked with yellow ochre and reddish bay; thighs, and feathered legs, pale buff; toes, covered to the claws, which are black, large, and sharp-pointed.

The bird, from which the foregoing figure and description were taken, was shot on the sea-shore, near Great Egg Harbour, in New Jersey, in the month of November, and, on dissection, was found to be a female. Turton describes a species called the white fronted owl (*S. albifrons,*) which, in everything except the size, agrees with this bird, and has, very probably, been taken from a young male, which is sometimes found considerably less than the female.

SEA-SIDE FINCH. (*Fringilla maritima.*)

PLATE XXXIV. Fig. 2.

AMMODRAMUS MARITIMUS.—Swainson.*

Ammodramus, *Swain. Zool. Journ.* No. 11. p. 348.—Fringilla maritima, *Bonap.
Synop.* p. 110.—The Sea-side Finch, *Aud. Orn. Biog.* i. p. 470, pl. 93, male and
female.

OF this bird I can find no description. It inhabits the low
rush-covered sea islands along our Atlantic coast, where I first
found it ; keeping almost continually within the boundaries
of tide water, except when long and violent east or north-
easterly storms, with high tides, compel it to seek the shore.
On these occasions it courses along the margin, and among
the holes and interstices of the weeds and sea-wrack, with a
rapidity equalled only by the nimblest of our sandpipers, and
very much in their manner.   At these times also it roosts on
the ground, and runs about after dusk.

* The sea-side and short-tailed finches constitute the genus *Ammodramus* of Swainson. The former was discovered by Wilson ; the latter
is the sharp-tailed oriole of Latham. They are both peculiar to North
America, and are nearly confined to the salt marshes on the coast. They
are very curious in their structure, combining, as remarked by our
author, properties for either running or climbing. The tail is truly
scansorial ; the feet partly so ; the hallux formed for running, having
the claw elongated, and of a flat bend, as among the larks.

Mr Audubon has figured this bird with the nest. He says it is placed
so near the ground, that one might suppose it sunk into it, although this
is not actually the case. It is composed externally of coarse grass, and
is lined with finer kinds, but exhibits little regularity. The eggs are
from four to six, elongated, grayish white, freckled with brown all over.
They build in elevated shrubby places, where many nests may be found
in the space of an acre. When the young are grown, they betake themselves to the ditches and sluices which intersect the salt marshes, and
find abundant food. They enter the larger holes of crabs, and every
crack and crevice of the drying mud. In this they much resemble the
wrens, who enjoy entering and prying into every chink or opening of
their own haunts. Mr Audubon had some dressed in a pie, but found
them quite unpalatable. —ED.

This species derives its whole subsistence from the sea. I examined a great number of individuals by dissection, and found their stomachs universally filled with fragments of shrimps, minute shell-fish, and broken limbs of small sea crabs. Its flesh, also, as was to be expected, tasted of fish, or was what is usually termed sedgy. Amidst the recesses of these wet sea marshes, it seeks the rankest growth of grass and sea weed, and climbs along the stalks of the rushes with as much dexterity as it runs along the ground, which is rather a singular circumstance, most of our climbers being rather awkward at running.

The sea-side finch is six inches and a quarter long, and eight and a quarter in extent; chin, pure white, bordered on each side by a stripe of dark ash, proceeding from each base of the lower mandible; above that is another slight streak of white; from the nostril over the eye extends another streak, which immediately over the lores is rich yellow, bordered above with white, and ending in yellow olive; crown, brownish olive, divided laterally by a stripe of slate blue, or fine light ash; breast, ash, streaked with buff; belly, white; vent, buff coloured, and streaked with black; upper parts of the back, wings, and tail, a yellowish brown olive, intermixed with very pale blue; greater and lesser coverts, tipt with dull white; edge of the bend of the wing, rich yellow; primaries edged with the same immediately below their coverts; tail, cuneiform, olive brown, centered with black; bill, dusky above, pale blue below, longer than is usual with finches; legs and feet, a pale bluish white; irides, hazel. Male and female nearly alike in colour.

SHARP-TAILED FINCH. (*Fringilla caudacuta.*)

PLATE XXXIV.—FIG. 3.

Sharp-tailed Oriole, *Lath. Gen. Synop.* ii. p. 448. pl. 17.—
Peale's Museum, No. 6442.

AMMODRAMUS CAUDACUTUS.—SWAINSON.*

Ammodramus, *Swain. Zool. Journ.* No. ii. p. 348.—Fringilla caudacuta, *Bonap.
Synop.* p. 110.

A BIRD of this denomination is described by Turton, Syst. p. 562, but which by no means agrees with the present. This, however, may be the fault of the describer, as it is said to be a bird of Georgia: unwilling, therefore, to multiply names unnecessarily, I have adopted his appellation. In some future part of the work I shall settle this matter with more precision.

This new (as I apprehend it) and beautiful species is an associate of the former, inhabits the same places, lives on the same food; and resembles it so much in manners, that but for their dissimilarity in some essential particulars, I would be disposed to consider them as the same in a different state of plumage. They are much less numerous than the preceding, and do not run with equal celerity.

The sharp-tailed finch is five inches and a quarter long, and seven inches and a quarter in extent; bill, dusky; auriculars, ash; from the bill over the eye, and also below it, run two broad stripes of brownish orange; chin, whitish; breast, pale buff, marked with small pointed spots of black; belly, white; vent, reddish buff; from the base of the upper mandible a broad stripe of pale ash runs along the crown and hind head,

---

* Mr Audubon has figured a bird, very closely allied in plumage, under the title of *Ammodramus Henslowii*, and, in the letter-press, has described it as Henslow's bunting, *Emberiza Henslowii*. It will evidently come under the first genus, and if new and distinct, will form a third North American species. It is named after Professor Henslow of Cambridge, and was obtained near Cincinnati. There is no account of its history and habits.—ED.

bordered on each side by one of blackish brown ; back, a yellowish brown olive, some of the feathers curiously edged with semicircles of white ; sides under the wings buff, spotted with black ; wing-coverts and tertials black, broadly edged with light reddish buff ; tail, cuneiform, short ; all the feathers sharp pointed ; legs, a yellow clay colour ; irides, hazel.

I examined many of these birds, and found but little difference in the colour and markings of their plumage.

Since writing the above, I have become convinced that the bird described by Mr Latham, under the name of sharp-tailed oriole, is the present species. Latham states, that his description and figure were taken from a specimen deposited in Mrs Blackburn's collection, and that it came from New York.

SAVANNAH FINCH. (*Fringilla Savanna.*)

PLATE XXXIV.—Fig. 4. Male.*

Peale's Museum, No. 6583.

ZONOTRICHIA ? SAVANNA.—Jardine.

Fringilla Savanna, *Bonap. Synop.* p. 108.

This delicately marked sparrow has been already taken notice of, in a preceding part of this work, where a figure of the female was introduced. The present figure was drawn from a very beautiful male, and is a faithful representation of the original.

The length is five and a half inches ; extent, eight and a half ; bill, pale brown ; eyebrows, Naples yellow ; breast and whole lower parts, pure white, the former marked with small pointed spots of brown ; upper parts, a pale whitish drab, mottled with reddish brown ; wing-coverts, edged and tipt with white ; tertials, black, edged with white and bay ; legs, pale clay ; ear-feathers, tinged with Naples yellow. The female and young males are less, and much darker.

* The female is described in Vol. I. p. 342.

Drawn from Nature by A. Wilson.   *Engraved by W.H.Lizars.*

1. Winter Falcon. 2. Magpie. 3. Crow.

This is, probably, the most timid of all our sparrows. In winter it frequents the sea-shores; but, as spring approaches, migrates to the interior, as I have lately discovered, building its nest in the grass nearly in the same form, though with fewer materials, as that of the bay-winged bunting. On the 23d of May, I found one of these at the root of a clump of rushes in a grass field, with three young, nearly ready to fly. The female counterfeited lameness, spreading her wings and tail, and using many affectionate stratagems to allure me from the place. The eggs I have never seen.

WINTER FALCON. *(Falco hyemalis.)*

PLATE XXXV.—Fig. 1.

Turton, Syst. p. 156.—*Arct. Zool.* p. 209, No. 107.—*Peale's Museum*, No. 272 and 273.

ASTUR? HYEMALIS.—Jardine.*

The Winter Hawk, *Aud.* pl. 71; *Orn. Biog.* p. 164.

This elegant and spirited hawk is represented in the plate of one-half its natural size; the other two figures are reduced in the same proportion. He visits us from the north early in November, and leaves us late in March.

* This species, with the *Falco lineatus* of our author, have been the subject of dispute, as to their identity. The Prince of Musignano thinks they are the same, but in different states of plumage, according to age. Audubon says they are decidedly distinct, and has given plates of each, with an account of the differences he observed in their habits. I have transcribed his observations at some length, that these distinctions may be seen and judged of individually. I am inclined to consider them distinct, and cannot reconcile the great difference of habit to birds of one species, particularly in the same country. With regard to their station, again, they present a most interesting form. They are intermediate, as it were, between *Buteo, Astur,* and *Circus.* The colours are those of *Buteo* and *Circus;* while the form and active habits of the one is that of *Astur;* those of the winter hawk more of *Circus;* the wings are short for a true Buzzard, and possess the proportional length of the feathers of the goshawks. The feet of both are decidedly *Astur,* running perhaps into the more slender form of *Circus;* and from the pre-

This is a dexterous frog catcher ; who, that he may pursue
his profession with full effect, takes up his winter residence
ponderance of their form to the goshawks, I have chosen that as their
present appellation, but certainly with a query.

I have transcribed the habits of both species as given by Audubon,
that the comparison may be the more easy, and at the description of *F.
lineatus* have referred to this page :—

" The winter hawk is not a constant resident in the United States,
but merely visits them, making its first appearance there at the approach
of winter.   The flight is smooth and light, although greatly protracted,
when necessity requires it to be so.   It sails, at times, at a considerable
elevation ; and, notwithstanding the comparative shortness of its wings,
performs this kind of motion with grace, and in circles of more than
ordinary diameter.   It is a remarkably silent bird, often spending the
greater part of the day without uttering its notes more than once or
twice, which it does just before it alights, to watch with great patience
and perseverance for the appearance of its prey.   Its haunts are the
extensive meadows and marshes which occur along our rivers.   There
it pounces with a rapid motion on the frogs, which it either devours on
the spot, or carries to the perch, or the top of the hay-stack, on which
it previously stood.   It generally rests at night on the ground, among
the tall sedges of the marshes.   I have never seen this hawk in pursuit
of any other birds than those of its own species, each individual chasing
the others from the district which it had selected for itself.   The cry of the
winter hawk is clear and prolonged, and resembles the syllables *kay-o.*"

" The red-shouldered hawk, or, as I would prefer calling it, the red-
breasted hawk, although dispersed over the greater part of the United
States, is rarely observed in the middle districts, where, on the contrary,
the winter falcon usually makes its appearance from the north at the
approach of every autumn, and is of more common occurrence.   This
bird is one of the most noisy of its genus, during spring especially,
when it would be difficult to approach the skirts of woods bordering a
large plantation, without hearing its discordant shrill notes, *ka-hee, ka-
hee,* as it is seen sailing in rapid circles at a very great elevation.   Its
ordinary flight is even and protracted.   It is a more general inhabitant of
the woods than most of our other species, particularly during the summer.

" The interior of woods seems, as I have said, the fittest haunts for
the red-shouldered hawk.   He sails through them a few yards above
the ground, and suddenly alights on the low branch of a tree, or the
top of a dead stump, from which he silently watches in an erect posture
for the appearance of squirrels, upon which he pounces directly, and
kills them in an instant, afterwards devouring them on the ground.

" At the approach of spring, this species begins to pair, and its flight
is accompanied with many circlings and zigzag motions, during which

almost entirely among our meadows and marshes. He sometimes stuffs himself so enormously with these reptiles, that the
prominency of his craw makes a large bunch, and he appears
to fly with difficulty. I have taken the broken fragments and
whole carcasses of ten frogs, of different dimensions, from
the crop of a single individual. Of his genius and other
exploits, I am unable to say much. He appears to be a
fearless and active bird, silent, and not very shy. One which
I kept for some time, and which was slightly wounded, dis-
dained all attempts made to reconcile him to confinement;
and would not suffer a person to approach without being
highly irritated, throwing himself backward, and striking,
with expanded talons, with great fury. Though shorter
winged than some of his tribe, yet I have no doubt but, with
proper care, he might be trained to strike nobler game, in a
bold style, and with great effect. But the education of hawks in
this country may well be postponed for a time, until fewer improvements remain to be made in that of the human subject.

Length of the winter hawk, twenty inches; extent, forty-one
inches, or nearly three feet six inches; cere and legs, yellow,
the latter long, and feathered for an inch below the knee;
bill, bluish black, small, furnished with a tooth in the upper
mandible; eye, bright amber, cartilage over the eye, very
prominent, and of a dull green; head, sides of the neck, and

it emits its shrill cries. The top of a tall tree seems to be preferred,
as I have found its nest most commonly placed there, not far from the
edges of woods bordering plantations; it is seated in the forks of a large
branch, towards its extremity, and is as bulky as that of the common
crow; it is formed externally of dry sticks and Spanish moss, and is
lined with withered grass and fibrous roots of different sorts, arranged
in a circular manner. The eggs are generally four, sometimes five, of
a broad oval form, granulated all over, pale blue, faintly blotched with
brownish red at the smaller end."

From the above account it is seen that the red-shouldered hawk has
much more the habits of an *Astur* than the other, which seems to lean
towards the *Circi;* the breeding places of the latter are, however, not
mentioned by any writer. The different states of plumage in these
birds are deserving of farther research.—Ed.

throat, dark brown, streaked with white; lesser coverts with a strong glow of ferruginous; secondaries, pale brown, indistinctly barred with darker; primaries, brownish orange, spotted with black, wholly black at the tips; tail long, slightly rounded, barred alternately with dark and pale brown; inner vanes, white; exterior feathers, brownish orange; wings, when closed, reach rather beyond the middle of the tail; tail-coverts, white, marked with heart-shaped spots of brown; breast and belly, white, with numerous long drops of brown, the shafts blackish; femoral feathers, large, pale yellow ochre, marked with numerous minute streaks of pale brown; claws, black. The legs of this bird are represented by different authors as slender; but I saw no appearance of this in those I examined.

The female is considerably darker above, and about two inches longer.

MAGPIE. *(Corvus pica.)*

PLATE XXXV.—FIG. 2.

Arct. Zool. No. 136.—*Lath.* i. 392.—*Buff.* iii. 85.—*Peale's Museum,* No. 1333.

PICA CAUDATA.—RAY.*

THIS bird is much better known in Europe than in this country, where it has not been long discovered; although it is now found to inhabit a wide extent of territory, and in

---

* The common magpie of Europe is typical of that section among the *Corvidæ*, to which the name of *Pica* has been given. They retain the form of the bill as in *Corvus;* their whole members are weaker; the feathers on the rump are more lax and puffy, and the tail is always very lengthened.

The Appendix to Captain Franklin's narrative, by Mr Sabine, first gave rise to the suspicion that two very nearly allied species of magpie were found in the northern parts of America; and that gentleman has accordingly described the specimens killed at Cumberland House, during the first Arctic expedition, under the name of *Corvus Hudsonicus*—of which the following are the principal distinctions—and he seems to consider that bird more particularly confined to the more northern parts of the

great numbers. The drawing was taken from a very beautiful specimen, sent from the Mandan nation, on the Missouri, to Mr Jefferson, and by that gentleman presented to Mr Peale of this city, in whose museum it lived for several months, and where I had an opportunity of examining it. On carefully comparing it with the European magpie in the same collection, no material difference could be perceived. The figure on the plate is reduced to exactly half the size of life.

This bird unites in its character courage and cunning, turbulency and rapacity. Not inelegantly formed, and distinguished by gay as well as splendid plumage, he has long been noted in those countries where he commonly resides, and his habits and manners are there familiarly known. He is

continent, while the other was met with in the United States and the Missouri country.

"The Hudson's Bay magpie is of less size in all its parts than the common magpie, except in its tail, which exceeds that of its congener in length ; but the most remarkable and obvious difference is, in a loose tuft of grayish and white feathers on the back. Length of the body, exclusive of the tail, seven inches, that of the tail from eleven to twelve inches, that of the common being from nine to ten."

In the "Northern Zoology," *Corvus Hudsonicus* is quoted as a synonym. The authors remark, " This bird, so common in Europe, is equally plentiful in the interior prairie lands of America ; but it is singular, that, though it abounds on the shores of Sweden, and other maritime parts of the Old World, it is very rare on the Alantic, eastward of the Mississippi, or Lake Winipeg." " The manners of the American bird are precisely what we have been accustomed to observe in the English one. On comparing its eggs with those of the European bird, they were found to be longer and narrower ; and though the colours are the same, the blotches are larger and more diffused."

The distinctions mentioned by Mr Sabine seem very trivial ; indeed they may be confined entirely to a less size. The grayish tuft of feathers on the rump is the same in the common magpie of Britain. I have had an opportunity of examining only one North American specimen, which is certainly smaller, but in no other respect different. The authors of the "Northern Zoology" mention their having compared Arctic specimens with one from the interior of China, and they found no difference. The geographical distribution may therefore extend to a greater range than was supposed,—Europe, China, and America.—ED.

particularly pernicious to plantations of young oaks, tearing up the acorns; and also to birds, destroying great numbers of their eggs and young, even young chickens, partridges, grouse and pheasants.   It is perhaps on this last account that the whole vengeance of the game laws has lately been let loose upon him in some parts of Britain, as appears, by accounts from that quarter, where premiums, it is said, are offered for his head, as an arch poacher; and penalties inflicted on all those who permit him to breed on their premises.   Under the lash of such rigorous persecution, a few years will probably exterminate the whole tribe from the island.   He is also destructive to gardens and orchards ; is noisy and restless, almost constantly flying from place to place ; alights on the backs of the cattle, to rid them of the larvæ that fester in the skin ; is content with carrion when nothing better offers ; eats various kinds of vegetables, and devours greedily grain, worms, and insects of almost every description.   When domesticated, he is easily taught to imitate the human voice, and to articulate words pretty distinctly; has all the pilfering habits of his tribe, filling every chink, nook, and crevice, with whatever he can carry off; is subject to the epilepsy, or some similar disorder ; and is, on the whole, a crafty, restless, and noisy bird.

He generally selects a tall tree, adjoining the farm house, for his nest, which is placed among the highest branches ; this is large, composed outwardly of sticks, roots, turf, and dry weeds, and well lined with wool, cow hair, and feathers ; the whole is surrounded, roofed, and barricaded with thorns, leaving only a narrow entrance.   The eggs are usually five of a greenish colour, marked with numerous black or dusky spots. In the northern parts of Europe, he migrates at the commencement of winter.

In this country, the magpie was first taken notice of at the factories, or trading houses, on Hudson's Bay, where the Indians used sometimes to bring it in, and gave it the name of heart-bird,—for what reason is uncertain.   It appears, however, to be rather rare in that quarter.   These circumstances are taken notice of by Mr Pennant and other British naturalists.

In 1804, an exploring party under the command of Captains Lewis and Clark, on their route to the Pacific Ocean across the continent, first met with the magpie somewhere near the great bend of the Missouri, and found that the number of these birds increased as they advanced. Here also the blue jay disappeared ; as if the territorial boundaries and jurisdiction of these two noisy and voracious families of the same tribe had been mutually agreed on, and distinctly settled. But the magpie was found to be far more daring than the jay, dashing into their very tents, and carrying off the meat from the dishes. One of the hunters who accompanied the expedition informed me, that they frequently attended him while he was engaged in skinning and cleaning the carcass of the deer, bear, or buffalo he had killed, often seizing the meat that hung within a foot or two of his head. On the shores of the Koos-koos-ke river, on the west side of the great range of Rocky Mountains, they were found to be equally numerous.

It is highly probable that those vast plains, or prairies, abounding with game and cattle, frequently killed for the mere hides, tallow, or even marrow bones, may be one great inducement for the residency of these birds, so fond of flesh and carrion. Even the rigorous severity of winter in the high regions along the head waters of Rio du Nord, the Arkansaw, and Red River, seems insufficient to force them from those favourite haunts ; though it appears to increase their natural voracity to a very uncommon degree. Colonel Pike relates, that in the month of December, in the neighbourhood of the North Mountain, N. lat. 41° W. long. 34°, Reaumur's thermometer standing at 17° below 0, these birds were seen in great numbers. " Our horses," says he, " were obliged to scrape the snow away to obtain their miserable pittance ; and, to increase their misfortunes, the poor animals were attacked by the magpies, who, attracted by the scent of their sore backs, alighted on them, and, in defiance of their wincing and

kicking, picked many places quite raw; the difficulty of procuring food rendering those birds so bold, as to alight on our
men's arms, and eat meat out of their hand." *

The magpie is eighteen inches in length; the head, neck,
upper part of the breast and back, are a deep velvety black;
primaries, brownish black, streaked along their inner vanes
with white; secondaries, rich purplish blue; greater coverts,
green blue; scapulars, lower part of the breast and belly,
white; thighs and vent, black; tail, long; the two exterior
feathers scarcely half the length of the longest, the others
increasing to the two middle ones, which taper towards their
extremities. The colour of this part of the plumage is very
splendid, being glossy green, dashed with blue and bright
purple; this last colour bounds the green; nostrils, covered
with a thick tuft of recumbent hairs, as are also the sides of
the mouth; bill, legs, and feet, glossy black. The female
differs only in the less brilliancy of her plumage.

CROW. (*Corvus corone.*†)

PLATE XXXV.—Fig. 3.

Peale's Museum, No. 1246.

CORVUS CORONE?—Linnæus.

This is perhaps the most generally known, and least beloved,
of all our land birds; having neither melody of song, nor

* Pike's Journal, p. 170.

† " The voice of this bird is so remarkably different from that of the
Corone of Europe, that I was at first led to believe it a distinct species ;
but the most scrupulous examination and comparison of European and
American specimens proved them to be the same," are the words of
Bonaparte in his Nomenclature to Wilson ; and *Corvus corone* is quoted
as the name and synonym to this species in the " Northern Zoology,"
from a male killed on the plains of the Saskatchewan.

This is one of the birds I have yet been unable to obtain for comparison with European specimens, and it may seem presumption to differ
from the above authorities, without ever having seen the bird in question.

beauty of plumage, nor excellence of flesh, nor civility of manners, to recommend him ; on the contrary, he is branded as a thief and a plunderer—a kind of black-coated vagabond, who hovers over the fields of the industrious, fattening on their labours, and, by his voracity, often blasting their expectations. Hated as he is by the farmer, watched and persecuted by almost every bearer of a gun, who all triumph in his destruction, had not Heaven bestowed on him intelligence and sagacity far beyond common, there is reason to believe that the whole tribe (in these parts at least) would long ago have ceased to exist.

The crow is a constant attendant on agriculture, and a general inhabitant of the cultivated parts of North America. In the interior of the forest he is more rare, unless during the season of breeding. He is particularly attached to low flat corn countries, lying in the neighbourhood of the sea, or of large rivers ; and more numerous in the northern than southern states, where vultures abound, with whom the crows are unable to contend. A strong antipathy, it is also said, prevails between the crow and the raven, insomuch, that where the latter is numerous, the former rarely resides. Many of the first settlers of the Gennesee country have informed me, that, for a long time, ravens were numerous with them, but no crows, and even now the latter are seldom observed in that country. In travelling from Nashville to Natchez, a

I cannot, nevertheless, reconcile Wilson's account of the difference of habits and cry to those of Britain and Europe. It seems a species more intermediate between the common rook, *C. frugilegus*, and the *C. corone ;* their gregarious habits, and feeding so much on grain, are quite at variance with the carrion crow ; Wilson's account of the crow roost on the Delaware is so different, that, as far as habit is concerned, it is impossible to refer them to one ; and though some allowance might be made for the diversity of habit in the two countries, I do not see in what manner the cry of the bird should be so distinctly affected as to be remarked by nearly all authors who have mentioned them.

Burns's line in the " Cottar's Saturday Night " alludes certainly to the common rook, and he, I am sure, knew the difference between a crow and a corbie.—ED.

distance of four hundred and seventy miles, I saw few or no crows, but ravens frequently, and vultures in great numbers.

The usual breeding time of the crow, in Pennsylvania, is in March, April, and May, during which season they are dispersed over the woods in pairs, and roost in the neighbourhood of the tree they have selected for their nest. About the middle of March they begin to build, generally choosing a high tree; though I have also known them prefer a middlesized cedar. One of their nests, now before me, is formed externally of sticks, wet moss, thin bark, mixed with mossy earth, and lined with large quantities of horse hair, to the amount of more than half a pound, some cow hair, and some wool, forming a very soft and elastic bed. The eggs are four, of a pale green colour, marked with numerous specks and blotches of olive.

During this interesting season, the male is extremely watchful, making frequent excursions of half a mile or so in circuit, to reconnoitre; and the instant he observes a person approaching, he gives the alarm, when both male and female retire to a distance till the intruder has gone past. He also regularly carries food to his mate, while she is sitting; occasionally relieves her; and, when she returns, again resigns up his post. At this time, also, as well as until the young are able to fly, they preserve uncommon silence, that their retreat may not be suspected.

It is in the month of May, and until the middle of June, that the crow is most destructive to the corn fields, digging up the newly planted grains of maize, pulling up by the roots those that have begun to vegetate, and thus frequently obliging the farmer to replant, or lose the benefit of the soil; and this sometimes twice, and even three times, occasioning a considerable additional expense, and inequality of harvest. No mercy is now shown him. The myriads of worms, moles, mice, caterpillars, grubs, and beetles, which he has destroyed, are altogether overlooked on these occasions. Detected in robbing the hens' nests, pulling up the corn, and killing the young chickens, he is considered as an outlaw, and sentenced to destruction. But the great difficulty is, how to put this sentence in execution. In vain the gunner skulks along the hedges and fences ; his faithful sentinels planted on some commanding point, raise the alarm, and disappoint vengeance of its object. The coast again clear, he returns once more in silence, to finish the repast he had begun. Sometimes he approaches the farm house by stealth, in search of young chickens, which he is in the habit of snatching off, when he can elude the vigilance of the mother hen, who often proves too formidable for him. A few days ago, a crow was observed eagerly attempting to seize some young chickens in an orchard, near the room where I write; but these clustering close round the hen, she resolutely defended them, drove the crow into an apple tree, whither she instantly pursued him with such spirit and intrepidity, that he was glad to make a speedy retreat, and abandon his design.

The crow himself sometimes falls a prey to the superior strength and rapacity of the great owl, whose weapons of offence are by far the more formidable of the two.*

* " A few years ago," says an obliging correspondent, " I resided on the banks of the Hudson, about seven miles from the city of New York. Not far from the place of my residence was a pretty thick wood or swamp, in which great numbers of crows, who used to cross the river from the opposite shore, were accustomed to roost. Returning homeward one afternoon, from a shooting excursion, I had occasion to pass through this swamp. It was near sunset, and troops of crows were flying in all directions over my head. While engaged in observing their flight, and endeavouring to select from among them an object to shoot at, my ears were suddenly assailed by the distressful cries of a crow, who was evidently struggling under the talons of a merciless and rapacious enemy. I hastened to the spot whence the sounds proceeded, and, to my great surprise, found a crow lying on the ground, just expiring, and seated upon the body of the yet warm and bleeding quarry, *a large brown owl*, who was beginning to make a meal of the unfortunate robber of corn fields. Perceiving my approach, he forsook his prey with evident reluctance, and flew into a tree at a little distance, where he sat watching all my movements, alternately regarding, with longing eyes, the victim he had been forced to leave, and darting at me no very

Towards the close of summer, the parent crows, with their new families, forsaking their solitary lodgings, collect together, as if by previous agreement, when evening approaches. About an hour before sunset, they are first observed, flying, somewhat in Indian file, in one direction, at a short height above the tops of the trees, silent and steady, keeping the general curvature of the ground, continuing to pass sometimes till after sunset, so that the whole line of march would extend for many miles. This circumstance, so familiar and picturesque, has not been overlooked by the poets, in their descriptions of a rural evening. Burns, in a single line, has finely sketched it :—

" The blackening trains of crows to their repose."

The most noted crow-roost with which I am acquainted is near Newcastle, on an island in the Delaware. It is there known by the name of the Pea Patch, and is a low, flat, alluvial spot, of a few acres, elevated but a little above high water mark, and covered with a thick growth of reeds. This

friendly looks, that seemed to reproach me for having deprived him of his expected regale. I confess that the scene before me was altogether novel and surprising. I am but little conversant with natural history ; but I had always understood, that the depredations of the owl were confined to the smaller birds and animals of the lesser kind, such as mice, young rabbits, &c., and that he obtained his prey rather by fraud and stratagem, than by open rapacity and violence. I was the more confirmed in this belief, from the recollection of a passage in Macbeth, which now forcibly recurred to my memory.—The courtiers of King Duncan are recounting to each other the various prodigies that preceded his death, and one of them relates to his wondering auditors, that

' An eagle, towering in his pride of place,
Was by a *mousing owl* hawk'd at and kill'd.'

But to resume my relation : That the owl was the murderer of the unfortunate crow, there could be no doubt. No other bird of prey was in sight ; I had not fired my gun since I entered the wood ; nor heard any one else shoot : besides, the unequivocal situation in which I found the parties would have been sufficient before any 'twelve good men and true,' or a jury of crows, to have convicted him of his guilt. It is proper to add, that I avenged the death of the hapless crow by a well-aimed shot at the felonious robber, that extended him breathless on the ground."

appears to be the grand rendezvous, or head-quarters, of the greater part of the crows within forty or fifty miles of the spot. It is entirely destitute of trees, the crows alighting and nestling among the reeds, which by these means are broken down and matted together. The noise created by those multitudes, both in their evening assembly and reascension in the morning, and the depredations they commit in the immediate neighbourhood of this great resort, are almost incredible. Whole fields of corn are sometimes laid waste by thousands alighting on it at once, with appetites whetted by the fast of the preceding night; and the utmost vigilance is unavailing to prevent, at least, a partial destruction of this their favourite grain. Like the stragglers of an immense, undisciplined, and rapacious army, they spread themselves over the fields, to plunder and destroy wherever they alight. It is here that the character of the crow is universally execrated; and to say to the man who has lost his crop of corn by these birds, that crows are exceedingly useful for destroying vermin, would be as consolatory as to tell him who had just lost his house and furniture by the flames, that fires are excellent for destroying bugs.

The strong attachment of the crows to this spot may be illustrated by the following circumstance: Some years ago, a sudden and violent north-east storm came on during the night, and the tide, rising to an uncommon height, inundated the whole island. The darkness of the night, the suddenness and violence of the storm, and the incessant torrents of rain that fell, it is supposed, so intimidated the crows, that they did not attempt to escape, and almost all perished. Thousands of them were next day seen floating in the river; and the wind, shifting to the north-west, drove their dead bodies to the Jersey side, where for miles they blackened the whole shore.

This disaster, however, seems long ago to have been repaired; for they now congregate on the Pea Patch in as immense multitudes as ever.*

* The following is extracted from a late number of a newspaper printed in that neighbourhood :—

So universal is the hatred to crows, that few States, either here or in Europe, have neglected to offer rewards for their destruction. In the United States, they have been repeatedly ranked in our laws with the wolves, the panthers, foxes, and squirrels, and a proportionable premium offered for their heads, to be paid by any justice of the peace to whom they are delivered. On all these accounts, various modes have been invented for capturing them. They have been taken in clap nets, commonly used for taking pigeons ; two or three live crows being previously procured as decoys, or, as they are called, *stool-crows.* Corn has been steeped in a strong decoction of hellebore, which, when eaten by them, produces giddiness, 'and finally it is said, death. Pieces of paper formed into the shape of a hollow cone, besmeared within with birdlime, and a grain or two of corn dropt on the bottom, have also been adopted. Numbers of these being placed on the ground, where corn has been planted, the crows, attempting to reach the grains, are instantly hoodwinked, fly directly upwards to a great height; but generally descend near the spot whence they rose, and are easily taken. The reeds of their roosting places are sometimes set on fire during a dark night, and the gunners having previously posted themselves around, the crows rise in great uproar, and, amidst the general consternation, by the light of the burnings, hundreds of them are shot down.

Crows have been employed to catch crows, by the following

"The farmers of Red Lion Hundred held a meeting at the village of St George's, in the state of Delaware, on Monday the 6th inst., to receive proposals of John Deputy, on a plan for banishing or destroying the crows. Mr Deputy's plan being heard and considered, was approved, and a committee appointed to contract with him, and to procure the necessary funds to carry the same into effect. Mr Deputy proposes, that for five hundred dollars he will engage to kill or banish the crows from their roost on the Pea Patch, and give security to return the money on failure.

"The sum of five hundred dollars being thus required, the committee beg leave to address the farmers and others of Newcastle county and elsewhere on the subject."

stratagem : A live crow is pinned by the wings down to the ground on his back, by means of two sharp, forked sticks. Thus situated, his cries are loud and incessant, particularly if any other crows are within view. These sweeping down about him, are instantly grappled by the prostrate prisoner, by the same instinctive impulse that urges a drowning person to grasp at everything within his reach. Having disengaged the game from his clutches, the trap is again ready for another experiment ; and by pinning down each captive, successively, as soon as taken, in a short time you will probably have a large flock screaming above you, in concert with the outrageous prisoners below. Many farmers, however, are content with hanging up the skins, or dead carcasses, of crows in their corn fields, *in terrorem ;* others depend altogether on the gun, keeping one of their people supplied with ammunition, and constantly on the look out. In hard winters the crows suffer severely ; so that they have been observed to fall down in the fields, and on the roads, exhausted with cold and hunger. In one of these winters, and during a long continued deep snow, more than six hundred crows were shot on the carcass of a dead horse, which was placed at a proper distance from the stable, from a hole of which the discharges were made. The preniums awarded for these, with the price paid for the quills, produced nearly as much as the original value of the horse, besides, as the man himself assured me, saving feathers sufficient for filling a bed.

The crow is easily raised and domesticated ; and it is only when thus rendered unsuspicious of, and placed on terms of familiarity with man, that the true traits of his genius and native disposition fully develop themselves. In this state he soon learns to distinguish all the members of the family ; flies towards the gate, screaming, at the approach of a stranger ; learns to open the door by alighting on the latch ; attends regularly at the stated hours of dinner and breakfast, which he appears punctually to recollect; is extremely noisy and loquacious; imitates the sound of various words pretty

distinctly ; is a great thief and hoarder of curiosities, hiding
in holes, corners, and crevices, every loose article he can
carry off, particularly small pieces of metal, corn, bread, and
food of all kinds ; is fond of the society of his master, and will
know him even after a long absence, of which the following is
a remarkable instance, and may be relied on as a fact :—A
very worthy gentleman, now [1811] living in the Gennesee
country, but who at the time alluded to resided on the
Delaware, a few miles below Easton, had raised a crow, with
whose tricks and society he used frequently to amuse himself.
This crow lived long in the family ; but at length disappeared,
having, as was then supposed, been shot by some vagrant
gunner, or destroyed by accident. About eleven months after
this, as the gentleman, one morning, in company with
several others, was standing on the river shore, a number of
crows happening to pass by, one of them left the flock, and
flying directly towards the company, alighted on the gentleman's shoulder, and began to gabble away with great
volubility, as one long absent friend naturally enough does
on meeting with another. On recovering from his surprise,
the gentleman instantly recognised his old acquaintance, and
endeavoured, by several civil but sly manœuvres, to lay hold
of him ; but the crow not altogether relishing quite so much
familiarity, having now had a taste of the sweets of liberty,
cautiously eluded all his attempts ; and suddenly glancing
his eye on his distant companions, mounted in the air after
them, soon overtook and mingled with them, and was never
afterwards seen to return.

The habits of the crow in his native state are so generally
known as to require little further illustration. His watchfulness, and jealous sagacity in distinguishing a person with a
gun, are notorious to every one. In spring, when he makes
his appearance among the groves and low thickets, the whole
feathered songsters are instantly alarmed, well knowing the
depredations and murders he commits on their nests, eggs, and
young. Few of them, however, have the courage to attack

him, except the king bird, who, on these occasions, teases and pursues him from place to place, diving on his back while high in air, and harassing him for a great distance. A single pair of these noble spirited birds, whose nest was built near, have been known to protect a whole field of corn from the depredations of the crows, not permitting one to approach it.

The crow is eighteen inches and a half long, and three feet two inches in extent; the general colour is a shining glossy blue black, with purplish reflections; the throat and lower parts are less glossy; the bill and legs, a shining black, the former two inches and a quarter long, very strong and covered at the base with thick tufts of recumbent feathers; the wings, when shut, reach within an inch and a quarter of the tip of the tail, which is rounded; fourth primary, the longest; secondaries scalloped at the ends, and minutely pointed, by the prolongation of the shaft; iris, dark hazel.

The above description agrees so nearly with the European species, as to satisfy me that they are the same; though the voice of ours is said to be less harsh, not unlike the barking of a small spaniel: the pointedness of the ends of the tail-feathers, mentioned by European naturalists, and occasioned by the extension of the shafts, is rarely observed in the present species, though always very observable in the secondaries.

The female differs from the male in being more dull coloured, and rather deficient in the glossy and purplish tints and reflections. The difference, however, is not great.

Besides grain, insects, and carrion, they feed on frogs, tadpoles, small fish, lizards, and shell fish; with the latter they frequently mount to a great height, dropping them on the rocks below, and descending after them to pick up the contents. The same habit is observable in the gull, the raven, and sea-side crow. Many other aquatic insects, as well as marine plants, furnish them with food; which accounts for their being so generally found, and so numerous, on the sea-shore, and along the banks of our large rivers.

WHITE-HEADED, OR BALD EAGLE.* (*Falco leucocephalus.*)

PLATE XXXVI.

Linn. Syst. 124.—*Lath.* i. 29.—Le pygargue à tête blanche, *Buff.* i. 99, *Pl. enl.* 411.—*Arct. Zool.* 196, No. 89.—Bald Eagle, *Catesby*, i. 1.—*Peale's Museum*, No. 78.

HALIÆETUS LEUCOCEPHALUS.—Savigny.†

Aigle á tête blanche, *Cuv. Regn. Anim.* i. p. 315.—*Temm. Man.* i. p. 52.—Falco leucocephalus (sub-gen. Haliæetus), *Bonap. Synop.* p. 26.—The White-headed Eagle, *Aud. Orn. Biog.* i. p. 160, pl. 31, male.—Aquila (Haliæetus) leucocephala, *North. Zool.* ii. p. 15.

This distinguished bird, as he is the most beautiful of his tribe in this part of the world, and the adopted emblem of our country, is entitled to particular notice. He is represented in the plate of one-third his natural size, and was drawn from one

* The epithet *bald* applied to this species, whose head is thickly covered with feathers, is equally improper and absurd with the titles goatsucker, kingfisher, &c., bestowed on others ; and seems to have been occasioned by the white appearance of the head, when contrasted with the dark colour of the rest of the plumage. The appellation, however, being now almost universal, is retained in the following pages.

† This species and the sea eagle of Europe have been thought to be the same by many ornithologists ; some of a latter date appear still to confound them, and to be unable to satisfy themselves regarding the distinction. The subject has even been left in doubt in a work which has been recommended as a text-book to the British student. They are decidedly distinct, the one being the representing form of the other in their respective countries. The common sea eagle, *Haliæetus albicilla*, is, I believe, exclusively European ; the *H. leucocephalus*, according to Temminck, is common to the northern hemispheres of both the Old and New World, though much more abundant in the latter. The adult birds may be at once distinguished, and the confusion can only have arisen from the similarity of the young : when closely compared, they will also be found to possess considerable distinctions.

In habit, too, there is a difference. I have had both species alive in my possession for several years ; that of America, more active and restless in disposition, is constantly in motion, and incessantly utters its shrill barking cry. Both species are difficult to be tamed, but the stranger will hardly allow his cage to be cleaned out. Though four years old, the head and tail have not attained their pure whiteness,

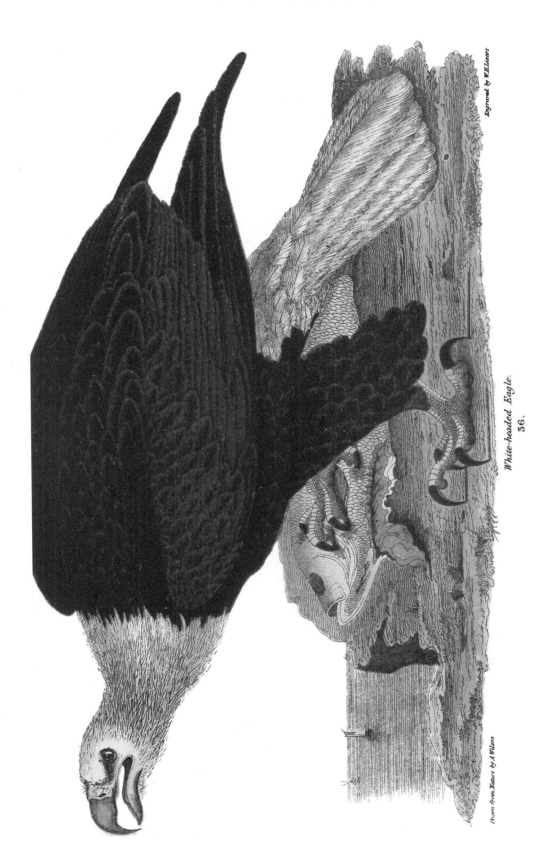

Drawn from Nature by A. Wilson

White-headed Eagle.
56.

Engraved by W.H.Lizars

of the largest and most perfect specimens I have yet met with. In the background is seen a distant view of the celebrated Cataract of Niagara, a noted place of resort for these birds,

being still marked with some patches of brown; but I have found this to be invariably the case with birds in confinement, from three to five years being then required to complete their perfect change,[1] whereas three years is the generally supposed time in a wild state. Fish is preferred to any other food by both, but nothing appears to come amiss to them.

Savigny established his genus for this form, or for the large bare-legged fishing eagles. They are not so powerfully formed, or so much adapted for rapid flight as the falcons and eagles. The tarsi are weaker —the tail more graduated—the whole form more inelegant; and when at rest, the secondaries hang in a drooping and sluggish manner over their wings; their habits, unless when in search of prey, or in the breeding season, much less daring and active. Such may be said to be the general characters of the group; our present species, however, seems to have a disposition more akin to the very fiercest : we have seen him to be very savage in his cage; in his native wilds he seems little less so. Fish is the favourite food, though they do not seem able to take them by plunging, but content themselves with either seizing from the ospreys what they have caught, or, where the water is so shallow as to allow them, clutch the fish without diving. Audubon says it only now and then procures fish for itself. He has seen them several times attempting to take red-fins by wading briskly through the water, and striking at them with their bill. When fish are not to be had, they appear hardly contented with the smaller animals or birds; pigs and sheep are a common fare, and our author has even mentioned one instance of a child being attacked. The male and female hunt in concert, and it must be when attacking some large-winged game, or water-fowl which have had recourse to the lake or river for safety, that their energies will be best observed. Audubon thus describes a swan hunt :—

" The next moment, however, the wild trumpet-like sound of a yet distant but approaching swan is heard : a shriek from the female eagle comes across the stream; for she is fully as alert as her mate. The snow-white bird is now in sight : her long neck is stretched forward; her eye is on the watch, vigilant as that of her enemy; her large wings seem with difficulty to support the weight of her body, although they flap incessantly. So irksome do her exertions seem, that her very legs are spread beneath her tail, to aid her in her flight. She approaches : the eagle has marked her for his prey. As the swan is passing the

---

[1] Mr Audubon mentions having known it six, and says in a wild state they breed the second year in full plumage.

as well on account of the fish procured there, as for the
numerous carcasses of squirrels, deer, bears, and various other
animals, that, in their attempts to. cross the river above the

dreaded pair, starts from his perch, in full preparation for the chase, the
male bird, with an awful scream.

" Now is the moment to witness a display of the eagle's powers. He
glides through the air like a falling star, and, like a flash of lightning,
comes upon the timorous quarry, which now, in agony and despair,
seeks, by various manœuvres, to elude the grasp of his cruel talons. It
mounts, doubles, and willingly would plunge into the stream, were it
not prevented by the eagle, which, long possessed of the knowledge that,
by such a stratagem, the swan might escape him, forces it to remain in
the air, by attempting to strike it with his talons from beneath. The
hope of escape is soon given up by the swan. It has already become
much weakened, and its strength fails at the sight of the courage and
swiftness of its antagonist. Its last gasp is about to escape, when the
ferocious eagle strikes with his talons the under side of its wing, and,
with unresisted power, forces the bird to fall in a slanting direction upon
the nearest shore."

And, again, when hunting in concert after some bird which has
alighted on the water :—

" At other times, when these eagles, sailing in search of prey, discover
a goose, a duck, or a swan, that has alighted on the water, they accom-
plish its destruction in a manner that is worthy of our attention. Well
aware that water-fowl have it in their power to dive at their approach,
and thereby elude their attempts upon them, they ascend in the air, in
opposite directions, over the lake or river on which the object which
they are desirous of possessing has been observed. Both reach a certain
height, immediately after which, one of them glides with great swiftness
towards the prey ; the latter, meantime, aware of the eagle's intention,
dives the moment before he reaches the spot. The pursuer then rises
in the air, and is met by its mate, which glides towards the water-bird,
that has just emerged to breathe, and forces it to plunge again beneath
the surface, to escape the talons of this second assailant. The first eagle
is now poising itself in the place where its mate formerly was, and rushes
anew, to force the quarry to make another plunge. By thus alternately
gliding, in rapid and often repeated rushes, over the ill-fated bird, they
soon fatigue it, when it stretches out its neck, swims deeply, and makes
for the shore, in the hope of concealing itself among the rank weeds.
But this is of no avail ; for the eagles follow it in all its motions ; and
the moment it approaches the margin, one of them darts upon it."

The bald eagle was met with in the Overland Arctic Expedition, but,
towards the north, was only a summer visitant ; in the Fur Countries,

Falls, have been dragged into the current, and precipitated down that tremendous gulf, where, among the rocks that bound the Rapids below, they furnish a rich repast for the

it is one of the earliest, arriving in the month of March, which has thence received the name of *Meekeeshew,* or *Eepeeshim,* or eagle month. It appears also migratory everywhere to the North ; it was not met with to the north of the Great Slave Lake, lat. 62° N. although it is common in the summer in the country lying between that and Lake Superior, and its breeding-places in the district are numerous. In the month of October, when the rivers are frozen over, it entirely quits Hudson's Bay lands ; and it is only on the sea-coasts that individuals can be then met with.

In this place we must introduce another splendid fishing eagle, which, if ultimately proved to be an undescribed species, will stand as the *Hæliæetus Washingtonii* of Audubon. It has been first beautifully figured and described by that gentleman, and a specimen of it exists in the Academy of Philadelphia. Its immense size, and some other differences, seem to keep it distinct from any species we are acquainted with, and it is most probably before this time proved to be new. We strongly suspect, however, that the state in which it is figured is not that of the adult plumage, and that this has yet to be found : we can only wish that its discoverer may be successful in his present arduous journey. It must be of very rare occurrence, three or four being all that Mr Audubon has ever found of it. We have transcribed the more essential parts of his description. From it there will be seen a difference in their habits from the white-headed bird, building and roosting on rocks ; and in their mode of fishing, which is performed like the osprey.

It was in February, 1814, that Mr Audubon first saw this bird, while on a trading voyage on the Upper Mississipi. He was assured that it was rare ; and, from the accounts he received, being convinced that it was unknown to naturalists, he felt anxious to learn its habits, and to discover in what particulars it differed from the rest of its genus. Mr Audubon did not again meet with it for some years, and his next meeting was partly accidental : he was engaged in collecting crayfish, and perceived, on the steep and rocky banks of the Ohio, the marks of the breeding-place of some bird of prey. His inquiries among the people in the neighbourhood led him to suppose that it was an eagle, different from any of those known in America. He resolved to watch the nest ; and the following is the result :—

" In high expectation I seated myself about a hundred yards from the foot of the rock. Never did time pass more slowly. I could not help betraying the most impatient curiosity, for my hopes whispered it was a sea eagle's nest. Two long hours had elapsed before the old bird made his appearance, which was announced to us by the loud hissings of the

vulture, the raven, and the bald eagle, the subject of the present account. This bird has been long known to naturalists, being common to both continents, and occasionally met with from a very high northern latitude, to the borders of the torrid zone, but chiefly in the vicinity of the sea, and along the shores and cliffs of our lakes and large rivers. Formed by nature for braving the severest cold; feeding equally on the produce of the sea and of the land; possessing

two young ones, which crawled to the extremity of the hole to receive a fine fish. I had a perfect view of this noble bird, as he held himself to the edging rock, hanging like the barn, bank, or social swallow, his tail spread, and his wings partly so. I trembled lest a word should escape my companions. The slightest murmur had been treason from them. They entered into my feelings, and, though little interested, joined with me. In a few minutes the other parent joined her mate. She glanced her quick and piercing eye around, and instantly perceived that her abode had been discovered. She dropped her prey, with a loud shriek, communicated the alarm to the male, and, hovering with him over our heads, kept up a growling cry." It was not till two years after that Mr Audubon had the good fortune to shoot this eagle; and the following description was then taken :—

"Bill, bluish black, the edges pale; the soft margin towards the commissure, and the base of the under mandible, yellow; cere, yellowish brown; lore, light greenish blue; iris, chestnut brown; feet, deep yellow; claws, bluish black; upper part of the head, hind neck, back scapulars, rump, tail-coverts, and posterior tibial feathers, blackish brown, glossed with a coppery tint; throat, foreneck, breast, and belly, light brownish yellow, each feather marked along the centre with blackish brown; wing-coverts, light grayish brown, those next the body becoming darker, and approaching the colour of the back; primary quills, dark brown, deeper on their inner webs; secondaries, lighter, and on their outer webs, of nearly the same light tint as their coverts; tail, uniform dark brown; anterior tibial feathers, grayish brown.

"Length, three feet seven inches; extent of wings, ten feet two inches; bill, three and a quarter inches along the back; along the gap, which commences directly under the eye, to the tip of the lower mandible, three and one-third, and one and three quarters deep; length of wing when folded, thirty-two inches; length of tail, fifteen inches; tarsus, four and a half; middle, four and three-quarters; hind claw, two and a half.

"The two stomachs, large and baggy; their contents in the individual described were fish, fishes' scales, and entrails of various kinds; intestines, large, but thin and transparent."—Ed.

powers of flight capable of outstripping even the tempests themselves; unawed by anything but man; and, from the ethereal heights to which he soars, looking abroad, at one glance, on an immeasurable expanse of forests, fields, lakes, and ocean, deep below him, he appears indifferent to the little localities of change of seasons; as, in a few minutes, he can pass from summer to winter, from the lower to the higher regions of the atmosphere, the abode of eternal cold, and thence descend, at will, to the torrid, or the arctic regions of the earth.   He is, therefore, found at all seasons in the countries he inhabits; but prefers such places as have been mentioned above, from the great partiality he has for fish.

In procuring these, he displays, in a very singular manner, the genius and energy of his character, which is fierce, contemplative, daring, and tyrannical,—attributes not exerted but on particular occasions, but, when put forth, overpowering all opposition.   Elevated on the high dead limb of some gigantic tree that commands a wide view of the neighbouring shore and ocean, he seems calmly to contemplate the motions of the various feathered tribes that pursue their busy avocations below,—the snow-white gulls slowly winnowing the air; the busy *Tringæ* coursing along the sands; trains of ducks streaming over the surface; silent and watchful cranes, intent and wading; clamorous crows; and all the winged multitudes that subsist by the bounty of this vast liquid magazine of nature.   High over all these hovers one, whose action instantly arrests his whole attention.   By his wide curvature of wing, and sudden suspension in air, he knows him to be the fish hawk, settling over some devoted victim of the deep.   His eye kindles at the sight, and, balancing himself, with half opened wings, on the branch, he watches the result.   Down, rapid as an arrow from heaven, descends the distant object of his attention, the roar of its wings reaching the ear as it disappears in the deep, making the surges foam around.   At this moment, the eager looks of the eagle are all ardour; and, levelling his neck for flight, he sees the fish hawk once more emerge, struggling with his prey, and mounting in the air with screams of exultation. These are the signal for our hero, who, launching into the air, instantly gives chase, and soon gains on the fish hawk; each exerts his utmost to mount above the other, displaying in these rencontres the most elegant and sublime aërial evolutions. The unencumbered eagle rapidly advances, and is just on the point of reaching his opponent, when, with a sudden scream, probably of despair and honest execration, the latter drops his fish; the eagle, poising himself for a moment, as if to take a more certain aim, descends like a whirlwind, snatches it in his grasp ere it reaches the water, and bears his ill-gotten booty silently away to the woods.

These predatory attacks, and defensive manœuvres of the eagle and the fish hawk, are matters of daily observation along the whole of our sea board, from Georgia to New England, and frequently excite great interest in the spectators. Sympathy, however, on this, as on most other occasions, generally sides with the honest and laborious sufferer, in opposition to the attacks of power, injustice, and rapacity—qualities for which our hero is so generally notorious, and which, in his superior, *man*, are certainly detestable. As for the feelings of the poor fish, they seem altogether out of the question.

When driven, as he sometimes is, by the combined courage and perseverance of the fish hawks, from their neighbourhood, and forced to hunt for himself, he retires more inland, in search of young pigs, of which he destroys great numbers. In the lower parts of Virginia and North Carolina, where the inhabitants raise vast herds of those animals, complaints of this kind are very general against him. He also destroys young lambs in the early part of spring; and will sometimes attack old sickly sheep, aiming furiously at their eyes.

In corroboration of the remarks I have myself made on the manners of the bald eagle, many accounts have reached me from various persons of respectability, living on or near our sea-coast: the substance of all these I shall endeavour to incorporate with the present account.

Mr John L. Gardiner, who resides on an island of three thousand acres, about three miles from the eastern point of Long Island, from which it is separated by Gardiner's Bay, and who has, consequently, many opportunities of observing the habits of these birds, has favoured me with a number of interesting particulars on this subject; for which I beg leave thus publicly to return my grateful acknowledgment.

"The bald eagles," says this gentleman, "remain on this island during the whole winter. They can be most easily discovered on evenings by their loud snoring while asleep on high oak trees; and, when awake, their hearing seems to be nearly as good as their sight. I think I mentioned to you, that I had myself seen one flying with a lamb ten days old, and which it dropped on the ground from about ten or twelve feet high. The struggling of the lamb, more than its weight, prevented its carrying it away. My running, hallooing, and being very near, might prevent its completing its design. It had broke the back in the act of seizing it; and I was under the necessity of killing it outright to prevent its misery. The lamb's dam seemed astonished to see its innocent offspring borne off into the air by a bird.

"I was lately told," continues Mr Gardiner, "by a man of truth, that he saw an eagle rob a hawk of its fish, and the hawk seemed so enraged as to fly down at the eagle, while the eagle very deliberately, in the air, threw himself partly over on his back, and, while he grasped with one foot the fish, extended the other to threaten or seize the hawk. I have known several hawks unite to attack the eagle; but never knew a single one to do it. The eagle seems to regard the hawks as the hawks do the king birds—only as teasing, troublesome fellows."

From the same intelligent and obliging friend, I lately received a well-preserved skin of the bald eagle, which, from its appearance, and the note that accompanied it, seems to have belonged to a very formidable individual. "It was shot," says Mr Gardiner, "last winter, on this island, and weighed thirteen pounds; measured three feet in length, and seven from tip to tip of the expanded wings; was extremely fierce looking; though wounded, would turn his back to no one; fastened his claws into the head of a dog, and was with difficulty disengaged. I have ridden on horseback within five or six rods of one, who, by his bold demeanour, raising his feathers, &c., seemed willing to dispute the ground with its owner. The crop of the present was full of mutton, from my part blood Merinos; and his intestines contained feathers, which he probably devoured with a duck, or winter gull, as I observed an entire foot and leg of some water fowl. I had two killed previous to this, which weighed ten pounds avoirdupois each."

The intrepidity of character, mentioned above, may be further illustrated by the following fact, which occurred a few years ago, near Great Egg Harbour, New Jersey. A woman, who happened to be weeding in the garden, had set her child down near, to amuse itself while she was at work; when a sudden and extraordinary rushing sound, and a scream from her child, alarmed her, and starting up, she beheld the infant thrown down, and dragged some few feet, and a large bald eagle bearing off the fragment of its frock, which being the only part seized, and giving way, providentially saved the life of the infant.

The appetite of the bald eagle, though habituated to long fasting, is of the most voracious, and often a most indelicate kind. Fish, when he can obtain them, are preferred to all other fare. Young lambs and pigs are dainty morsels, and made free with on all favourable occasions. Ducks, geese, gulls, and other sea fowl, are also seized with avidity. The most putrid carrion, when nothing better can be had, is acceptable; and the collected groups of gormandising vultures, on the approach of this dignified personage, instantly disperse, and make way for their master, waiting his departure in sullen silence, and at a respectful distance, on the adjacent trees.

In one of those partial migrations of tree squirrels that

sometimes take place in our western forests, many thousands of them were drowned in attempting to cross the Ohio; and at a certain place, not far from Wheeling, a prodigious number of their dead bodies were floated to the shore by an eddy. Here the vultures assembled in great force, and had regaled themselves for some time, when a bald eagle made his appearance, and took sole possession of the premises, keeping the whole vultures at their proper distance for several days. He has also been seen navigating the same river on a floating carrion, though scarcely raised above the surface of the water, and tugging at the carcase, regardless of snags, sawyers, planters, or shallows. He sometimes carries his tyranny to great extremes against the vultures. In hard times, when food happens to be scarce, should he accidentally meet with one of these who has its craw crammed with carrion, he attacks it fiercely in the air; the cowardly vulture instantly disgorges, and the delicious contents are snatched up by the eagle before they reach the ground.

The nest of this species is generally fixed on a very large and lofty tree, often in a swamp or morass, and difficult to be ascended. On some noted tree of this description, often a pine or cypress, the bald eagle builds, year after year, for a long series of years. When both male and female have been shot from the nest, another pair has soon after taken possession. The nest is large, being added to and repaired every season, until it becomes a black, prominent mass, observable at a considerable distance. It is formed of large sticks, sods, earthy rubbish, hay, moss, &c. Many have stated to me that the female lays first a single egg, and that, after having sat on it for some time, she lays another; when the first is hatched, the warmth of that, it is pretended, hatches the other. Whether this be correct or not, I cannot determine; but a very respectable gentleman of Virginia assured me, that he saw a large tree cut down, containing the nest of a bald eagle, in which were two young, one of which appeared nearly three times as large as the other. As a proof of their attachment to their young, a person near Norfolk informed me, that, in clearing a piece of wood on his place, they met with a large dead pine tree, on which was a bald eagle's nest and young. The tree being on fire more than half way up, and the flames rapidly ascending, the parent eagle darted around and among the flames, until her plumage was so much injured that it was with difficulty she could make her escape, and even then, she several times attempted to return to relieve her offspring.

No bird provides more abundantly for its young than the bald eagle. Fish are daily carried thither in numbers, so that they sometimes lie scattered round the tree, and the putrid smell of the nest may be distinguished at the distance of several hundred yards. The young are at first covered with a thick whitish or cream coloured cottony down; they gradually become of a gray colour as their plumage developes itself; continue of the brown gray until the third year, when the white begins to make its appearance on the head, neck, tail-coverts, and tail; these, by the end of the fourth year, are completely white, or very slightly tinged with cream; the eye also is at first hazel, but gradually brightens into a brilliant straw colour, with the white plumage of the head. Such at least was the gradual progress of this change, witnessed by myself, on a very fine specimen brought up by a gentleman, a friend of mine, who, for a considerable time, believed it to be what is usually called the gray eagle, and was much surprised at the gradual metamorphosis. This will account for the circumstance, so frequently observed, of the gray and white-headed eagle being seen together, both being, in fact, the same species, in different stages of colour, according to their difference of age.

The flight of the bald eagle, when taken into consideration with the ardour and energy of his character, is noble and interesting. Sometimes the human eye can just discern him, like a minute speck, moving in slow curvatures along the face of the heavens, as if reconnoitring the earth at that immense distance. Sometimes he glides along in a direct horizontal line, at a vast height, with expanded and unmoving wings, till he gradually disappears in the distant blue ether. Seen gliding in easy circles over the high shores and mountainous cliffs that tower above the Hudson and Susquehanna, he attracts the eye of the intelligent voyager, and adds great interest to the scenery. At the great Cataract of Niagara, already mentioned, there rises from the gulf into which the Fall of the Horse-shoe descends, a stupendous column of smoke, or spray, reaching to the heavens, and moving off in large black clouds, according to the direction of the wind, forming a very striking and majestic appearance. The eagles are here seen sailing about, sometimes losing themselves in this thick column, and again reappearing in another place, with such ease and elegance of motion, as renders the whole truly sublime.

> High o'er the watery uproar, silent seen,
> Sailing sedate in majesty serene,
> Now midst the pillar'd spray sublimely lost,
> And now, emerging, down the Rapids tost,
> Glides the bald eagle, gazing calm, and slow,
> O'er all the horrors of the scene below ;
> Intent alone to sate himself with blood,
> From the torn victims of the raging flood.

The white-headed eagle is three feet long, and seven feet in extent ; the bill is of a rich yellow ; cere, the same, slightly tinged with green ; mouth, flesh-coloured ; tip of the tongue, bluish black ; the head, chief part of the neck, vent, tail-coverts, and tail, are white in the perfect, or old birds of both sexes,—in those under three years of age these parts are of a gray brown ; the rest of the plumage is deep dark brown, each feather tipt with pale brown, lightest on the shoulder of the wing, and darkest towards its extremities. The conformation of the wing is admirably adapted for the support of so large a bird ; it measures two feet in breadth· on the greater quills, and sixteen inches on the lesser ; the longest primaries are twenty inches in length, and upwards of one inch in circumference where they enter the skin ;

the broadest secondaries are three inches in breadth across
the vane; the scapulars are very large and broad, spreading
from the back to the wing, to prevent the air from passing
through; another range of broad flat feathers, from three to
ten inches in length, also extends from the lower part of the
breast to the wing below, for the same purpose; between
these lies a deep triangular cavity; the thighs are remarkably
thick, strong, and muscular, covered with long feathers pointing backwards, usually called the femoral feathers; the legs,
which are covered half way below the knee, before, with dark
brown downy feathers, are of a rich yellow, the colour of ripe
Indian corn; feet the same; claws, blue black, very large
and strong, particularly the inner one, which is considerably
the largest; soles, very rough and warty; the eye is sunk under
a bony, or cartilaginous projection, of a pale yellow colour, and
is turned considerably forwards, not standing parallel with the
cheeks; the iris is of a bright straw colour, pupil black.

The male is generally two or three inches shorter than the
female; the white on the head, neck, and tail being more
tinged with yellowish, and its whole appearance less formidable; the brown plumage is also lighter, and the bird itself
less daring than the female,—a circumstance common to
almost all birds of prey.

The bird from which the foregoing drawing and description
were taken, was shot near Great Egg Harbour, in the month
of January. It was in excellent order, and weighed about
eleven pounds. Dr Samuel B. Smith, of this city, obliged
me with a minute and careful dissection of it; from whose
copious and very interesting notes on the subject, I shall
extract such remarks as are suited to the general reader.

" The eagle you sent me for dissection was a beautiful
female. It had two expansions of the gullet. The first principally composed of longitudinal bundles of fibre, in which
(as the bird is ravenous and without teeth) large portions of
unmasticated meats are suffered to dissolve before they pass
to the lower or proper stomach, which is membranous. I did
not receive the bird time enough to ascertain whether any

chilification was effected by the juices from the vessels of this enlargement of the œsophagus. I think it probable, that it also has a regurgitating, or vomiting power, as the bird constantly swallows large quantities of indigestible substances, such as quills, hairs, &c. In this sac of the eagle, I found the quill-feathers of the small white gull ; and in the true stomach, the tail and some of the breast-feathers of the same bird, and the dorsal vertebræ of a large fish. This excited some surprise, until you made me acquainted with the fact of its watching the fish hawks, and robbing them of their prey. Thus we see, throughout the whole empire of animal life, power is almost always in a state of hostility to justice ; and of the Deity only can it truly be said, that *justice* is commensurate with *power !*

" The eagle has the several auxiliaries to digestion and assimilation in common with man. The liver was unusually large in your specimen. It secretes bile, which stimulates the intestines, prepares the chyle for blood, and by this very secretion of bile (as it is a deeply respiring animal), separates or removes some obnoxious principles from the blood. (See Dr Rush's admirable lecture on this important viscus in the human subject.) The intestines were also large, long, convolute, and supplied with numerous lacteal vessels, which differ little from those of men, except in colour, which was transparent. The kidneys were large, and seated on each side the vertebræ, near the anus. They are also destined to secrete some offensive principles from the blood.

" The eggs were small and numerous ; and, after a careful examination, I concluded that no sensible increase takes place in them till the *particular* season. This may account for the unusual excitement which prevails in these birds in the sexual intercourse. Why there are so many eggs, is a mystery. It is, perhaps, consistent with natural law, that everything should be abundant ; but, from this bird, it is said, no more than two young are hatched in a season, consequently, no more eggs are wanted than a sufficiency to produce that effect. Are the eggs numbered originally, and is there no increase of number, but a gradual loss, till all are deposited ? If so, the number may correspond to the long life and vigorous health of this noble bird. Why there are but two young in a season, is easily explained. Nature has been studiously parsimonious of her physical strength, from whence the tribes of animals incapable to resist, derive security and confidence."

The eagle is said to live to a great age,—sixty, eighty, and, as some assert, one hundred years. This circumstance is remarkable, when we consider the seeming intemperate habits of the bird. Sometimes fasting, through necessity, for several days, and at other times gorging itself with animal food till its craw swells out the plumage of that part, forming a large protuberance on the breast. This, however, is its natural food, and for these habits its whole organisation is particularly adapted. It has not, like men, invented rich wines, ardent spirits, and a thousand artificial poisons, in the form of soups, sauces, and sweetmeats. Its food is simple, it indulges freely, uses great exercise, breathes the purest air, is healthy, vigorous, and long lived. The lords of the creation themselves might derive some useful hints from these facts, were they not already, in general, too wise, or too proud, to learn from their *inferiors*, the fowls of the air and beasts of the field.

FISH HAWK, OR OSPREY. *(Falco haliœtus.)*

PLATE XXXVII.—Fig. 1.

Carolina Osprey, *Lath. Syn.* i. p. 46.—26. A.—Falco piscator, *Briss.* i. p. 361. 14. 362. 15.—Faucon Pêcheur de la Caroline, *Buff.* i. p. 142.—Fishing Hawk, *Catesby, Car.* i. p. 2.—*Turt. Syst.* i. 149.—*Peale's Museum*, No. 144.

PANDION HALIÆETUS.—Savigny.[*]

Le Balbuzard, *Cuv. Regn. Anim.* i. p. 316.—Aigle Balbuzard, *Temm. Man.* i. p. 47.—Balbusardus haliætus, *Flem. Br. Anim.* p. 51.—Osprey, Falco haliæetus, *Selby, Illust. Br. Ornith.* i. p. 12, pl. 4.—Falco haliætus (sub-gen. *Pandion*), *Bonap. Synop.* p. 26.—The Fish Hawk, or Osprey, *Aud.* pl. 81. male *Orn. Biog.* i. 415.—Aquila (*Pandion*) haliæeta, *North. Zool.* ii. p. 20.

This formidable, vigorous-winged, and well known bird, subsists altogether on the finny tribes that swarm in our bays,

---

[*] This is the type of another aquatic group, and a real fisher. It does not, like the white-headed eagle, though fond of fish, subsist only upon

1. Fish-Hawk. 2. Fish-Crow. 3. Ring Plover. 4. Least Snipe.

creeks, and rivers ; procuring his prey by his own active skill and industry ; and seeming no farther dependent on the land than as a mere resting-place, or, in the usual season, a spot of deposit for his nest, eggs, and young.   The figure here given is

the plunder of others, but labours for itself in the most dexterous manner ; and for this, the beautiful adaptation of its form renders every assistance.   The body is very strongly built, but is rather of a narrow and elongated shape ; the head is less than the ordinary proportional dimensions ; and the wings are expansive, powerful, and sharp-pointed. The manner of seizing their prey is by soaring above the surface of the sea, or lake, and, when in sight of a fish, closing the wings, and darting, as it were, by the weight of the body, which, in the descent, may be perceived to be directed by the motion of the tail.   For this purpose, those parts which we have mentioned are finely framed, and for the remainder of the operation, the legs and feet are no less beautifully modelled.   The thighs, instead of being clothed with finely lengthened plumes, as in most of the other falcons, and which, when wet, would prove a great encumbrance, are covered with a thick downy plumage ; the tarsi are short and very strong ; the toes have the same advantages ; and underneath, at the junction of each joint, have a large protuberance, covered, as are the other parts of the sole, with a thick and strong array of hard jagged scales, which are sufficient, by the roughness, to prevent any escape of their slippery prey when it is once fairly clutched ; the claws are also very strong, and hooked, and are round as a cylinder, both above and beneath, which will ensure an easy, piercing, or quick retraction from any body at which they may be struck.   The outer toe is also capable of being turned either way—a most essential assistance in grasping.   In striking their prey they do not appear to dive deep ; indeed, their feet, by which alone it is taken, could not then be brought into action, but they are often concealed in the spray occasioned by their rapid descent.

The size of a fish they are able to bear away is very great, and sometimes exceeds their own weight.   That of the female is little more than five pounds, and Mr Audubon has figured his specimen with a *weak fish* more than that weight ; while our author mentions a shad that, when partly eaten, weighed more than six pounds.   These authenticated accounts lead us almost to credit the more marvellous stories of that amusing sporting writer, Mr Loyd.

That gentleman relates, that in Sweden the eagle sometimes strikes so large a pike, that not being able to disengage his talons, he is carried under water and drowned.   Dr Mullenborg vouched for this, by the fact of having himself seen an enormous pike, with an eagle fastened to his

reduced to one-third the size of life, to correspond with that of the bald eagle, his common attendant and constant plunderer.

The fish hawk is migratory, arriving on the coasts of New York and New Jersey about the twenty-first of March, and retiring to the south about the twenty-second of September. Heavy equinoctial storms may vary these periods of arrival and departure a few days ; but long observation has ascertained that they are kept with remarkable regularity. On the arrival of these birds in the northern parts of the United States, in March, they sometimes find the bays and ponds frozen, and experience a difficulty in procuring fish for many days. Yet there is no instance on record of their attacking birds, or inferior land animals, with intent to feed on them ; though

back, lying dead on a piece of ground which had been overflowed, but from whence the water had retreated.

He mentions also an account of a struggle between an eagle and a pike, witnessed by a gentleman, on the Gotha river, at no great distance from Wenersborg. In this instance, when the eagle first seized the pike, he was enabled to lift him a short distance into the air, but the weight of the fish, together with its struggles, soon carried them back again to the water, under which for a while they both disappeared. Presently, however, the eagle again came to the surface, uttering the most piercing cries, and making apparently every endeavour to extricate his talons, but all in vain ; and after struggling, he was carried under water.

Savigny formed his well marked genus *Pandion* from this species, which we now adopt. The osprey is common to both continents, and I possess one from New Holland in no way different. It is met with in England occasionally, but, according to Montague, is particularly plentiful in Devonshire. In Scotland, a pair or two may be found about most of the Highland lochs, where they fish, and, during the breeding season, build on the ruined towers so common on the edges or insulated rocks of these wild waters. The nest is an immense fabric of rotten sticks—

Itself a burden for the tallest tree,

and is generally placed, if such exists, on the top of the chimney, and if this be wanting, on the highest summit of the building. An aged tree may sometimes be chosen, but ruins are always preferred, if near. They have the same propensity of returning to an old station with those of America ; and if one is shot, a mate is soon found, and brought to the ancient abode. Loch Lomond, Loch Awe and Kilchurn Castle, and Loch Menteith, have been long breeding places.—Ed.

their great strength of flight, as well as of feet and claws, would seem to render this no difficult matter. But they no sooner arrive, than they wage war on the bald eagles, as against a horde of robbers and banditti; sometimes succeeding, by force of numbers and perseverance, in driving them from their haunts, but seldom or never attacking them in single combat.

The first appearance of the fish hawk in spring is welcomed by the fishermen, as the happy signal of the approach of those vast shoals of herring, shad, &c., that regularly arrive on our coasts, and enter our rivers in such prodigious multitudes. Two of a trade, it is said, seldom agree; the adage, however, will not hold good in the present case, for such is the respect paid the fish hawk, not only by this class of men, but, generally, by the whole neighbourhood where it resides, that a person who should attempt to shoot one of them would stand a fair chance of being insulted. This prepossession in favour of the fish hawk is honourable to their feelings. They associate, with its first appearance, ideas of plenty, and all the gaiety of business; they see it active and industrious like themselves; inoffensive to the productions of their farms; building with confidence, and without the least disposition to concealment, in the middle of their fields, and along their fences; and returning, year after year, regularly to its former abode.

The nest of the fish hawk is usually built on the top of a dead, or decaying tree, sometimes not more than fifteen, often upwards of fifty feet, from the ground. It has been remarked by the people of the sea coasts, that the most thriving tree will die in a few years after being taken possession of by the fish hawk. This is attributed to the fish oil, and to the excrements of the bird; but is more probably occasioned by the large heap of wet salt materials of which the nest is usually composed. In my late excursions to the sea shore, I ascended to several of these nests that had been built in from year to year, and found them constructed as follows:—Externally, large sticks, from half an inch to an inch and a half in diameter, and two or three feet in length, piled to the height of four or five feet, and from two to three feet in breadth ; these were intermixed with corn stalks, seaweed, pieces of wet turf, in large quantities, mullein stalks, and lined with dry sea-grass ; the whole forming a mass very observable at half a mile's distance, and large enough to fill a cart, and be no inconsiderable load for a horse. These materials are so well put together, as often to adhere, in large fragments, after being blown down by the wind. My learned and obliging correspondent of New York, Dr Samuel L. Mitchill, observes, that "A sort of superstition is entertained in regard to the fish hawk. It has been considered a fortunate incident to have a nest, and a pair of these birds, on one's farm. They have, therefore, been generally respected ; and neither the axe nor the gun has been lifted against them. Their nest continues from year to year. The same couple, or another, as the case may be, occupies it, season after season. Repairs are duly made, or, when demolished by storms, it is industriously rebuilt. There was one of these nests, formerly, upon the leafless summit of a venerable chestnut tree on our farm, directly in front of the house, at the distance of less than half a mile. The withered trunk and boughs, surmounted by the coarse wrought and capacious nest, was a more picturesque object than an obelisk : and the flights of the hawks, as they went forth to hunt—returned with their game—exercised themselves in wheeling round and round, and circling about it—were amusing to the beholder, almost from morning to night. The family of these hawks, old and young, was killed by the Hessian *Jagers.* A succeeding pair took possession of the nest ; but in the course of time, the prongs of the trunk so rotted away, that the nest could no longer be supported. The hawks have been obliged to seek new quarters. We have lost this part of our prospect ; and our trees have not afforded a convenient site for one of their habitations since."

About the first of May, the female fish hawk begins to lay her eggs, which are commonly three in number, sometimes only two, and rarely four. They are somewhat larger than those of the common hen, and nearly of the same shape. The ground colour varies, in different eggs, from a reddish cream, to nearly a white, splashed and daubed all over with dark Spanish brown, as if done by art.* During the time the female is sitting, the male frequently supplies her with fish ; though she occasionally takes a short circuit to sea herself, but quickly returns again. The attention of the male, on such occasions, is regulated by the circumstances of the case. A pair of these birds, on the south side of Great Egg Harbour river, and near its mouth, was noted for several years. The female, having but one leg, was regularly furnished, while sitting, with fish in such abundance, that she seldom left the nest, and never to seek for food. This kindness was continued both before and after incubation. Some animals, who claim the name and rationality of man, might blush at the recital of this fact.

On the appearance of the young, which is usually about the last of June, the zeal and watchfulness of the parents are extreme. They stand guard, and go off to fish, alternately : one parent being always within a short distance of the nest. On the near approach of any person, the hawk utters a plaintive whistling note, which becomes shriller as she takes to wing, and sails around, sometimes making a rapid descent, as if aiming directly for you ; but checking her course, and

* Of the palatableness of these eggs I cannot speak from personal experience ; but the following incident will show that the experiment has actually been made :—A country fellow, near Cape May, on his way to a neighbouring tavern, passing a tree, on which was a fish hawk's nest, immediately mounted, and robbed it of the only egg it contained, which he carried with him to the tavern, and desired the landlord to make it into egg-nogg. The tavern keeper, after a few wry faces, complied with his request, and the fellow swallowed the cordial. Whether from its effects on the olfactory nerves (for he said it smelt abominably), on the imagination, or on the stomach alone, is uncertain, but it operated as a most outrageous emetic, and cured the man, for that time at least, of his thirst for egg-nogg. What is rather extraordinary, the landlord (Mr Beasley) assured me, that, to all appearance, the egg was perfectly fresh.

sweeping past, at a short distance overhead, her wings making a loud whizzing in the air. My worthy friend Mr Gardiner informs me, that they have even been known to fix their claws in a negro's head, who was attempting to climb to their nest; and I had lately a proof of their daring spirit in this way, through the kindness of a friend, resident, for a few weeks, at Great Egg Harbour. I had requested of him the favour to transmit me, if possible, a live fish hawk, for the purpose of making a drawing of it, which commission he very faithfully executed; and I think I cannot better illustrate this part of the bird's character, than by quoting his letter at large:—

" Beasley's, Great Egg Harbour, 30th *June* 1811.

" Sir,—Mr Beasley and I went to reconnoitre a fish hawk's nest on Thursday afternoon. When I was at the nest, I was struck with so great violence on the crown of the hat, that I thought a hole was made in it. I had ascended fearlessly, and never dreamt of being attacked. I came down quickly. There were in the nest three young ones, about the size of pullets, which, though full feathered, were unable to fly. On Friday morning, I went again to the nest to get a young one, which I thought I could nurse to a considerable growth, sufficient to answer your purpose, if I should fail to procure an old one, which was represented to me as almost impossible, on account of his shyness, and the danger from his dreadful claws. On taking a young one, I intended to lay a couple of snares in the nest, for which purpose I had a strong cord in my pocket. The old birds were on the tree when Captain H. and I approached it. As a defence, profiting by the experience of yesterday, I took a walking stick with me. When I was about half up the tree, the bird I send you struck at me repeatedly with violence; he flew round, in a small circle, darting at me at every circuit, and I striking at him. Observing that he always described a circle in the air, before he came at me, I kept a *hawk's eye* upon him, and the moment he passed me, I availed myself of the opportunity to ascend.

When immediately under the nest, I hesitated at the formidable opposition I met, as his rage appeared to increase with my presumption in invading his premises. But I mounted to the nest. At that moment he darted directly at me with all his force, whizzing through the air, his choler apparently redoubled. Fortunately for me, I struck him on the extreme joint of the right wing with my stick, which brought him to the ground. During this contest, the female was flying round and round at a respectful distance. Captain H. held him till I tied my handkerchief about his legs: the captain felt the effect of his claws. I brought away a young one to keep the old one in a good humour. I put them in a very large coop; the young one ate some fish, when broken and put into its throat; but the old one would not eat for two days. He continued sullen and obstinate, hardly changing his position. He walks about now, and is approached without danger. He takes very little notice of the young one. A Joseph Smith, working in the field where this nest is, had the curiosity to go up and look at the eggs: the bird clawed his face in a shocking manner; his eye had a narrow escape. I am told that it has never been considered dangerous to approach a hawk's nest. If this be so, this bird's character is peculiar; his affection for his young, and his valiant opposition to an invasion of his nest, entitle him to conspicuous notice. He is the *prince* of fish hawks; his character and his portrait seem worthy of being handed to the historic muse. A hawk more worthy of the honour which awaits him could not have been found. I hope no accident will happen to him, and that he may fully answer your purpose,—Yours,

"Thomas Smith."

" This morning the female was flying to and fro, making a mournful noise."

The young of the fish hawk are remarkable for remaining long in the nest before they attempt to fly. Mr Smith's letter is dated June 30th, at which time, he observes, they were as

large as pullets, and full feathered. Seventeen days after, I myself ascended to this same hawk's nest, where I found the two
remaining young ones seemingly full grown. They made no
attempts to fly, though they both placed themselves in a stern
posture of defence as I examined them at my leisure. The
female had procured a *second* helpmate ; but he did not seem
to inherit the spirit of his predecessor, for, like a true step-
father, he left the nest at my approach, and sailed about at a
safe distance with his mate, who showed great anxiety and
distress during the whole of my visit. It is universally asserted,
by the people of the neighbourhood where these birds breed,
that the young remain so long before they fly, that the parents
are obliged at last to compel them to shift for themselves,
beating them with their wings, and driving them from the
nest.   But that they continue to assist them even after this,
I know to be a fact, from my own observation, as I have seen
the young bird meet its parent in the air, and receive from
him the fish he carried in his claws.

The flight of the fish hawk, his manœuvres while in search
of fish, and his manner of seizing his prey, are deserving of
particular notice. In leaving the nest, he usually flies direct
till he comes to the sea, then sails around, in easy curving
lines, turning sometimes in the air as on a pivot, apparently
without the least exertion, rarely moving the wings, his legs
extended in a straight line behind, and his remarkable length,
and curvature, or bend of wing, distinguishing him from all
other hawks. The height at which he thus elegantly glides
is various, from one hundred to one hundred and fifty, and
two hundred feet, sometimes much higher, all the while calmly
reconnoitering the face of the deep below. Suddenly he is
seen to check his course, as if struck by a particular object,
which he seems to survey for a few moments with such steadiness, that he appears fixed in the air, flapping his wings.
This object, however, he abandons, or rather the fish he had
in his eye has disappeared, and he is again seen sailing around
as before. Now his attention is again arrested, and he descends

with great rapidity ; but ere he reaches the surface, shoots off on another course, as if ashamed that a second victim had escaped him. He now sails at a short height above the surface, and by a zigzag descent, and without seeming to dip his feet in the water, seizes a fish, which, after carrying a short distance, he probably drops, or yields up to the bald eagle, and again ascends, by easy spiral circles, to the higher regions of the air, where he glides about in all the ease and majesty of his species. At once, from this sublime aërial height, he descends like a perpendicular torrent, plunging into the sea with a loud rushing sound, and with the certainty of a rifle. In a few moments he emerges, bearing in his claws his struggling prey, which he always carries head foremost, and, having risen a few feet above the surface, shakes himself as a water spaniel would do, and directs his heavy and laborious course directly for the land. If the wind blow hard, and his nest lie in the quarter from whence it comes, it is amusing to observe with what judgment and exertion he beats to windward, not in a direct line, that is, *in the wind's eye*, but making several successive tacks to gain his purpose. This will appear the more striking, when we consider the size of the fish which he sometimes bears along. A shad was taken from a fish hawk near Great Egg Harbour, on which he had begun to regale himself, and had already ate a considerable portion of it ; the remainder weighed six pounds. Another fish hawk was passing Mr Beasley's, at the same place, with a large flounder in his grasp, which struggled and shook him so, that he dropt it on the shore. The flounder was picked up, and served the whole family for dinner. It is singular that the hawk never descends to pick up a fish which he happens to drop, either on the land or on the water. There is a kind of abstemious dignity in this habit of the hawk, superior to the gluttonous voracity displayed by most other birds of prey, particularly by the bald eagle, whose piratical robberies committed on the present species, have been already fully detailed in treating of his history. The hawk, however, in his fishing pursuits, sometimes mistakes his mark, or over-rates his strength, by striking fish too large and powerful for him to manage, by whom he is suddenly dragged under; and, though he sometimes succeeds in extricating himself, after being taken three or four times down, yet oftener both parties perish. The bodies of sturgeon, and several other large fish, with that of a fish hawk fast grappled in them, have, at different times, been found dead on the shore, cast up by the waves.

The fish hawk is doubtless the most numerous of all its genus within the United States. It penetrates far into the interior of the country up our large rivers, and their head waters. It may be said to line the sea-coast from Georgia to Canada. In some parts I have counted, at one view, more than twenty of their nests within half a mile. Mr Gardiner informs me, that, on the small island on which he resides, there are at least "three hundred nests of fish hawks that have young, which, on an average, consume probably not less than six hundred fish daily." Before they depart in the autumn, they regularly repair their nests, carrying up sticks, sods, &c., fortifying them against the violence of the winter storms, which, from this circumstance, they would seem to foresee and expect. But, notwithstanding all their precautions, they frequently, on their return in spring, find them lying in ruins around the roots of the tree; and sometimes the tree itself has shared the same fate. When a number of hawks, to the amount of twenty or upwards, collect together on one tree, making a loud squealing noise, there is generally a nest built soon after on the same tree. Probably this congressional assembly were settling the right of the new pair to the premises; or it might be a kind of wedding, or joyous festive meeting on the occasion. They are naturally of a mild and peaceable disposition, living together in great peace and harmony; for though with them, as in the best regulated communities, instances of attack and robbery occur among themselves, yet these instances are extremely rare. Mr Gardiner observes, that they are sometimes seen high in the air, sailing and cutting strange gambols, with loud vociferations, darting down several hundred feet perpendicular, frequently with part of a fish in one claw, which they seem proud of, and to claim *high hook*, as the fishermen call *him* who takes the greatest number. On these occasions, they serve as a barometer to foretell the changes of the atmosphere; for, when the fish hawks are seen thus sailing high in air, in circles, it is universally believed to prognosticate a change of weather, often a thunder storm, in a few hours. On the faith of the certainty of these signs, the experienced coaster wisely prepares for the expected storm, and is rarely mistaken.

There is one singular trait in the character of this bird, which is mentioned in treating of the purple grakle, and which I have since had many opportunities of witnessing. The grakles, or crow blackbirds, are permitted by the fish hawk to build their nests among the interstices of the sticks of which his own is constructed,—several pairs of grakles taking up their abode there, like humble vassals around the castle of their chief, laying, hatching their young, and living together in mutual harmony. I have found no less than four of these nests clustered around the sides of the former, and a fifth fixed on the nearest branch of the adjoining tree; as if the proprietor of this last, unable to find an unoccupied corner on the premises, had been anxious to share, as much as possible, the company and protection of this generous bird.

The fish hawk is twenty-two inches in length, and five feet three inches in extent; the bill is deep black, the upper as well as lower cere (for the base of the lower mandible has a loose moveable skin), and also the sides of the mouth, from the nostrils backwards, are light blue; crown and hind head pure white, front streaked with brown; through the eye, a bar of dark blackish brown passes to the neck behind, which, as well as the whole upper parts, is deep brown, the edges of the feathers lighter; shafts of the wing-quills, brownish white; tail, slightly rounded, of rather a paler brown than the body, crossed with eight bars of very dark brown; the wings, when shut, extend about an inch beyond the tail, and are nearly black towards the tips; the inner vanes of both quill and tail-feathers are whitish, barred with brown; whole lower parts, pure white, except the thighs, which are covered with short plumage, and streaked down the fore part with pale brown; the legs and feet are a very pale light blue, prodigiously strong and disproportionably large; they are covered with flat scales of remarkable strength and thickness, resembling, when dry, the teeth of a large rasp, particularly on the soles, intended, no doubt, to enable the bird to seize with more security his slippery prey; the thighs are long, the legs short, feathered a little below the knee, and, as well as the feet and claws, large; the latter hooked into semicircles, black, and very sharp pointed; the iris of the eye, a fiery yellow orange.

The female is full two inches longer; the upper part of the head, of a less pure white, and the brown streaks on the front spreading more over the crown; the throat and upper part of the breast are also dashed with large blotches of a pale brown, and the bar passing through the eye, not of so dark a brown. The toes of both are exceedingly strong and warty, and the hind claw a full inch and a quarter in diameter. The feathers on the neck and hind head are long and narrow, and generally erected when the bird is irritated, resembling those of the eagle. The eye is destitute of the projecting bone common to most of the falcon tribe; the nostril, large, and of a curving triangular shape. On dissection, the two glands on the rump, which supply the bird with oil for lubricating its feathers to protect them from the wet, were found to be remarkably large, capable, when opened, of admitting the end of the finger, and contained a large quantity of white greasy matter, and some pure yellow oil; the gall was in small quantity. The numerous convolutions and length of the intestines surprised me; when carefully extended, they measured within an inch or two of nine feet, and were no thicker than those of a robin! The crop, or craw, was middle sized, and contained a nearly dissolved fish; the stomach was a large oblong pouch, capable of considerable distension, and was also filled with half-digested fish : no appearance of a muscular gizzard.

By the descriptions of European naturalists, it would appear that this bird, or one near akin to it, is a native of the eastern continent in summer, as far north as Siberia ; the bald buzzard of Turton almost exactly agreeing with the present species in size, colour, and manners, with the exception of its breeding or making its nest among the reeds, instead of on trees. Mr Bewick, who has figured and described the female of this bird under the appellation of the osprey, says, that "it builds on the ground, among reeds, and lays three or four eggs, of an elliptical form, rather less than those of a hen." This difference of habit may be owing to particular local circumstances, such deviations being usual among many of our native birds. The Italians are said to compare its descent upon the water to a piece of lead falling upon that element ; and distinguish it by the name of *Aquila plumbina,* or the leaden eagle. In the United States it is everywhere denominated the fish hawk, or fishing hawk, a name truly expressive of its habits.

The regular arrival of this noted bird at the vernal equinox when the busy season of fishing commences, adds peculiar interest to its first appearance, and procures it many a benediction from the fisherman. With the following lines, illustrative of these circumstances, I shall conclude its history :—

Soon as the sun, great ruler of the year,
Bends to our northern climes his bright career,
And from the caves of ocean calls from sleep
The finny shoals and myriads of the deep ;
When freezing tempests back to Greenland ride,
And day and night the equal hours divide ;
True to the season, o'er our sea-beat shore,
The sailing osprey high is seen to soar,
With broad unmoving wing ; and, circling slow,
Marks each loose straggler in the deep below ·

Sweeps down like lightning ! plunges with a roar !
And bears his struggling victim to the shore.

The long-housed fisherman beholds with joy,
The well-known signals of his rough employ ;
And, as he bears his nets and oars along,
Thus hails the welcome season with a song :—

THE FISHERMAN'S HYMN.

The osprey sails above the sound,
  The geese are gone, the gulls are flying;
The herring shoals swarm thick around,
  The nets are launch'd, the boats are plying;
    Yo, ho, my hearts ! let's seek the deep,
      Raise high the song, and cheerly wish her,
    Still as the bending net we sweep,
      " God bless the fish hawk and the fisher ! "

She brings us fish—she brings us spring,
  Good times, fair weather, warmth, and plenty,
Fine store of shad, trout, herring, ling,
  Sheepshead, and drum, and old-wives dainty.
    Yo, ho, my heart ! let's seek the deep,
      Ply every oar, and cheerly wish her,
    Still as the bending net we sweep,
      " God bless the fish hawk and the fisher ! "

She rears her young on yonder tree,
  She leaves her faithful mate to mind 'em ;
Like us, for fish, she sails to sea,
  And, plunging, shews us where to find 'em.
    Yo, ho, my hearts ! let's seek the deep,
      Ply every oar, and cheerly wish her,
    While the slow bending net we sweep,
      " God bless the fish hawk and the fisher ! "

FISH CROW. (*Corvus ossifragus.*)

PLATE XXXVII.—Fig 2.

Peale's Museum, No. 1369.

CORVUS OSSIFRAGUS.—Wilson.*

Corvus ossifragus, *Bonap. Synop.* p. 57.

This is another roving inhabitant of our sea-coasts, ponds, and river shores, though a much less distinguished one than the preceding, this being the first time, as far as I can learn, that he has ever been introduced to the notice of the world.

I first met with this species on the sea-coasts of Georgia, and observed that they regularly retired to the interior as evening approached, and came down to the shores of the river Savannah by the first appearance of day. Their voice first attracted my notice, being very different from that of the common crow, more hoarse and guttural, uttered as if something stuck in their throat, and varied into several modulations as they flew along. Their manner of flying was also unlike the others, as they frequently sailed about, without flapping the wings, something in the manner of the raven; and I soon perceived that their food, and their mode of procuring it, were also both different: their favourite haunts being about the banks of the river, along which they usually sailed, dexterously

* This is a very curious bird, first named and described by our author. It is one of the predacious species, with the nostrils clothed with feathers, and seems to feed nearly alone on fish or reptiles, doing almost no harm to the husbandman. In the latter circumstance, it resembles also our carrion crow, which often kills the common frog; and last summer I observed one flying with an adder in his bill. He had caught it on a detached piece of muir, and, on my approach, rose, taking the prey along with him, most probably before it was sufficiently despatched, as the writhings of the reptile caused him to alight several times at short distances, before being perfectly at ease. Being on horseback, I could not follow to see the end of the engagement. The species seems peculiar to the coast of North America, and does not extend very far northward.— ED.

snatching up, with their claws, dead fish, or other garbage, that floated on the surface. At the country seat of Stephen Elliot, Esq., near the Ogechee river, I took notice of these crows frequently perching on the backs of the cattle, like the magpie and jackdaw of Britain ; but never mingling with the common crows and differing from them in this particular, that the latter generally retire to the shore, the reeds, and marshes, to roost, while the fish crow always, a little before sunset, seeks the interior high woods to repose in.

On my journey through the Mississippi territory last year, I resided for some time at the seat of my hospitable friend Dr Samuel Brown, a few miles from Fort Adams, on the Mississippi. In my various excursions there, among the lofty fragrance-breathing magnolia woods and magnificent scenery that adorn the luxuriant face of nature in those southern regions, this species of crow frequently made its appearance, distinguished by the same voice and habits it had in Georgia. There is, in many of the ponds there, a singular kind of lizard, that swims about with its head above the surface, making a loud sound, not unlike the harsh jarring of a door. These the crow now before us would frequently seize with his claws, as he flew along the surface, and retire to the summit of a dead tree to enjoy his repast. Here I also observed him a pretty constant attendant at the pens where the cows were usually milked, and much less shy, less suspicious, and more solitary than the common crow. In the county of Cape May, New Jersey, I again met with these crows, particularly along Egg Harbour river ; and latterly on the Schuylkill and Delaware, near Philadelphia, during the season of shad and herring fishing, viz., from the middle of March till the beginning of June. A small party of these crows, during this period, regularly passed Mr Bartram's gardens to the high woods to roost, every evening a little before sunset, and as regularly returned, at or before sunrise every morning, directing their course towards the river. The fishermen along these rivers also inform me, that they have particularly remarked this crow by his croaking voice, and his fondness for fish; almost always hovering about their fishing places to glean up the refuse. Of their manner of breeding I can only say, that they separate into pairs, and build in tall trees near the sea or river shore; one of their nests having been built this season in a piece of tall woods near Mr Beasley's at Great Egg Harbour. The male of this nest furnished me with the figure in the plate, which was drawn of full size, and afterwards reduced to one-third the size of life, to correspond with the rest of the figures on the same plate. From the circumstance of six or seven being usually seen here together in the month of July, it is probable that they have at least four or five young at a time.

I can find no description of this species by any former writer. Mr Bartram mentions a bird of this tribe which he calls the *great sea-side crow ;* but the present species is considerably inferior in size to the common crow, and having myself seen and examined it in so many and remotely situated parts of the country, and found it in all these places alike, I have no hesitation in pronouncing it to be a new and hitherto undescribed species.

The fish crow is sixteen inches long, and thirty-three in extent; black all over, with reflections of steel-blue and purple; the chin is bare of feathers around the base of the lower mandible; upper mandible notched near the tip, the edges of both turned inwards about the middle; eye, very small, placed near the corner of the mouth, and of a dark hazel colour; recumbent hairs or bristles, large and long; ear-feathers, prominent; first primary, little more than half the length, fourth the longest; wings, when shut, reach within two inches of the tip of the tail; tail, rounded, and seven inches long from its insertion; thighs, very long; legs, stout; claws, sharp, long and hooked, hind one the largest, all jet black. Male and female much alike.

I would beg leave to recommend to the watchful farmers of the United States, that, in their honest indignation against the common crow, they would spare the present species, and not shower destruction indiscriminately on their black friends and enemies; at least on those who *sometimes* plunder them, and those who never molest or injure their property.

RINGED PLOVER. (*Charadrius hiaticula.*)

PLATE XXXVII.—FIG. 3.

Lath. Syn. v. p. 201. 8.—*Arct. Zool.* ii. No. 401.—Petit pluvier, à Collier, *Buff.* viii. p. 90. 6, *Pl. enl.* 921.—Pluvialis torquato minor, *Briss.* v. p. 63. 8. t. 5. f. 2.—*Turt. Syst.* p. 411. 2.—*Peale's Museum,* No. 4150.

CHARADRIUS MELODUS.—ORD.*

Charadrius melodus, *Bonap. Synop.* p. 296.—Charadrius Okenii? *Wagl. Syst. Av.* No. 24.

IT was not altogether consistent with my original plan, to introduce any of the grallæ, or waders, until I had advanced nearer to a close with the land birds; but as the scenery here seemed somewhat appropriate, I have taken the liberty of placing in it two birds, reduced to one-third of their natural size, both being *varieties* of their respective species, each of which will appear in their proper places, in some future part of this work, in full size, and in their complete plumage.

* This little plover has proved to be one of those very closely allied species so difficult of distinction, without a comparison with its congeners. The present figure is in the adult spring dress, and will be again represented by Bonaparte in that of autumn, in our third volume. The synonyms of Wilson are, of course, erroneous. Those also of Temminck, quoted in his Manual, and the observations on Wilson's plate and description, must share a similar fate. The observations in the nomenclature of Wilson, by the Prince of Musignano, will best explain how this species ought to stand. " *C. hiaticula* was at first given by Wilson as a variety, of which he intended to describe the type in a future volume; but when he did so in his seventh volume, he clearly and positively pointed out the difference in markings, habits, migration, voice, &c., between the two, which he then considered as distinct species, but without applying a new name; and we have no doubt that, if he had made out the index himself, he would then have supplied the deficiency, as he had before done in respect to some land birds. Mr Ord supplied this void, by calling it *C. melodus.*"—ED.

The ringed plover is very abundant on the low sandy shores of our whole sea-coast during summer. They run, or rather seem to glide, rapidly along the surface of the flat sands, frequently spreading out their wings and tail like a fan, and fluttering along, to draw or entice one away from their nests. These are formed with little art, being merely shallow concavities dug in the sand, in which the eggs are laid, and, during the day at least, left to the influence of the sun to hatch them. The parents, however, always remain near the spot to protect them from injury, and probably, in cold, rainy, or stormy weather, to shelter them with their bodies. The eggs are three, sometimes four, large for the bird, of a dun clay colour, and marked with numerous small spots of reddish purple.

The voice of these little birds, as they move along the sand, is soft and musical, consisting of a single plaintive note occasionally repeated. As you approach near their nests, they seem to court your attention, and, the moment they think you observe them, they spread out their wings and tail, dragging themselves along, and imitating the squeaking of young birds; if you turn from them, they immediately resume their proper posture, until they have again caught your eye, when they display the same attempts at deception as before. A flat, dry, sandy beach, just beyond the reach of the summer tides, is their favourite place for breeding.

This species is subject to great variety of change in its plumage. In the month of July, I found most of those that were breeding on Summers's Beach, at the mouth of Great Egg Harbour, such as I have here figured; but, about the beginning or middle of October, they had become much darker above, and their plumage otherwise varied. They were then collected in flocks; their former theatrical and deceptive manœuvres seemed all forgotten. They appeared more active than before, as well as more silent, alighting within a short distance of one, and feeding about without the least appearance of suspicion. At the commencement of winter, they all go off towards the south.

This variety of the ringed plover is seven inches long, and fourteen in extent; the bill is reddish yellow for half its length, and black at the extremity; the front and whole lower parts, pure white, except the side of the breast, which is marked with a curving streak of black, another spot of black bounding the front above; back and upper parts, very pale brown, inclining to ashy white, and intermixed with white; wings, pale brown; greater coverts, broadly tipt with white; interior edges of the secondaries, and outer edges of the primaries, white, and tipt with brown; tail, nearly even, the lower half white, brown towards the extremity, the outer feather pure white, the next white, with a single spot of black; eye, black and full, surrounded by a narrow ring of yellow; legs, reddish yellow; claws, black; lower side of the wings, pure white.

LITTLE SANDPIPER. (*Tringa pusilla.*)

PLATE XXXVII.—Fig. 4.

Lath. Syn. v. p. 184. 32.—*Arct. Zool.* ii. No. 397.—Cinclus dominicensis minor. *Briss.* v. p. 222. 13. t. 25. f. 2.—*Turt. Syst.* p. 410.—*Peale's Museum*, No. 4138.

TRINGA MINUTILLA ?—Vieillot.[*]

Tringa pusilla, *Bonap. Synop.* p. 319.

This is the least of its tribe in this part of the world, and in its mode of flight has much more resemblance to the snipe than to the sandpiper. It is migratory, departing early in October for the south. It resides chiefly among the sea marshes, and feeds among the mud at low water; springs with a zigzag irregular flight, and a feeble twit. It is not altogether confined to the neighbourhood of the sea, for I have found several of them on the shores of the Schuylkill, in the month of August. In October, immediately before they go

[*] The Prince of Musignano considers this species peculiar to America; that it is different from the *T. minuta* and *Temminckii* of Europe, and that it is not the Linnæan *T. pusilla*. If the latter opinion be correct, *pusilla* cannot be retained, and I have added with a query the name given to it by Vieillot.—Ed.

1. Barn Swallow. 2. Female. 3. White-bellied S. 4. Bank S.

away, they are usually very fat. Their nests or particular breeding places I have not been able to discover.

This minute species is found in Europe, and also at Nootka Sound, on the western coast of America. Length, five inches and a half; extent, eleven inches; bill and legs, brownish black; upper part of the breast, gray brown, mixed with white; back and upper parts, black; the whole plumage above, broadly edged with bright bay and yellow ochre; primaries, black; greater coverts, the same, tipt with white; eye, small, dark hazel; tail, rounded, the four exterior feathers on each side, dull white, the rest, dark brown; tertials, as long as the primaries; head above, dark brown, with paler edges; over the eye, a streak of whitish; belly and vent, white; the bill is thick at the base, and very slender towards the point; the hind toe, small. In some specimens the legs were of a dirty yellowish colour. Sides of the rump, white; just below the greater coverts, the primaries are crossed with white.

Very little difference could be perceived between the plumage of the males and females. The bay on the edges of the back and scapulars was rather brighter in the male, and the brown deeper.

BARN SWALLOW. (*Hirundo Americana.*)

PLATE XXXVIII.—Fig. 1. Male; Fig. 2. Female.

Peale's Museum, No. 7609.

HIRUNDO AMERICANA ?—Wilson.*

Hirundo rufa, *Bonap. Synop.* p. 64. — Hirundo Americana, *North. Zool.* ii. p. 329.

There are but few persons in the United States unacquainted with this gay, innocent, and active little bird. Indeed the whole tribe are so distinguished from the rest of small birds

---

* Wilson at once perceived the difference between the present species, and, as it is commonly called, the "chimney swallow" of Europe, though many of his contemporaries considered them only as varieties. The Prince of Musignano has, however, considered it as previously described

by their sweeping rapidity of flight, their peculiar aërial evolutions of wing over our fields and rivers, and through our very streets, from morning to night, that the light of heaven itself,

by Latham under the title of *H. rufa,* and again figured as the same by Vieillot.

The authors of the "Northern Zoology" have again appended the following note to their notice of the bird ; and, in the uncertainty, we have chosen to retain Wilson's original name, until the species is really determined from authentic specimens.

"It appears to us very doubtful whether the *Hirondelle à ventre roux de Cayenne* of Buffon (*Ed. Sonn.* xix. p. 35), of which methodists have made their *Hirundo rufa,* is really the same as the *H. Americana* of Wilson. From the evidence we at present have, we are disposed to consider them distinct. The only authentic account of the cayenne species is that given by Buffon, which all the compilers have since copied. From this, it appears to be only *five inches and a half long* (French measure ?), ours is fully seven. The front is whitish (*le front blanchâtre*); ours is very deep rufous. But the most remarkable difference between the birds is in the construction of their nests,—the cayenne bird building one *without* mud, and so long as sometimes to measure a foot and a half, with an opening *near the bottom ;* the *Americana* of Wilson, on the contrary, using a good deal of mud ; the length is only seven inches, and the opening *at top,* with an external rim, for the parents occasionally to sit upon. Until this matter is investigated, we cannot suppose that individuals of the *same species* would, in different countries, build their nests in such very dissimilar ways."

It appears to be exclusively American, and migrates from north to south, and the reverse. There is a great resemblance between the two species ; but they may be at once distinguished by the pure white and the rich chestnut which clothes the under parts of each, and they would seem to be another of those representing forms which are so frequent, and run so closely in colour and habits through both continents.

Wilson, when mentioning the distinction of this species, includes a difference in habit, from our species building in chimneys, and not in barns, like the American. Chimneys are by no means the common building place of the British swallow, although those in the neighbourhood of towns may use that resort for want of another, in the same way that those in a mining country use the neglected shafts. In the country, barns, shades of thrashing mills, or any outhouse with an open door or window, under the portico of a front door, are their constant building-place ; and although houses in the country have chimneys as well as those in town, they are very seldom, if ever, resorted to. Their nests are also of the same structure and materials, built with clay mingled

the sky, the trees, or any other common objects of Nature, are not better known than the swallows. We welcome their first appearance with delight, as the faithful harbingers and

with straw, and lined with feathers, placed against a rafter, beam, or wall, and open at top.[1] The eggs also very similar.

Bewick mentions a curious instance of variation, which may be also taken as a strong proof of the annual return of birds to the same building-places. "At Cameston Hall, near Bath, a pair of swallows built their nests on the upper part of the frame of an old picture over the chimney—coming in through a broken pane in the window of the room. They came three years successively, and, in all probability, would have continued to do so, if the room had not been put into repair, which prevented their access to it."

Swallows have been divided into various genera, as might be supposed from their being commonly indicated swallows, swifts, or martins. Some form among these are found in almost every country, except as we approach the poles; and in North America, where the whole *Hirundinidæ* will be comprised in six individuals, we have two real swallows, two martins, the very strongly formed purple swallow, and the representative of the swifts in *Chætura pelasgica.* These will come under observation as we proceed. The present, with the republican, or cliff swallow, figured by Bonaparte in his continuation, with that of Europe, are true forms of *Hirundo,* one which possesses great activity, though not so much strength in flight as the swifts, but which will show the more exact relative proportion of power between the members. They are very generally distributed, have the wings long, and the tail forked; the only form where these members are more extended, is in the genus *Macropterix,* lately formed by Mr Swainson from an Indian group, which will perhaps show the farthest development of the wings and tail, but which bear the same disproportion as in the broad-shaped and sickle-winged humming birds. In all their various flights, the motions are conducted with great celerity and elegance, and are directed by the rapid motion of the tail.

The subject of their migrations, which I believe takes place with all species, and in all countries, has occupied much speculation; of the fact there can now be no doubt, and the collection of vast crowds together before departure, seems more confined to this form than to any of the others; so far, at least, as my own observation has extended. The American species congregate; so do the republican swallows; and towards the end of August, our own may be seen daily in flocks, on the

---

[1] According to Professor Rennie, it is called, in Sweden, *Ladu Swala*, barn swallow; while, in the south of Europe, where chimneys are rare, it builds in gateways, porches, and galleries.

companions of flowery spring and ruddy summer; and when, after a long, frost-bound, and boisterous winter, we hear it announced, that "the swallows are come," what a train of charming ideas are associated with the simple tidings!

house tops or cornices, on railings, or on a bare tree, where the later broods are still fed and exercised by the parents, and the southern journey of the whole mass, as it were, delayed until all had required sufficient strength.

At times, these congregations are much greater than at others, or like some great assemblage from the neighbouring country. One of these took place in 1815 near Rotherham, and has been made the subject of an anonymous pamphlet, by a clergyman in that neighbourhood. The assemblage and departure is thus described in it:—"Early in the month of September 1815, the swallows, that beautiful and social tribe of the feathered race, began to assemble in the neighbourhood of Rotherham, at the willow ground, on the banks of the canal, preparatory to their migration to a warmer climate; and their numbers were daily augmented, until they became a vast flock, which no man could easily number. Thousands upon thousands—tens of thousands—and myriads; so great, indeed, that the spectator would almost have concluded, the whole swallow race were there collected in one huge host.

"It was their manner, while there, to rise from the willows in the morning, a little before six o'clock, when their thick columns literally darkened the sky. Their divisions were then into four, five, and sometimes into six grand wings, each of these filing and taking a different route,—one east, another west, another south; as if not only to be equally dispersed throughout the country, to provide food for their numerous troops, but also to collect with them whatever of their fellows, or straggling parties, might still be left behind.

"In the evening, about five o'clock, they began to return to their station, and continued coming in from all quarters, until nearly dark. It was here that you might see them go through their various aerial evolutions, in many a sportive ring and airy gambol, strengthening their pinions in these playful feats, for their long ethereal journey, as they cut the air and frolicked in the last beams of the setting sun, or lightly skimmed the surface of the glassy pool.

"The verdant enamel of summer had given place to the warm and mellow tints of autumn. The leaves were now fast falling from their branches, while the naked tops of many of the trees appeared. The golden sheaves were safely lodged in the barns, and the reapers had shouted their harvest-home. Frosty and misty mornings succeeded, the certain presages of the approach of winter. They were omens understood by the swallows, as signals for their march; and on the morning

The wonderful activity displayed by these birds forms a striking constrast to the slow habits of most other animals. It may be fairly questioned whether, among the whole feathered tribes which Heaven has formed to adorn this part of creation, there be any that, in the same space of time, pass over an equal extent of surface with the swallow. Let a person take his stand, on a fine summer evening, by a new-mown field, meadow, or river shore, for a short time, and among the numerous individuals of this tribe that flit before him, fix his eye on a particular one, and follow for a while all its circuitous labyrinths—its extensive sweeps—its sudden, rapidly reiterated zigzag excursions, little inferior to the lightning itself,—and then attempt, by the powers of mathematics, to calculate the length of the various lines it describes. Alas! even his omnipotent fluxions would avail him little here, and he would soon abandon the task in despair. Yet, that some definite conception may be formed of this extent, let us suppose that this little bird flies, in his usual way, at the rate of one mile in a minute, which, from the many experiments I have made, I believe to be within the truth ; and that he is so engaged for ten hours every day; and further, that this active life is extended to ten years (many of our small birds being known to live much longer, even in a state of domestication), the amount of all these, allowing three hundred and sixty-five days to a year, would give us two million one hundred and ninety thousand miles; upwards of eighty-seven times the circumference of the globe ! Yet this little winged seraph, if I may so speak, who, in a few days, and at will, can pass from the borders of the arctic regions to the torrid zone, is forced, when winter approaches, to descend to the bottoms of lakes, rivers, and mill-ponds, to bury itself in the mud with eels and snapping turtles, or to creep ingloriously into a cavern, a rat-hole, or a hollow tree, there to doze, with snakes, toads, and

of the 7th of October, their mighty army broke up their encampment, debouched from their retreat, rising, covered the heavens with their legions, and, directed by an unerring guide, took their trackless way."
—Ed.

other reptiles, until the return of spring! Is not this true, ye
wise men of Europe and America, who have published so many
credible narratives on this subject? The geese, the ducks,
the cat bird, and even the wren, which creeps about our outhouses in summer like a mouse, are all acknowledged to be
migratory, and to pass to southern regions at the approach of
winter: the swallow alone, on whom Heaven has conferred
superior powers of wing, must sink in torpidity at the bottom
of our rivers, or doze all winter in the caverns of the earth. I
am myself something of a traveller, and foreign countries afford
many novel sights: should I assert, that in some of my pere-
grinations I had met with a nation of Indians, all of whom, old
and young, at the commencement of cold weather, descend to
the bottom of their lakes and rivers, and there remain until the
breaking up of frost; nay, should I affirm, that thousands of
people, in the neighbourhood of this city, regularly undergo
the same semi-annual submersion—that I myself had fished
up a whole family of these from the bottom of Schuylkill,
where they had lain *torpid* all winter, carried them home, and
brought them all comfortably to themselves again; should I
even publish this in the learned pages of the *Transactions* of
our Philosophical Society,—who would believe me? Is, then,
the organisation of a swallow less delicate than that of a man?
Can a bird, whose vital functions are destroyed by a short pri-
vation of pure air and its usual food, sustain, for six months,
a situation where the most robust man would perish in a few
hours or minutes? Away with such absurdities! they are
unworthy of a serious refutation. I should be pleased to meet
with a man who has been personally more conversant with
birds than myself, who has followed them in their wide and
devious routes—studied their various manners—mingled with
and marked their peculiarities more than I have done; yet
the miracle of a resuscitated swallow, in the depth of winter,
from the bottom of a mill-pond, is, I confess, a phenomenon in
ornithology that I have never met with.

What better evidence have we that these fleet-winged tribes,

instead of following the natural and acknowledged migrations
of many other birds, lie torpid all winter in hollow trees, caves,
and other subterraneous recesses ? That the chimney swallow,
in the early part of summer, may have been found in a
hollow tree, and in great numbers too, is not denied; such
being, in some places of the country (as will be shown in the
history of that species), their actual places of rendezvous on
their first arrival, and their common roosting place long
after : or that the bank swallows, also, soon after their
arrival, in the early part of spring, may be chilled by the
cold mornings which we frequently experience at that season,
and be found in this state in their holes, I would as little
dispute ; but that either the one or the other has ever been
found, *in the midst of winter*, in a state of *torpidity*, I
do not—cannot believe.   Millions of trees, of all dimensions,
are cut down every fall and winter of this country, where, in
their proper season, swallows swarm around us.   Is it, therefore, in the least probable that we should, only once or twice
in an age, have no other evidence than one or two solitary and
very suspicious reports of a Mr Somebody having made a
discovery of this kind ?   If caves were their places of winter
retreat, perhaps no country on earth could supply them with
a greater choice.   I have myself explored many of these, in
various parts of the United States, both in winter and in spring,
particularly in that singular tract of country in Kentucky
called the Barrens, where some of these subterraneous caverns
are several miles in length, lofty and capacious, and pass
under a large and deep river—have conversed with the salt-
petre workers by whom they are tenanted ; but never heard
or met with one instance of a swallow having been found there
in winter.   These people treated such reports with ridicule.

It is to be regretted that a greater number of experiments
have not been made, by keeping live swallows through the
winter, to convince these believers in the torpidity of birds of
their mistake.   That class of cold-blooded animals which are
known to become torpid during winter, and of which hun-

dreds and thousands are found every season, are subject to the same when kept in a suitable room for experiment. How is it with the swallows in this respect? Much powerful testimony might be produced on this point: the following experiments, recently made by Mr James Pearson of London, and communicated by Sir John Trevelyan, Bart., to Mr Bewick, the celebrated engraver in wood, will be sufficient for our present purpose, and throw great light on this part of the subject.*

" Five or six of these birds were taken about the latter end of August 1784, in a bat fowling-net at night. They were put separately into small cages, and fed with nightingale's food : in about a week or ten days, they took food of themselves ; they were then put all together into a deep cage, four feet long, with gravel at the bottom ; a broad shallow pan, with water, was placed in it, in which they sometimes washed themselves, and seemed much strengthened by it. One day Mr Pearson observed that they went into the water with unusual eagerness, hurrying in and out again repeatedly with such swiftness as if they had been suddenly seized with a frenzy. Being anxious to see the result, he left them to themselves about half an hour, and, going to the cage again, found them all huddled together in a corner, apparently dead ; the cage was then placed at a proper distance from the fire, when only two of them recovered, and were as healthy as before : the rest died. The two remaining ones were allowed to wash themselves occasionally for a short time only ; but their feet soon after became swelled and inflamed, which Mr Pearson attributed to their perching, and they died about Christmas. Thus the first year's experiment was in some measure lost. Not discouraged by the failure of this, Mr Pearson determined to make a second trial the succeeding year, from a strong desire of being convinced of the truth of their going into a state of torpidity. Accordingly, the next season, having taken some more birds, he put them into the cage, and in every respect pursued the same methods as with

* See Bewick's British Birds, vol. i. p. 254.

the last; but, to guard their feet from the bad effects of the damp and cold, he covered the perches with flannel, and had the pleasure to observe that the birds throve extremely well; they sang their song during the winter, and, soon after Christmas, began to moult, which they got through without any difficulty, and lived three or four years, regularly moulting every year at the usual time. On the renewal of their feathers, it appeared that their tails were forked exactly the same as in those birds which return hither in the spring, and in every respect their appearance was the same. These birds, says Mr Pearson, were exhibited to the Society for Promoting Natural History, on the 14th day of February 1786, at the time they were in a deep moult, during a severe frost, when the snow was on the ground. Minutes of this circumstance were entered in the books of the society. These birds died at last from neglect, during a long illness which Mr Pearson had: they died in the summer. Mr Pearson concludes his very interesting account in these words :—'20th January 1797—I have now in my house, No. 21 Great Newport Street, Long Acre, four swallows in moult, in as perfect health as any bird ever appeared to be when moulting.' "

The barn swallow of the United States has hitherto been considered by many writers as the same with the common chimney swallow of Europe. They differ, however, considerably in colour, as well as in habits; the European species having the belly and vent white, the American species those parts of a bright chestnut; the former building in the corners of chimneys, near the top, the latter never in such places; but usually in barns, sheds, and other outhouses, on beams, braces, rafters, &c. It is difficult to reconcile these constant differences of manners and markings in one and the same bird; I shall therefore take the liberty of considering the present as a separate and distinct species.

The barn swallow arrives in this part of Pennsylvania from the south on the last week in March or the first week in April, and passes on to the north, as far, at least, as the

river St Lawrence. On the east side of the great range of
the Alleghany, they are dispersed very generally over the
country, wherever there are habitations, even to the summit
of high mountains; but, on account of the greater coldness
of such situations, are usually a week or two later in
making their appearance there. On the 16th of May, being
on a shooting expedition on the top of Pocano Mountain,
Northampton, when the ice on that and on several successive
mornings was more than a quarter of an inch thick, I
observed with surprise a pair of these swallows which had
taken up their abode on a miserable cabin there. It was then
about sunrise, the ground white with hoar frost, and the male
was twittering on the roof by the side of his mate with great
sprightliness. The man of the house told me that a single
pair came regularly there every season, and built their nest
on a projecting beam under the eaves, about six or seven feet
from the ground. At the bottom of the mountain, in a large
barn belonging to the tavern there, I counted upwards of
twenty nests, all seemingly occupied. In the woods they are
never met with; but, as you approach a farm, they soon catch
the eye, cutting their gambols in the air. Scarcely a barn,
to which these birds can find access, is without them; and, as
public feeling is universally in their favour, they are seldom
or never disturbed. The proprietor of the barn last mentioned, a German, assured me that if a man permitted the
swallows to be shot, his cows would give bloody milk, and
also that no barn where swallows frequented would ever be
struck with lightning; and I nodded assent. When the
tenets of superstition "lean to the side of humanity," one
can readily respect them. On the west side of the Alleghany
these birds become more rare. In travelling through the
States of Kentucky and Tennessee, from Lexington to the
Tennessee river, in the months of April and May, I did not see
a single individual of this species; though the purple martin,
and, in some places, the bank swallow, was numerous.

Early in May they begin to build. From the size and

structure of the nest, it is nearly a week before it is completely finished. One of these nests, taken on the 21st of June from the rafter to which it was closely attached, is now lying before me. It is in the form of an inverted cone, with a perpendicular section cut off on that side by which it adhered to the wood. At the top it has an extension of the edge or offset, for the male or female to sit on occasionally, as appeared by the dung ; the upper diameter was about six inches by five, the height externally seven inches. This shell is formed of mud, mixed with fine hay, as plasterers do their mortar with hair, to make it adhere the better ; the mud seems to have been placed in regular strata or layers, from side to side ; the hollow of this cone (the shell of which is about an inch in thickness) is filled with fine hay, well stuffed in ; above that is laid a handful of very large downy geese feathers. The eggs are five, white, specked, and spotted all over with reddish brown. Owing to the semi-transparency of the shell, the eggs have a slight tinge of flesh colour. The whole weighs about two pounds.

They have generally two broods in the season. The first make their appearance about the second week in June ; and the last brood leave the nest about the 10th of August. Though it is not uncommon for twenty, and even thirty, pair to build in the same barn, yet everything seems to be conducted with great order and affection ; all seems harmony among them, as if the interest of each were that of all. Several nests are often within a few inches of each other ; yet no appearance of discord or quarrelling takes place in this peaceful and affectionate community.

When the young are fit to leave the nest, the old ones entice them out by fluttering backwards and forwards, twittering and calling to them every time they pass ; and the young exercise themselves for several days in short essays of this kind within doors before they first venture abroad. As soon as they leave the barn, they are conducted by their parents to the trees or bushes by the pond, creek, or river shore, or other suitable situation, where their proper food is most abundant, and where they can be fed with the greatest convenience to both parties. Now and then they take a short excursion themselves, and are also frequently fed while on wing by an almost instantaneous motion of both parties, rising perpendicularly in air, and meeting each other. About the middle of August they seem to begin to prepare for their departure. They assemble on the roof in great numbers, dressing and arranging their plumage, and making occasional essays, twittering with great cheerfulness. Their song is a kind of sprightly warble, sometimes continued for a considerable time. From this period to the 8th of September, they are seen near the Schuylkill and Delaware every afternoon, for two or three hours before sunset, passing along to the south in great numbers, feeding as they skim along. I have counted several hundreds pass within sight in less than a quarter of an hour, all directing their course towards the south. The reeds are now their regular roosting places; and about the middle of September there is scarcely an individual of them to be seen. How far south they continue their route is uncertain; none of them remain in the United States. Mr Bartram informs me, that, during his residence in Florida, he often saw vast flocks of this and our other swallows passing from the peninsula towards the south in September and October, and also on their return to the north about the middle of March. It is highly probable that, were the countries to the south of the Gulf of Mexico, and as far south as the great river Maranon, visited and explored by a competent naturalist, these regions would be found to be the winter rendezvous of the very birds now before us, and most of our other migratory tribes.

In a small volume which I have lately met with, entitled, "An Account of the British Settlement of Honduras," by Captain George Henderson, of the 5th West India Regiment, published in London in 1809, the writer, in treating of that part of its natural history which relates to birds, gives the following particulars :—" Myriads of swallows," says he, " are also the occasional inhabitants of Honduras. The time of their residence is generally confined to the period of the rains [that is, from October to February], after which they totally disappear. There is something remarkably curious and deserving of notice in the ascent of these birds. As soon as the dawn appears, they quit their place of rest, which is usually chosen amid the rushes of some watery savannah ; and invariably rise to a certain height, in a compact spiral form, and which at a distance often occasions them to be taken for an immense column of smoke. This attained, they are then seen separately to disperse in search of food, the occupation of their day. To those who may have had the opportunity of observing the phenomenon of a waterspout, the similarity of evolution in the ascent of these birds will be thought surprisingly striking. The descent, which regularly takes place at sunset, is conducted much in the same way, but with inconceivable rapidity ; and the noise which accompanies this can only be compared to the falling of an immense torrent, or the rushing of a violent gust of wind. Indeed, to an observer, it seems wonderful that thousands of these birds are not destroyed, in being thus propelled to the earth with such irresistible force." *

How devoutly it is to be wished that the natural history of those regions were more precisely known, so absolutely necessary as it is to the perfect understanding of this department of our own !

The barn swallow is seven inches long, and thirteen inches in extent ; bill, black ; upper part of the head, neck, back, rump, and tail-coverts, steel blue, which descends rounding on the breast ; front and chin, deep chestnut ; belly, vent, and lining of the wing, light chestnut ; wings and tail, brown black, slightly glossed with reflections of green ; tail, greatly forked, the exterior feather on each side an inch and a half longer than the next, and tapering towards the extremity, each

* Henderson's Honduras, p. 119.

feather, except the two middle ones, marked on its inner vane with an oblong spot of white ; lores, black ; eye, dark hazel ; sides of the mouth, yellow ; legs, dark purple.

The female differs from the male in having the belly and vent rufous white, instead of light chestnut; these parts are also slightly clouded with rufous ; and the exterior tail-feathers are shorter.

These birds are easily tamed, and soon become exceedingly gentle and familiar. I have frequently kept them in my room for several days at a time, where they employed themselves in catching flies, picking them from my clothes, hair, &c., calling out occasionally as they observed some of their old companions passing the windows.

GREEN-BLUE, OR WHITE-BELLIED SWALLOW.
(*Hirundo viridis.*)

PLATE XXXVIII.—Fig. 3.

Peale's Museum, No. 7707.

HIRUNDO BICOLOR.—Vieillot.*

Hirundo viridis, *Aud. Ann. Lyc. of New York*, i. p. 166.—The White-bellied Swallow, *Aud. Orn. Biog.* i. p. 491, pl. 98.—Hirundo bicolor, *Bonap. Synop.* p. 65.—*North. Zool.* ii. p. 328.

This is the species hitherto supposed by Europeans to be the same with their common martin, *Hirundo urbica*, a bird nowhere to be found within the United States. The English

* This beautiful and highly curious little bird has, like the last, been confused with a European species, *H. urbica*. Gmelin and Latham esteem it only a variety, while other writers make it identical. From the European martin it may always at once be distinguished by wanting the purely white rump, so conspicuous during the flight of the former. The priority of the name will be in favour of Vieillot, and it should stand as *H. bicolor* of that naturalist.

The martins possess a greater preponderance of power in the wings over the tail than the swallows ; and their flight, as our author remarks, is consequently more like sailing than flying. All their turns are round and free, and performed most frequently in large sweeps, without any

martin is blue black above, the present species greenish blue; the former has the whole rump white, and the legs and feet are covered with short white downy feathers; the latter has nothing of either. That ridiculous propensity in foreign writers to consider most of our birds as varieties of their own, has led them into many mistakes, which it shall be the business of the author of the present work to point out decisively, wherever he may meet with them.

The white-bellied swallow arrives in Pennsylvania a few days

motion of the wings. In their other forms they hardly differ, though almost any one will say this is a martin, that a swallow. I am inclined to keep them as a subordinate group, and there also would be placed the water martins, which have already been made into a genus by Boje. They are all nearly of the same form, are gregarious, and build and feed in large companies.

The white-bellied swallow bears more analogy to the water martins than that of Europe, or those which frequent inland districts. According to Audubon, they sit and roost on the sedges and tall water plants, as well as upon the bushes; and they sometimes in the beginning of autumn, as mentioned by our author, collect on the shores or sandbanks of rivers, in considerable numbers. About the end of July, in the present year, I had an opportunity of seeing the latter incident take place with our common sand martin (*H. riparia*), one very hot evening, when residing on the shores of the Solway Frith, where the beach is unusually flat and sandy. Several hundreds of these were collected upon a space not exceeding two acres, most of them were upon the ground, a few occasionally rising and making a short circuit. At this part, a small stream entered the sea, and they seemed partly resting and washing, and partly feeding on a small fly that had apparently come newly to existence, and covered the sands in immense profusion. None of our other species mingled, though they were abundant in the neighbourhood.

The American bird is also remarkable as being a berry eater, an occurrence nearly unknown among the *Hirundinidæ*. Neither is their breeding in holes of trees frequent among them. The only instance of a similar propensity is one related of the common swift, in " Loudon's Magazine of Natural History," which, however, is a species more likely to suit itself to circumstances of the kind, as it appears to have done in this instance, where it formed its breeding place in the deserted holes of woodpeckers. Audubon has traced their migrations through the year, and has proved that they winter in Louisiana. I believe they belong exclusively to the New World.—ED.

later than the preceding species. It often takes possession of an apartment in the boxes appropriated to the purple martin ; and also frequently builds and hatches in a hollow tree. The nest consists of fine loose dry grass, lined with large downy feathers, rising above its surface, and so placed as to curl inwards, and completely conceal the eggs. These last are usually four or five in number, and pure white. They also have two broods in the season.

The voice of this species is low and guttural ; they are more disposed to quarrel than the barn swallows, frequently fighting in the air for a quarter of an hour at a time, particularly in spring, all the while keeping up a low rapid chatter. They also sail more in flying ; but, during the breeding season, frequent the same situations in quest of similar food. They inhabit the northern Atlantic States as far as the district of Maine, where I have myself seen them ; and my friend Mr Gardiner informs me that they are found on the coast of Long Island and its neighbourhood. About the middle of July, I observed many hundreds of these birds sitting on a flat sandy beach near the entrance of Great Egg Harbour. They were also very numerous among the myrtles of these low islands, completely covering some of the bushes. One man told me that he saw one hundred and two shot at a single discharge. For sometime before their departure, they subsist principally on the myrtle berries (*Myrica cerifera*), and become extremely fat. They leave us early in September.

This species appears to have remained hitherto undescribed, owing to the misapprehension before mentioned. It is not perhaps quite so numerous as the preceding, and rarely associates with it to breed, never using mud of any kind in the construction of its nest.

The white-bellied swallow is five inches and three quarters long, and twelve inches in extent ; bill and eye, black ; upper parts, a light glossy greenish blue ; wings, brown black, with slight reflections of green ; tail, forked, the two exterior feathers being about a quarter of an inch longer than the middle ones, and all of a uniform brown black ; lores, black ; whole lower parts pure white ; wings, when shut, extend about a quarter of an inch beyond the tail ; legs, naked, short, and strong, and, as well as the feet, of a dark purplish flesh colour ; claws, stout.

The female has much less of the greenish gloss than the male, the colours being less brilliant ; otherwise alike.

BANK SWALLOW, OR SAND MARTIN.
(*Hirundo riparia.*)

PLATE XXXVIII.—Fig. 4.

Lath. Syn. iv. p. 568, 10.—*Arct. Zool.* ii. No. 332.—L'Hirondelle de rivage, *Buff.* vi. 632, *Pl. enl.* 543, f. 2.—*Turt. Syst.* 629.—*Peale's Museum*, No. 7637.

HIRUNDO? RIPARIA ?—Linnæus.[*]

Hirundo riparia, *Bonap. Synop.* p. 65.—Cotile riparia, *Boje.*

This appears to be the most sociable with its kind, and the least intimate with man, of all our swallows, living together in large communities of sometimes three or four hundred.

[*] I have been unable to compare specimens of these birds from both countries, but from the best authorities, I am induced to consider them identical. A doubt has been expressed by Vieillot, who considered the American bird as possessing a greater length of tarsus, and having that part also clothed with short plumes. Bonaparte has, again, from actual comparison, said they were entirely similar.

As in America, they are the first swallow which appears in this country, arriving soon after the commencement of March. Their breeding-places are in the same situations, but often pierced into the banks for a much greater length. If the bank is sandy and easily scratched, seven or eight feet will scarcely reach the extremity, a wonderful length, if we consider the powers of the worker.

They are abundant over every part of North America, and were met by Dr Richardson in the 68th parallel. "We observed," says that naturalist, "thousands of these sand martins fluttering at the entrance of their burrows, near the mouth of the Mackenzie, in the 68th parallel, on the 4th of July. They are equally numerous in every district of the Fur Countries, wherein banks suitable for burrowing exist ; but it is not likely that they ever rear more than one brood north of the Lake Superior."—Ed.

On the high sandy bank of a river, quarry, or gravel pit, at a foot or two from the surface, they commonly scratch out holes for their nests, running them in a horizontal direction to the depth of two and sometimes three feet. Several of these holes are often within a few inches of each other, and extend in various strata along the front of the precipice, sometimes for eighty or one hundred yards. At the extremity of this hole, a little fine dry grass, with a few large downy feathers, form the bed on which their eggs, generally five in number, and pure white, are deposited. The young are hatched late in May ; and here I have taken notice of the common crow, in parties of four or five, watching at the entrance of these holes, to seize the first straggling young that should make its appearance. From the clouds of swallows that usually play round these breeding places, they remind one at a distance of a swarm of bees.

The bank swallow arrives here earlier than either of the preceding ; begins to build in April, and has commonly two broods in the season. Their voice is a low mutter. They are particularly fond of the shores of rivers, and, in several places along the Ohio, they congregate in immense multitudes. We have sometimes several days of cold rain and severe weather after their arrival in spring, from which they take refuge in their holes, clustering together for warmth, and have been frequently found at such times in almost a lifeless state with the cold ; which circumstance has contributed to the belief that they lie torpid all winter in these recesses. I have searched hundreds of these holes in the months of December and January, but never found a single swallow, dead, living, or torpid. I met with this bird in considerable numbers on the shores of the Kentucky river, between Lexington and Danville. They likewise visit the seashore in great numbers previous to their departure, which continues from the last of September to the middle of October.

The bank swallow is five inches long, and ten inches in extent ; upper parts mouse coloured, lower white, with a

1. Chimney Swallow. 2. Purple Martin. 3. Female. 4. Connecticut Warbler.

39.

band of dusky brownish across the upper part of the breast; tail, forked, the exterior feather slightly edged with whitish; lores and bill, black; legs, with a few tufts of downy feathers behind; claws, fine pointed and very sharp; over the eye, a streak of whitish; lower side of the shafts, white; wings and tail, darker than the body. The female differs very little from the male.

This bird appears to be in nothing different from the European species; from which circumstance, and its early arrival here, I would conjecture that it passes to a high northern latitude on both continents.

CHIMNEY SWALLOW. (*Hirundo pelasgia.*)

PLATE XXXIX.—Fig. 1.

Lath. Syn. v. p. 583, 32.—*Catesb. Car. App.* t. 8.—Hirondelle de la Caroline, *Buff.* vi. p. 700.—Hirundo Carolinensis, *Briss.* ii. p. 501, 9.—Aculeated Swallow, *Arct. Zool.* ii. No. 335, 18.—*Turt. Syst.* p. 630.—*Peale's Museum,* No. 7663.
CHÆTURA PELASGIA.—Stephens.*

Chætura pelasgia, *Steph. Cont. Sh. Zool. Sup.* p. 76.—Cypselus pelasgius, *Bonap. Synop.* p. 63.

This species is peculiarly our own, and strongly distinguished from all the rest of our swallows by its figure, flight, and manners. Of the first of these, the representation in the plate will give a correct idea; its other peculiarities shall be detailed as fully as the nature of the subject requires.

---

* This species has been taken as the type of Mr Stephens' genus *Chætura.* In form they resemble the swifts; and the first observed distinction will be the structure of the tail, where the quills of the feathers are elongated, and run to a sharp or subulated point. The bill is more compressed laterally; the legs and feet possess very great muscularity; the toes alone are scaled, and the tarsi are covered with a naked skin, through which the form of the muscles is plainly visible; the claws are much hooked. All these provisions are necessary to their mode of life. Without some strong support, they could not cling for a great length of time in the hollows of trees, or in chimneys; and their tails are used, in the manner of a woodpecker, to assist the power of

This swallow, like all the rest of its tribe in the United States, is migratory, arriving in Pennsylvania late in April or early in May, and dispersing themselves over the whole country wherever there are vacant chimneys in summer sufficiently high and convenient for their accommodation. In no other situation with us are they observed at present to build. This circumstance naturally suggests the query, Where did these birds construct their nests before the arrival of Europeans in this country, when there were no such places for their accommodation? I would answer, Probably in the same situations in which they still continue to build in the remote regions of our western forests, where European improvements of this kind are scarcely to be found, namely, in the hollow of a tree, which in some cases has the nearest resemblance to their present choice, of any other. One of the first settlers in the State of Kentucky informed me, that he cut down a large hollow beech tree, which contained forty or fifty nests of the chimney swallow, most of which, by the fall of the tree, or by the weather, were lying at the bottom of the hollow; but sufficient fragments remained adhering to the sides of the tree to enable him to number them. They appeared, he said, to be of many years' standing. The present site which they have chosen must, however, hold out many more advantages than the former, since we see that, in the whole thickly settled parts of the United States, these birds have uniformly adopted this new convenience, not a single pair being observed to prefer the woods. Security from birds of prey and other animals—from storms that frequently overthrow the timber, and the numerous ready conveniences which these new situations afford, are doubtless some of the advan-

the strong feet. They present, in a beautiful manner, the scansorial form among the *Fissirostres;* one species, the *Ch. senex* (*Cypselus senex,* Temm.), even feeds in the manner of the true climbers, running up the steep rocks, assisted by its tail, in search of food.

The group will contain a considerable number. We have them from India, North and South America, and New Holland, but I am not aware that Africa has yet produced any species.—ED.

tages. The choice they have made certainly bespeaks something more than mere unreasoning instinct, and does honour to their discernment.

The nest of this bird is of singular construction, being formed of very small twigs, fastened together with a strong adhesive glue or gum, which is secreted by two glands, one on each side of the hind head, and mixes with the saliva. With this glue, which becomes hard as the twigs themselves, the whole nest is thickly besmeared. The nest itself is small and shallow, and attached by one side or edge to the wall, and is totally destitute of the soft lining with which the others are so plentifully supplied. The eggs are generally four, and white. This swallow has two broods in the season. The young are fed at intervals during the greater part of the night, a fact which I have had frequent opportunities of remarking both here and in the Mississippi territory. The noise which the old ones make in passing up and down the funnel has some resemblance to distant thunder. When heavy and long-continued rains occur, the nest, losing its hold, is precipitated to the bottom. This disaster frequently happens. The eggs are destroyed; but the young, though blind (which they are for a considerable time), sometimes scramble up along the vent, to which they cling like squirrels, the muscularity of their feet, and the sharpness of their claws, at this tender age, being remarkable. In this situation they continue to be fed for perhaps a week or more. Nay, it is not uncommon for them voluntarily to leave the nest long before they are able to fly, and to fix themselves on the wall, where they are fed until able to hunt for themselves.

When these birds first arrive in spring, and for a considerable time after, they associate together every evening in one general rendezvous; those of a whole district roosting together. This place of repose, in the more unsettled parts of the country, is usually a large hollow tree, open at top; trees of that kind, or *swallow trees* as they are usually called, having been noticed in various parts of the country, and generally believed to be the winter quarters of these birds, where, heaps upon heaps, they dozed away the winter in a state of torpidity. Here they have been seen on their resurrection in spring, and here they have again been remarked descending to their deathlike sleep in autumn.

Among the various accounts of these trees that might be quoted, the following are selected as bearing the marks of authenticity. "At Middlebury, in this State," says Mr Williams, "History of Vermont," p. 16, "there was a large hollow elm, called by the people in the vicinity, the swallow tree. From a man who for several years lived within twenty rods of it, I procured this information. He always thought the swallows tarried in the tree through the winter, and avoided cutting it down on that account. About the first of May the swallows came out of it in large numbers, about the middle of the day, and soon returned. As the weather grew warmer, they came out in the morning, with a loud noise, or roar, and were soon dispersed. About half an hour before sundown, they returned in millions, circulating two or three times round the tree, and then descending like a stream into a hole about sixty feet from the ground. It was customary for persons in the vicinity to visit this tree to observe the motions of these birds: and when any person disturbed their operations, by striking violently against the tree with their axes, the swallows would rush out in millions, and with a great noise. In November 1791, the top of this tree was blown down twenty feet below where the swallows entered : there has been no appearance of the swallows since. Upon cutting down the remainder, an immense quantity of excrements, quills, and feathers were found, but no appearance or relics of any nests.

"Another of these swallow trees was at Bridport. The man who lived the nearest to it gave this account : The swallows were first observed to come out of the tree in the spring, about the time that the leaves first began to appear on the trees ; from that season they came out in the morning about half an hour after sunrise. They rushed out like a stream, as big as the hole in the tree would admit, and ascended in a perpendicular line, until they were above the height of the adjacent trees ; then assumed a circular motion, performing their evolutions two or three times, but always in a larger circle, and then dispersed in every direction. A little before sundown, they returned in immense numbers, forming several circular motions, and then descended like a stream into the hole, from whence they came out in the morning. About the middle of September, they were seen entering the tree for the last time. These birds were all of the species called the house, or chimney swallow. The tree was a large hollow elm ; the hole at which they entered was about forty feet above the ground, and about nine inches in diameter. The swallows made their first appearance in the spring, and their last appearance in the fall, in the vicinity of this tree ; and the neighbouring inhabitants had no doubt but that the swallows continued in it during the winter. A few years ago a hole was cut at the bottom of the tree : from that time the swallows have been gradually forsaking the tree, and have now almost deserted it."

Though Mr Williams himself, as he informs us, is led to believe, from these and some other particulars which he details, " that the house swallow, in this part of America, generally resides during the winter in the hollow of trees ; and the ground swallows (bank swallows) find security in the mud at the bottom of lakes, rivers, and ponds ;" yet I cannot, in the cases just cited, see any sufficient cause for such a belief. The birds were seen to pass out on the 1st of May, or in the spring, when the leaves began to appear on the trees, and, about the middle of September, they were seen entering the tree for the last time ; but there is no information here of their being seen at any time during winter, either within or around the tree. This most important part of the matter is taken for granted without the least examination, and, as will be presently shown, without foundation. I shall, I think, also prove that, if these trees had been cut down in the depth of winter, not a single swallow would have been found either in a living or torpid state ! And that this was merely a place of rendezvous for *active living birds* is evident from the "immense quantity of excrements" found within it, which birds in a state of *torpidity* are not supposed to produce. The total absence of the relics of nests is a proof that it was not a breeding place, and that the whole was nothing more than one of those places to which this singular bird resorts immediately on its arrival in May, in which, also, many of the males continue to roost during the whole summer, and from which they regularly depart about the middle of September. From other circumstances, it appears probable that some of these trees have been for ages the summer rendezvous or general roosting place of the whole chimney swallows of an extensive district. Of this sort I conceive the following to be one, which is thus described by a late traveller to the westward :—

Speaking of the curiosities of the State of Ohio, the writer observes :—" In connection with this, I may mention a large collection of feathers found within a hollow tree which I examined, with the Rev. Mr Story, May 18th, 1803. It is in the upper part of Waterford, about two miles distant from the Muskingum. A very large sycamore, which, through age, had decayed and fallen down, contained in its hollow trunk, five and a half feet in diameter, and for nearly fifteen feet upwards, a mass of decayed feathers, with a small admixture of brownish dust, and the exuviæ of various insects. The feathers were so rotten, that it was impossible to determine to what kinds of birds they belonged. They were less than those of the pigeon ; and the largest of them were like the pinion and tail feathers of the swallow. I examined carefully this astonishing collection, in the hope of finding the bones and bills, but could not distinguish any. The tree, with some remains of its ancient companions lying around, was of a growth preceding that of the neighbouring forest. Near it, and even out of its mouldering ruins, grow thrifty trees of a size which indicate two or three hundred years of age." *

* Harris's Journal, p. 180.

Such are the usual roosting places of the chimney swallow in the more thinly settled parts of the country. In towns, however, they are differently situated, and it is matter of curiosity to observe, that they frequently select the court-house chimney for their general place of rendezvous, as being usually more central, and less liable to interruption during the night. I might enumerate many places where this is their practice. Being in the town of Reading, Pennsylvania, in the month of August, I took notice of sixty or eighty of these birds, a little before evening, amusing themselves by ascending and descending the chimney of the court-house there. I was told that, in the early part of summer, they were far more numerous at that particular spot. On the 20th of May, in returning from an excursion to the Great Pine Swamp, I spent part of the day in the town of Easton, where I was informed by my respected friend, Mordecai Churchman, cashier of the bank there, and one of the people called Quakers, that the chimney swallows of Easton had selected the like situation; and that, from the windows of his house, which stands nearly opposite to the court-house, I might, in an hour or two, witness their whole manœuvres.

I accepted the invitation with pleasure. Accordingly, a short time after sunset, the chimney swallows, which were generally dispersed about town, began to collect around the court-house, their numbers every moment increasing, till, like motes in the sunbeams, the air seemed full of them. These, while they mingled amongst each other seemingly in every direction, uttering their peculiar note with great sprightliness, kept a regular circuitous sweep around the top of the court-house, and about fourteen or fifteen feet above it, revolving with great rapidity for the space of at least ten minutes. There could not be less than four or five hundred of them. They now gradually varied their line of motion, until one part of its circumference passed immediately over the chimney, and about five or six feet above it. Some as they passed made a slight feint of entering, which was repeated by those immedi-

ately after, and by the whole circling multitude in succession:
in this feint they approached nearer and nearer at every
revolution, dropping perpendicularly, but still passing over;
the circle meantime becoming more and more contracted, and
the rapidity of its revolution greater, as the dusk of evening
increased, until, at length, one, and then another, dropped in,
another and another followed, the circle still revolving, until
the whole multitude had descended, except one or two.   These
flew off, as if to collect the stragglers, and, in a few seconds,
returned, with six or eight more, which, after one or two rounds,
dropped in one by one, and all was silence for the night.   It
seemed to me hardly possible that the internal surface of the
vent could accommodate them all, without clustering on one
another, which I am informed they never do; and I was very
desirous of observing their ascension in the morning, but having
to set off before day, I had not that gratification.   Mr Churchman, however, to whom I have since transmitted a few queries,
has been so obliging as to inform me, that towards the beginning of June the number of those that regularly retired to the
court-house to roost was not more than one-fourth of the
former; that on the morning of the 23d of June, he particularly observed their reascension, which took place at a quarter-
past four, or twenty minutes before sunrise, and that they
passed out in less than three minutes; that at my request the
chimney had been examined from above; but that, as far down
at least as nine feet, it contained no nests; though at a former
period it is certain that their nests were very numerous there,
so that the chimney was almost choked, and a sweep could
with difficulty get up it.   But then it was observed that their
place of nocturnal retirement was in another quarter of the
town.   "On the whole," continues Mr Churchman, "I am of
opinion that those who continue to roost at the court-house
are male birds, or such as are not engaged in the business of
incubation, as that operation is going on in almost every
unoccupied chimney in town.   It is reasonable to suppose,
if they made use of that at the court-house for this purpose,

at least some of their nests would appear towards the top, as we find such is the case where but few nests are in a place."

In a subsequent letter Mr Churchman writes as follows:— "After the young brood produced in the different chimneys in Easton had taken wing, and a week or ten days previous to their total disappearance, they entirely forsook the court-house chimney, and rendezvoused in accumulated numbers in the southernmost chimney of John Ross's mansion, situated perhaps one hundred feet north-eastward of the court-house. In this last retreat I several times counted more than two hundred go in of an evening, when I could not perceive a single bird enter the court-house chimney. I was much diverted one evening on seeing a cat, which came upon the roof of the house, and placed herself near the chimney, where she strove to arrest the birds as they entered without success: she at length ascended to the chimney top and took her station, and the birds descended in gyrations without seeming to regard grimalkin, who made frequent attempts to grab them. I was pleased to see that they all escaped her fangs. About the first week in the ninth month [September], the birds quite disappeared; since which I have not observed a single individual. Though I was not so fortunate as to be present at their general assembly and council, when they concluded to take their departure, nor did I see them commence their flight, yet I am fully persuaded that none of them remain in any of our chimneys here. I have had access to Ross's chimney, where they last resorted, and could see the lights out from bottom to top, without the least vestige or appearance of any birds. Mary Ross also informed me that they have had their chimneys swept previous to their making fires, and, though late in autumn, no birds have been found there. Chimneys, also, which have not been used, have been ascended by sweeps in the winter without discovering any. Indeed, all of them are swept every fall and winter, and I have never heard of the swallows being found, in either a dead, living, or torpid state. As to the court-house, it has been occupied as a place of worship two or three times a-week for several weeks past, and at those times there has been fire in the stoves, the pipes of them both going into the chimney, which is shut up at bottom by brick work : and, as the birds had forsaken that place, it remains pretty certain that they did not return there ; and, if they did, the smoke, I think, would be deleterious to their existence, especially as I never knew them to resort to kitchen chimneys where fire was kept in the summer. I think I have noticed them enter such chimneys for the purpose of exploring; but I have also noticed that they immediately ascended, and went off, on finding fire and smoke."

The chimney swallow is easily distinguished in air from the rest of its tribe here by its long wings, its short body, the quick and slight vibrations of its wings, and its wide unexpected diving rapidity of flight ; shooting swiftly in various directions without any apparent motion of the wings, and uttering the sounds *tsip tsip tsip tsee tsee* in a hurried manner. In roosting, the thorny extremities of its tail are thrown in for its support. It is never seen to alight but in hollow trees or chimneys ; is always most gay and active in wet and gloomy weather ; and is the earliest abroad in the morning, and latest out in evening, of all our swallows. About the first or second week in September, they move off to the south, being often observed on their route, accompanied by the purple martins.

When we compare the manners of these birds, while here, with the account given by Captain Henderson of those that winter in such multitudes at Honduras, it is impossible not to be struck with the resemblance, or to suppress our strong suspicions that they may probably be the very same.

This species is four inches and a half in length, and twelve inches in extent; altogether of a deep sooty brown, except the chin and line over the eye, which are of a dull white ; the lores, as in all the rest, are black ; bill, extremely short, hard, and black ; nostrils, placed in a slightly elevated membrane ; legs, covered with a loose purplish skin ; thighs, naked, and of the same tint; feet, extremely muscular; the three fore toes nearly of a length; claws, very sharp; the wing, when closed, extends an inch and a half beyond the tip of the tail, which is rounded, and consists of *ten* feathers, scarcely longer than their coverts; their shafts extend beyond the vanes, are sharp-pointed, strong, and very elastic, and of a deep black colour; the shafts of the wing-quills are also remarkably strong; eye, black, surrounded by a bare blackish skin, or orbit.

The female can scarcely be distinguished from the male by her plumage.

PURPLE MARTIN. (*Hirundo purpurea.*)

PLATE XXXIX.—FIG. 2, MALE; FIG. 3, FEMALE.

Lath. Syn. iv. p. 574, 21 ; *Ibid.* iv. p. 575, 23. --*Catesb. Car.* i. 51.—*Arct. Zool.* ii. No. 333.—Hirondelle bleu de la Caroline, *Buff.* vi. p. 674, *Pl. enl.* 722.— Le Martinet couleur de pourpre, *Buff.* vi. p. 676.—*Turt. Syst.* 629.—*Edw.* 120.—Hirundo subis, *Lath.* iv. p. 575, 24.—*Peale's Museum*, Nos. 7645, 7646.

HIRUNDO PURPUREA.—LINNÆUS.*

Hirundo purpurea, *Bonap. Synop.* p. 64.—*North. Zool.* ii. p. 335.—The Purple Martin, *Aud. Orn. Biog.* i. p. 114, pl. 22, male and female.

THIS well-known bird is a general inhabitant of the United States, and a particular favourite wherever he takes up his

---

* This bird, at first sight, almost presents a different appearance from a swallow ; but, upon examination, all the members are truly that of *Hirundo*, developed, particularly the bill, to an extraordinary extent. The bill is very nearly that of a *Procnias* or *Ptiliogonys;* but the economy of the bird presents no affinity to the berry-eaters ; and the only difference in its feeding seems the preference to larger beetles, wasps, or bees, which its strength enables it to despatch without any danger to itself.

This bird exclusively belongs to the New World, and its migrations have a very extensive range. It makes its first appearance at Great Bear Lake on the 17th May, at which time the snow still partially covers the ground, and the rivers and lakes are fast bound in ice. In the middle of August, it retires again with its young brood from the Fur Countries. In a southern direction, Mr Swainson observed numbers round Pernambuco, 8½ degrees south of the line. They migrate in flocks, and at a very slow rate. The account of Mr Audubon,

abode. I never met with more than one man who disliked the martins, and would not permit them to settle about his house. This was a penurious close-fisted German, who hated them, because, as he said, "they ate his *peas*." I told him he must certainly be mistaken, as I never knew an instance of martins eating *peas ;* but he replied with coolness, that he had many times seen them himself "blaying near the hife, and going *schnip, schnap ;*" by which I understood that it was his *bees* that had been the sufferers ; and the charge could not be denied.

This sociable and half-domesticated bird arrives in the southern frontiers of the United States late in February or early in March; reaches Pennsylvania about the 1st of April, and extends his migrations as far north as the country round Hudson's Bay, where he is first seen in May, and disappears in August ; so, according to the doctrine of torpidity, has, consequently, a pretty long annual nap, in those frozen regions, of eight or nine months under the ice!   We, however, choose to consider him as advancing northerly with the gradual approach of spring, and retiring with his young family, on the first decline of summer, to a more congenial climate.

The summer residence of this agreeable bird is universally among the habitations of man, who, having no interest in his

who witnessed them, will show the possibility of much less powerful birds performing an immense distance, especially where every mile brings them an additional supply of food and a more genial climate.   I give his own words :—"I have had several opportunities, at the period of their arrival, of seeing prodigious flocks moving over that city (New Orleans) or its vicinity, at a considerable height, each bird performing circular sweeps as it proceeded, for the purpose of procuring food. These flocks were loose, and moved either westward, or towards the north-west, at a rate not exceeding four miles in the hour, as I walked under one of them, with ease, for upwards of two miles, at that rate, on the 4th of February 1821, on the bank of the river below the city, constantly looking up at the birds, to the great astonishment of many passengers, who were bent on far different pursuits.   My Fahrenheit's thermometer stood at 68°, the weather being calm and drizzly.   This flock extended about a mile and a half in length, by a quarter of a mile in breadth."—ED.

destruction, and deriving considerable advantage, as well as amusement, from his company, is generally his friend and protector. Wherever he comes, he finds some hospitable retreat fitted up for his accommodation, and that of his young, either in the projecting wooden cornice, on the top of the roof, or sign-post, in the box appropriated to the blue bird ; or, if all these be wanting, in the dove-house among the pigeons. In this last case, he sometimes takes possession of one quarter, or tier, of the premises, in which not a pigeon dare for a moment set its foot. Some people have large conveniencies formed for the martins, with many apartments, which are usually full tenanted, and occupied regularly every spring ; and, in such places, particular individuals have been noted to return to the same box for several successive years. Even the solitary Indian seems to have a particular respect for this bird. The Choctaws and Chickasaws cut off all the top branches from a sapling near their cabins, leaving the prongs a foot or two in length, on each of which they hang a gourd, or calabash, properly hollowed out for their convenience. On the banks of the Mississippi, the negroes stick up long canes, with the same species of apartment fixed to their tops, in which the martins regularly breed. Wherever I have travelled in this country, I have seen with pleasure the hospitality of the inhabitants to this favourite bird.

As superseding the necessity of many of my own observations on this species, I beg leave to introduce in this place an extract of a letter from the late learned and venerable John Joseph Henry, Esq., judge of the supreme court of Pennsylvania, a man of most amiable manners, which was written to me but a few months before his death, and with which I am happy to honour my performance :—" The history of the purple martin of America," says he, " which is indigenous in Pennsylvania, and countries very far north of our latitude, will, under your control, become extremely interesting. We know its manners, habitudes, and useful qualities here ; but we are not generally acquainted with some traits in its character, which, in my mind, rank it in the class of the most remarkable birds of passage. Somewhere (I cannot now refer to book and page) in Anson's Voyage, or in Dampier, or some other southern voyager, I recollect that the martin is named as an inhabitant of the regions of Southern America, particularly of Chili; and in consequence, from the knowledge we have of its immense emigration northward in our own country, we may fairly presume that its flight extends to the south as far as Tierra del Fuego. If the conjecture be well founded, we may, with some certainty, place this useful and delightful companion and friend of the human race as the first in the order of birds of passage. Nature has furnished it with a long, strong, and nervous pinion; its legs are short, too, so as not to impede its passage; the head and body are flattish; in short, it has every indication, from bodily formation, that Providence intended it as a bird of the longest flight. Belknap speaks of it as a visitant of New Hampshire. I have seen it in great numbers at Quebec. Hearne speaks of it in lat. sixty degrees north. To ascertain the times of the coming of the martin to New Orleans, and its migration to and from Mexico, Quito, and Chili, are desirable data in the history of this bird; but it is probable that the state of science in those countries renders this wish hopeless.

" Relative to the domestic history, if it may be so called, of the blue bird (of which you have given so correct and charming a description) and the martin, permit me to give you an anecdote:—In 1800 I removed from Lancaster to a farm a few miles above Harrisburg. Knowing the benefit derivable to a farmer from the neighbourhood of the martin, in preventing the depredations of the bald eagle, the hawks, and even the crows, my carpenter was employed to form a large box, with a number of apartments for the martin. The box was put up in the autumn. Near and around the house were a number of well-grown apple-trees and much shrubbery,—a very fit haunt for the feathered race. About the middle of February, the blue birds came; in a short time they were very familiar, and took possession of the box: these consisted of two or three pairs. By the 15th of May, the blue birds had eggs, if not young. Now the martins arrived in numbers, visited the box, and a severe conflict ensued. The blue birds, seemingly animated by their right of possession, or for the protection of their young, were victorious. The martins regularly arrived about the middle of May, for the eight following years, examined the apartments of the box, in the absence of the blue birds, but were uniformly compelled to fly upon the return of the latter.

" The trouble caused you by reading this note you will be pleased to charge to the martin. A box replete with that beautiful traveller is not very distant from my bed-head. Their notes seem discordant because of their numbers; yet to me they are pleasing. The industrious farmer and mechanic would do well to have a box fixed near the apartments of their drowsy labourers. Just as the dawn approaches, the martin begins its notes, which last half a minute or more, and then subside until the twilight is fairly broken. An animated and incessant musical chattering now ensues, sufficient to arouse the most sleepy person. Perhaps chanticleer is not their superior in this beneficial qualification; and he is far beneath the martin in his powers of annoying birds of prey."

I shall add a few particulars to this faithful and interesting sketch by my deceased friend :—About the middle, or 20th, of April, the martins first begin to prepare their nest. The last of these which I examined was formed of dry leaves of the weeping willow, slender straws, hay, and feathers in considerable quantity. The eggs were four, very small for the size of the bird, and pure white, without any spots. The first brood appears in May, the second late in July. During the period in which the female is laying, and before she commences incubation, they are both from home the greater part of the day. When the female is sitting, she is frequently visited by the male, who also occupies her place while she takes a short recreation abroad. He also often passes a quarter of an hour in the apartment beside her, and has become quite domesticated since her confinement. He sits on the outside dressing and arranging his plumage, occasionally passing to the door of the apartment as if to inquire how she does. His notes, at this time, seem to have assumed a peculiar softness, and his gratulations are expressive of much tenderness. Conjugal fidelity, even where there is a number together, seems to be faithfully preserved by these birds. On the 25th of May, a male and female martin took possession of a box in Mr Bartram's garden. A day or two after, a second female made her appearance, and stayed for several days; but, from the cold reception she met with, being frequently beat off by the male, she finally abandoned the place, and set off, no doubt, to seek for a more sociable companion.

The purple martin, like his half-cousin the king bird, is the terror of crows, hawks, and eagles. These he attacks whenever they make their appearance, and with such vigour and rapidity, that they instantly have recourse to flight. So well known is this to the lesser birds, and to the domestic poultry, that, as soon as they hear the martin's voice engaged in fight, all is alarm and consternation. To observe with what spirit and audacity this bird dives and sweeps upon and around the hawk or the eagle is astonishing. He also bestows an occasional bastinading on the king bird when he finds him too near his premises; though he will, at any time, instantly co-operate with him in attacking the common enemy.

The martin differs from all the rest of our swallows in the particular prey which he selects. Wasps, bees, large beetles, particularly those called by the boys *goldsmiths*, seem his favourite game. I have taken four of these large beetles from the stomach of a purple martin, each of which seemed entire, and even unbruised.

The flight of the purple martin unites in it all the swiftness, ease, rapidity of turning, and gracefulness of motion of its tribe. Like the swift of Europe, he sails much with little action of the wings. He passes through the most crowded parts of our streets, eluding the passengers with the quickness of thought; or plays among the clouds, gliding about at a vast height, like an aerial being. His usual note, *peuo, peuo, peuo,* is loud and musical; but is frequently succeeded by others more low and guttural. Soon after the 20th of August he leaves Pennsylvania for the south.

This bird has been described three or four different times, by European writers, as so many different species,—the Canadian swallow of Turton, and the great American martin of Edwards, being evidently the female of the present species. The violet swallow of the former author, said to inhabit Louisiana, differs in no respect from the present. Deceived by the appearance of the flight of this bird, and its similarity to that of the swift of Europe, strangers from that country have also asserted that the swift is common to North America and the United States. No such bird, however, inhabits any part of this continent that I have as yet visited.

The purple martin is eight inches in length, and sixteen inches in extent; except the lores, which are black, and the wings and tail, which are of a brownish black, he is of a rich and deep purplish blue, with strong violet reflections; the bill is strong, the gap very large; the legs also short, stout, and of a dark dirty purple; the tail consists of twelve feathers, is considerably forked, and edged with purple blue; the eye full and dark.

The female measures nearly as large as the male; the upper parts are blackish brown, with blue and violet reflections thinly scattered; chin and breast, grayish brown; sides under the wings, darker; belly and vent, whitish, not pure, with stains of dusky and yellow ochre; wings and tail, blackish brown.

CONNECTICUT WARBLER. (*Sylvia agilis.*)

PLATE XXXIX.—FIG. 4.

SYLVICOLA AGILIS.—JARDINE.*

Sylvia agilis, *Bonap. Synop.* p. 84 ; *Nomenclature,* p. 163.

THIS is a new species, first discovered in the State of Connecticut, and twice since met with in the neighbourhood of Philadelphia. As to its notes or nest, I am altogether unacquainted with them. The different specimens I have shot corresponded very nearly in their markings ; two of these were males, and the other undetermined, but conjectured also to be a male. It was found in every case among low thickets, but seemed more than commonly active, not remaining for a moment in the same position. In some of my future rambles I may learn more of this solitary species.

Length, five inches and three quarters ; extent, eight inches ; whole upper parts, a rich yellow olive ; wings, dusky brown, edged with olive ; throat, dirty white or pale ash ; upper part of the breast, dull greenish yellow ; rest of the lower parts, a pure rich yellow ; legs, long, slender, and of a pale flesh colour ; round the eye, a narrow ring of yellowish white ; upper mandible, pale brown ; lower, whitish ; eye, dark hazel.

Since writing the above, I have shot two specimens of a bird, which in every particular agrees with the above, except in having the throat of a dull buff colour, instead of pale ash ; both of these were females ; and I have little doubt but they are of the same species with the present, as their peculiar activity seemed exactly similar to the males above described.

These birds do not breed in the lower parts of Pennsylvania, though they probably may be found in summer in the alpine swamps and northern regions, in company with a numerous class of the same tribe that breed in these unfrequented solitudes.

* According to Bonaparte, this is a new species discovered by Wilson. Comparatively little is known regarding it.—ED.

1. Night Hawk. 2. Female.

40.

NIGHT HAWK. *(Caprimulgus Americanus.)*

PLATE XL.—FIG. 1, MALE; FIG. 2, FEMALE.

Long-winged Goatsucker, *Arct. Zool.* No. 337.—*Peale's Museum*, No. 7723, male ; 7724, female.

CAPRIMULGUS AMERICANUS ?—WILSON.*

Caprimulgus Virginianus, *Bonap. Synop.* p. 62.—Chordeiles Virgimorus, *Sw. North. Zool.* ii. p. 337.

THIS bird, in Virginia and some of the southern districts, is called a bat; the name night hawk is usually given it in the middle and northern States, probably on account of its appearance, when on wing, very much resembling some of our

* North America appears to contain three species of this curious genus, —the present one, with the following, and *C. Carolinensis*, afterwards described. The whole are nearly of like size, and, from the general similarity of marking which runs through the group, will somewhat resemble each other. Wilson may, therefore, claim the first merit of clearly distinguishing them, although he remained in uncertainty regarding the descriptions and synonyms of other authors. Vieillot appears to have described this species under the name of *C. popetue ;* but, notwithstanding, I cannot help preferring that given by Wilson, particularly as it seems confined to the New World.

Bonaparte remarks that the night hawks are among the swallows what the owls are among the *Falconidæ ;* and, if we may be allowed the expression, the *C. Americanus* has more of the hirundine look than the others. The whole plumage is harder, the ends of the quills are more pointed, the tail is forked, and the rictus wants the strong array of bristles which we consider one of the essentials in the most perfect form of *Caprimulgus.* We may here remark (although we know that there are exceptions), that we have generally observed in those having the tail forked, and, consequently, with a greater power of quick flight and rapid turnings, that the plumage is more rigid, and the flight occasionally diurnal. This is borne out, also, in our present species, which play "about in the air, over the breeding-place, even during the day ;" and, in their migrations, "may be seen almost everywhere, from five o'clock until after sunset, passing along the Schuylkill and the adjacent shores."

The truly night-feeding species have the plumage loose and downy, as in the nocturnal owls ; the wings more blunted, and the plumules coming to a slender point, and unconnected ; the tail rounded, and the

small hawks, and from its habit of flying chiefly in the evening. Though it is a bird universally known in the United States, and inhabits North America in summer from

rictus armed, in some instances, with very powerful bristles. Their organs of sight are also fitted only for a more gloomy light. They appear only at twilight, reposing during the day among furze or brake, or sitting, in their own peculiar manner, on a branch ; but if inactive amidst the clearer light, they are all energy and action when their own day has arrived. To these last will belong the common night hawk of Europe ; and a detail, in comparison of its manners with those of our author, may assist in giving some idea of the truly nocturnal species, which are similar, so far as variation of country and circumstances will allow. They are thus, in a few lines, accurately described by a poet whom Wilson would have admired :—

> Hark ! from yon quivering branch your direst foe,
> Insects of night, its whirring note prolongs,
> Loud as the sound of busy maiden's wheel:
> Then with expanded beak, and throat enlarged
> Even to its utmost stretch, its 'custom'd food
> Pursues voracious.

It frequents extensive moors and commons, perhaps more abundantly if they are either interspersed or bordered with brush or wood. At the commencement of twilight, when they are first roused from their daily slumber, they perch on some bare elevation of the ground, an old wall or fence, or heap of stones, in a moss country on a *peat stack*, and commence their monotonous *drum* or *whirr*, closely resembling the dull sound produced by a spinning-wheel ; and possessing the same variation of apparent distance in the sound, a modification of ventriloquism, which is perceived in the croak of the land rail, or the cry of the coot and water rail, or croaking of frogs ; at one time it is so near as to cause an alarm that you will disturb the utterer ; at another as if the bird had removed to the extreme limit of the listener's organs, while it remains unseen at a distance of perhaps not more than forty or fifty yards. At the commencement, this drumming sound seems to be continued for about ten or fifteen minutes, and occasionally during the night in the intervals of relaxation ; it is only, however, when perched that it is uttered, and never for so great a length of time as at the first. Their flight is never high, and is performed without any regularity; sometimes straight forward and in gliding circles, with a slow, steady clap of the wings, in the middle of which they will abruptly start into the air for thirty or forty feet, resuming their former line by a gradual fall ; at other times it will be performed in sudden jerks upwards, in the fall keeping the wings steady and closed over the back, skimming in the intervals near

Florida to Hudson's Bay, yet its history has been involved in considerable obscurity by foreign writers, as well as by some of our own country. Of this I shall endeavour to divest it in the present account.

the ground, and still retaining the wings like some gulls or terns, or a swallow dipping in the water, until they are again required to give the stroke upwards ; all the while the tail is much expanded, and is a conspicuous object in the male, from the white spots on the outer feathers. When in woods, or hawking near trees, the flight is made in glides among the branches, or it flutters close to the summits, and seizes the various *Phalænæ* which play around them. I once observed three or four of these birds hawking in this manner, on the confines of a spruce fir plantation, and, after various evolutions, they balanced themselves for a few seconds on the very summits of the leading shoots. This was frequently repeated while I looked on. During the whole of their flight, a short snap of the bill is heard, and a sort of *click, click,* with the distinct sound of the monosyllable *whip,* or, to convey the idea better, the sound of a whip suddenly lashed without cracking. The female, when disturbed from her nest, flits or skims along the surface for a short distance ; but I have never seen the young or eggs removed in the manner related of the American species, even after frequent annoyance. When the young are approached at night, before they are perfectly fledged, the old birds fly in circles round, approach very near, uttering incessantly their clicking cry, and make frequent dashes at the intruder, like a lapwing.

Among the night hawks, taking the form as understood to rank under *Caprimulgus* of Linnæus, we have a close resemblance of general form and characters, though there are one or two modifications which fully entitle the species to separation, and which work beautifully in the system of affinities or gradual development of form.* From these circumstances, Mr Swainson has formed a new genus for our present species.

In colour, the whole of *Caprimulgus* is very closely allied ; " drest, but with Nature's tenderest pencil touched," in various shades of brown, white, and russet ; the delicate blending of the markings produce an

---

* In some the mouth is furnished with very strong bristles, and in others it is entirely destitute of them, as may be seen in the species of North America. Again, the tail is square, round, or forked, sometimes to an extraordinary extent, as in the *C. psalurus* of Azara, and in *C. acutus* the shafts of the feathers project beyond the webs, and remind us of the genus *Chætura*. In some the tarsus is extremely short and weak, and covered with plumes to the very toes, in others long and naked. The wings are rounded or sharp-pointed ; and in the Sierra goatsucker we have the shaft of one of the secondaries running out to the length of twenty inches, with the web much expanded at the extremity, and presenting no doubt during flight a most unique appearance.—Ed.

Three species only of this genus are found within the United States,—the chuck-will's-widow, the whip-poor-will, and the night hawk. The first of these is confined to those States lying south of Maryland; the other two are found generally over the Union, but are frequently confounded one with the other, and by some supposed to be one and the same bird. A comparison of this with the succeeding plate, which contains the figure of the whip-poor-will, will satisfy those who still have their doubts on this subject; and the great difference of manners which distinguishes each will render this still more striking and satisfactory.

On the last week in April, the night hawk commonly makes its first appearance in this part of Pennsylvania. At what particular period they enter Georgia, I am unable to say; but I find, by my notes, that in passing to New Orleans by land, I first observed this bird in Kentucky on the 21st of April. They soon after disperse generally over the country, from the seashore to the mountains, even to the heights of the Alleghany; and are seen towards evening, in pairs, playing

effect always pleasing—often more so than in those which can boast of a more gorgeous apparel.

There is another structure in this bird which has given rise to much conjecture among naturalists, particularly those whose opportunities of observation have been comparatively limited, and has been looked upon as a peculiarity existing in this genus only,—I allude to the serrature of the centre claw. This structure we also find in many other genera, totally different from the present in almost every particular, and where the uses of combing its bristles or freeing itself from the vermin that persons have been willing to afflict this species with in more than ordinary proportions, could not be in any way applied. We find it among the *Ardeadæ, Platalea, Ibis, Phalacracorax,* and *Cursorius,* all widely differing in habit: the only assimilating form among them is the generally loose plumage. I have no hesitation in saying that the use of this structure has not yet been ascertained, and that, when found out, it will be different from any that has yet been suggested. The very variety of forms among which we find it will bear this out, and the presence of it in *Caprimulgus* will more likely turn out the extreme limit of the structure than that from which we should draw our conclusions. It is much more prevalent among the *Grallatores,* and our present form is the only one in any other division where it is at all found.—ED.

about, high in air, pursuing their prey, wasps, flies, beetles, and various other winged insects of the larger sort. About the middle of May the female begins to lay. No previous preparation or construction of a nest is made, though doubtless the particular spot has been reconnoitred and determined on. This is sometimes in an open space in the woods, frequently in a ploughed field, or in the corner of a cornfield. The eggs are placed on the bare ground, in all cases on a dry situation, where the colour of the leaves, ground, stones, or other circumjacent parts of the surface, may resemble the general tint of the eggs, and thereby render them less easy to be discovered. The eggs are most commonly two, rather oblong, equally thick at both ends, of a dirty bluish white, and marked with innumerable touches of dark olive brown. To the immediate neighbourhood of this spot the male and female confine themselves, roosting on the high trees adjoining during the greater part of the day, seldom, however, together, and almost always on separate trees. They also sit lengthwise on the branch, fence, or limb on which they roost, and never across, like most other birds : this seems occasioned by the shortness and slender form of their legs and feet, which are not at all calculated to grasp the branch with sufficient firmness to balance their bodies.

As soon as incubation commences, the male keeps a most vigilant watch around. He is then more frequently seen playing about in the air over the place, even during the day, mounting by several quick vibrations of the wings, then a few slower, uttering all the while a sharp, harsh squeak, till having gained the highest point, he suddenly precipitates himself, head foremost, and with great rapidity, down sixty or eighty feet, wheeling up again as suddenly ; at which instant is heard a loud booming sound, very much resembling that produced by blowing strongly into the bunghole of an empty hogshead, and which is doubtless produced by the sudden expansion of his capacious mouth while he passes through the air, as exhibited in the figure on the plate   He again mounts by alternate quick and leisurely motions of the wings, playing about as he ascends, uttering his usual hoarse squeak, till, in a few minutes, he again dives with the same impetuosity and violent sound as before. Some are of opinion that this is done to intimidate man or beast from approaching his nest, and he is particularly observed to repeat these divings most frequently around those who come near the spot, sweeping down past them, sometimes so near and so suddenly as to startle and alarm them. The same individual is, however, often seen performing these manœuvres over the river, the hill, the meadow, and the marsh, in the space of a quarter of an hour, and also towards the fall, when he has no nest. This singular habit belongs peculiarly to the male. The female has, indeed, the common hoarse note, and much the same mode of flight ; but never precipitates herself in the manner of the male. During the time she is sitting, she will suffer you to approach within a foot or two before she attempts to stir, and, when she does, it is in such a fluttering, tumbling manner, and with such appearance of a lame and wounded bird, as nine times in ten to deceive the person, and induce him to pursue her. This "pious fraud," as the poet Thomson calls it, is kept up until the person is sufficiently removed from the nest, when she immediately mounts and disappears. When the young are first hatched, it is difficult to distinguish them from the surface of the ground, their down being of a pale brownish colour, and they are altogether destitute of the common shape of birds, sitting so fixed and so squat as to be easily mistaken for a slight prominent mouldiness lying on the ground. I cannot say whether they have two broods in the season ; I rather conjecture that they have generally but one.

The night hawk is a bird of strong and vigorous flight, and of large volume of wing. It often visits the city, darting and squeaking over the streets at a great height, diving perpendicularly with the same hollow sound as before described. I have also seen them sitting on chimney tops in some of the most busy parts of the city, occasionally uttering their common note.

When the weather happens to be wet and gloomy, the night hawks are seen abroad at all times of the day, generally at a considerable height; their favourite time, however, is from two hours before sunset until dusk. At such times they seem all vivacity, darting about in the air in every direction, making frequent short sudden turnings, as if busily engaged in catching insects. Even in the hottest, clearest weather, they are occasionally seen abroad, squeaking at short intervals. They are also often found sitting along the fences, basking themselves in the sun. Near the seashore, in the vicinity of extensive salt marshes, they are likewise very numerous, skimming over the meadows, in the manner of swallows, until it is so dark that the eye can no longer follow them.

When wounded and taken, they attempt to intimidate you by opening their mouth to its utmost stretch, throwing the head forward, and uttering a kind of guttural whizzing sound, striking also violently with their wings, which seem to be their only offensive weapons, for they never attempt to strike with the bill or claws.

About the middle of August they begin to move off towards the south, at which season they may be seen almost every evening, from five o'clock until after sunset, passing along the Schuylkill and the adjacent shores, in widely scattered multitudes, all steering towards the south. I have counted several hundreds within sight at the same time, dispersed through the air, and darting after insects as they advanced. These occasional processions continue for two or three weeks; none are seen travelling in the opposite direction. Sometimes they are accompanied by at least twice as many barn swallows, some chimney swallows and purple martins. They are also most numerous immediately preceding a north-east storm. At this time also they abound in the extensive meadows on the Schuylkill and Delaware, where I have counted fifteen skimming over a single field in an evening. On shooting some of these, on the 14th of August, their stomachs were almost exclusively filled with crickets.

From one of them I took nearly a common snuff-box full of these insects, all seemingly fresh swallowed.

By the middle or 20th of September, very few of these birds are to be seen in Pennsylvania ; how far south they go, or at what particular time they pass the southern boundaries of the United States, I am unable to say. None of them winter in Georgia.

The ridiculous name goatsucker,—which was first bestowed on the European species, from a foolish notion that it sucked the teats of the goats, because, probably, it inhabited the solitary heights where they fed, which nickname has been since applied to the whole genus,—I have thought proper to omit. There is something worse than absurd in continuing to brand a whole family of birds with a knavish name, after they are universally known to be innocent of the charge. It is not only unjust, but tends to encourage the belief in an idle fable that is totally destitute of all foundation.

The night hawk is nine inches and a half in length, and twenty-three inches in extent; the upper parts are of a very deep blackish brown, unmixed on the primaries, but thickly sprinkled or powdered on the back scapulars and head with innumerable minute spots and streaks of a pale cream colour, interspersed with specks of reddish ; the scapulars are barred with the same, also the tail-coverts and tail, the inner edges of which are barred with white and deep brownish black for an inch and a half from the tip, where they are crossed broadly with a band of white, the two middle ones excepted, which are plain deep brown, barred and sprinkled with light clay ; a spot of pure white extends over the five first primaries, the outer edge of the exterior feather excepted, and about the middle of the wing; a triangular spot of white also marks the throat, bending up on each side of the neck ; the bill is exceedingly small, scarcely one-eighth of an inch in length, and of a black colour ; the nostrils circular, and surrounded with a prominent rim ; eye, large and full, of a deep bluish black ; the legs are short, feathered a little below the

1 Whip-poor will. 2 Female.

41.

knees, and, as well as the toes, of a purplish flesh colour, seamed with white; the middle claw is pectinated on its inner edge, to serve as a comb to clear the bird of vermin; the whole lower parts of the body are marked with transverse lines of dusky and yellowish. The tail is somewhat shorter than the wings when shut, is handsomely forked, and consists of ten broad feathers; the mouth is extremely large, and of a reddish flesh colour within; there are no bristles about the bill; the tongue is very small, and attached to the inner surface of the mouth.

The female measures about nine inches in length, and twenty-two in breadth; differs in having no white band on the tail, but has the spot of white on the wing; wants the triangular spot of white on the throat, instead of which there is a dully defined mark of a reddish cream-colour; the wings are nearly black, all the quills being slightly tipt with white; the tail is as in the male, and minutely tipt with white; all the scapulars, and whole upper parts, are powdered with a much lighter gray.

There is no description of the present species in Turton's translation of Linnæus. The characters of the genus given in the same work are also in this case incorrect, viz., "mouth furnished with a series of bristles; tail not forked,"—the night hawk having nothing of the former, and its tail being largely forked.

WHIP-POOR-WILL. *(Caprimulgus vociferus.)*

PLATE XLI.—Fig. 1, Male; Fig. 2, Female; Fig. 3, Young.

Peale's Museum, No. 7721, male; 7722, female.

CAPRIMULGUS VOCIFERUS.—Wilson.

Caprimulgus vociferus, *Bonap. Synop.* p. 61.—*North. Zool.* ii. p. 336.—Whip-poor-will, *Aud. Orn. Biog.* i. p. 422, pl. 32.

This is a singular and very celebrated species, universally noted over the greater part of the United States for the loud

reiterations of his favourite call in spring ; and yet personally
he is but little known, most people being unable to distinguish
this from the preceding species when both are placed before
them, and some insisting that they are the same.   This
being the case, it becomes the duty of his historian to give a
full and faithful delineation of his character and peculiarity
of manners, that his existence as a distinct and independent
species may no longer be doubted, nor his story mingled confusedly with that of another.   I trust that those best acquainted
with him will bear witness to the fidelity of the portrait.

On or about the 25th of April, if the season be not uncommonly cold, the whip-poor-will is first heard in this part
of Pennsylvania, in the evening as the dusk of twilight commences, or in the morning as soon as dawn has broke.   In
the State of Kentucky I first heard this bird on the 14th of
April, near the town of Danville.   The notes of this solitary
bird, from the ideas which are naturally associated with them,
seem like the voice of an old friend, and are listened to by
almost all with great interest.   At first they issued from some
retired part of the woods, the glen, or mountain ; in a few
evenings, perhaps, we hear them from the adjoining coppice,
the garden fence, the road before the door, and even from the
roof of the dwelling-house, long after the family have retired to
rest. Some of the more ignorant and superstitious consider this
near approach as foreboding no good to the family,—nothing
less than sickness, misfortune, or death, to some of its members.   These visits, however, so often occur without any bad
consequences, that this superstitious dread seems on the decline.

He is now a regular acquaintance.   Every morning and
evening his shrill and rapid repetitions are heard from the
adjoining woods, and when two or more are calling out at
the same time, as is often the case in the pairing season, and
at no great distance from each other, the noise, mingling with
the echoes from the mountains, is really surprising. Strangers,
in parts of the country where these birds are numerous, find
it almost impossible for some time to sleep ; while to those

long acquainted with them, the sound often serves as a lullaby to assist their repose.

These notes seem pretty plainly to articulate the words which have been generally applied to them, *whip-poor-will*, the first and last syllables being uttered with great emphasis, and the whole in about a second to each repetition; but when two or more males meet, their whip-poor-will altercations become much more rapid and incessant, as if each were straining to overpower or silence the other. When near, you often hear an introductory cluck between the notes. At these times, as well as at almost all others, they fly low, not more than a few feet from the surface, skimming about the house and before the door, alighting on the wood pile, or settling on the roof. Towards midnight they generally become silent, unless in clear moonlight, when they are heard with little intermission till morning. If there be a creek near, with high precipitous bushy banks, they are sure to be found in such situations. During the day they sit in the most retired, solitary, and deep shaded parts of the woods, generally on high ground, where they repose in silence. When disturbed, they rise within a few feet, sail low and slowly through the woods for thirty or forty yards, and generally settle on a low branch or on the ground. Their sight appears deficient during the day, as, like owls, they seem then to want that vivacity for which they are distinguished in the morning and evening twilight. They are rarely shot at or molested; and from being thus transiently seen in the obscurity of dusk, or in the deep umbrage of the woods, no wonder their particular markings of plumage should be so little known, or that they should be confounded with the night hawk, whom in general appearance they so much resemble. The female begins to lay about the second week in May, selecting for this purpose the most unfrequented part of the wood, often where some brush, old logs, heaps of leaves, &c., had been lying, and always on a dry situation. The eggs are deposited on the ground or on the leaves, not the slightest appearance of a nest being visible. These are

usually two in number, in shape much resembling those of
the night hawk, but having the ground colour much darker,
and more thickly marbled with dark olive. The precise
period of incubation I am unable to say.

In traversing the woods one day, in the early part of June,
along the brow of a rocky declivity, a whip-poor-will rose from
my feet, and fluttered along, sometimes prostrating herself,
and beating the ground with her wings, as if just expiring.
Aware of her purpose, I stood still, and began to examine the
space immediately around me for the eggs or young, one or
other of which I was certain must be near. After a long
search, to my mortification, I could find neither; and was just
going to abandon the spot, when I perceived somewhat like a
slight mouldiness among the withered leaves, and, on stooping
down, discovered it to be a young whip-poor-will, seemingly
asleep, as its eyelids were nearly closed; or perhaps this might
only be to protect its tender eyes from the glare of day. I sat
down by it on the leaves, and drew it as it then appeared.
(See fig. 3.) It was probably not a week old. All the while
I was thus engaged, it neither moved its body, nor opened its
eyes more than half; and I left it as I found it. After I had
walked about a quarter of a mile from the spot, recollecting
that I had left a pencil behind, I returned and found my
pencil, but the young bird was gone.

Early in June, as soon as the young appear, the notes of the
male usually cease, or are heard but rarely. Towards the
latter part of summer, a short time before these birds leave
us, they are again occasionally heard; but their call is then
not so loud—much less emphatical, and more interrupted than
in spring. Early in September they move off towards the south.

The favourite places of resort for these birds are on high,
dry situations; in low, marshy tracts of country they are
seldom heard. It is probably on this account that they are
scarce on the sea-coast and its immediate neighbourhood,
while towards the mountains they are very numerous. The
night hawks, on the contrary, delight in these extensive sea

marshes, and are much more numerous there than in the interior and higher parts of the country. But nowhere in the United States have I found the whip-poor-will in such numbers as in that tract of country in the State of Kentucky called the Barrens. This appears to be their most congenial climate and place of residence. There, from the middle of April to the 1st of June, as soon as the evening twilight draws on, the shrill and confused clamours of these birds are incessant, and very surprising to a stranger. They soon, however, become extremely agreeable; the inhabitants lie down at night lulled by their whistlings, and the first approach of dawn is announced by a general and lively chorus of the same music; while the full-toned *tooting*, as it is called, of the pinnated grouse forms a very pleasing bass to the whole.

I shall not, in the manner of some, attempt to amuse the reader with a repetition of the unintelligible names given to this bird by the Indians, or the superstitious notions generally entertained of it by the same people. These seem as various as the tribes, or even families, with which you converse; scarcely two of them will tell you the same story. It is easy, however, to observe, that this, like the owl, and other nocturnal birds, is held by them in a kind of suspicious awe, as a bird with which they wish to have as little to do as possible. The superstition of the Indian differs very little from that of an illiterate German or Scots Highlander, or the less informed of any other nation. It suggests ten thousand fantastic notions to each, and these, instead of being recorded with all the punctilio of the most important truths, seem only fit to be forgotten. Whatever, among either of these people, is strange and not comprehended, is usually attributed to supernatural agency; and an unexpected sight or uncommon incident is often ominous of good, but more generally of bad, fortune to the parties. Night, to minds of this complexion, brings with it its kindred horrors, its apparitions, strange sounds, and awful sights; and this solitary and inoffensive bird, being a frequent wanderer in these hours of ghosts and hobgoblins, is considered by the Indians as being, by habit and repute, little better than one of them. All these people, however, are not so credulous; I have conversed with Indians who treated these silly notions with contempt.

The whip-poor-will is never seen during the day, unless in circumstances such as have been described. Their food appears to be large moths, grasshoppers, pismires, and such insects as frequent the bark of old rotten and decaying timber. They are also expert in darting after winged insects. They will sometimes skim in the dusk within a few feet of a person, uttering a kind of low chatter as they pass. In their migrations north, and on their return, they probably stop a day or two at some of their former stages, and do not advance in one continued flight. The whip-poor-will was first heard this season (1811) on the 2d day of May, in a corner of Mr Bartram's woods, not far from the house, and for two or three mornings after in the same place, where I also saw it. From this time until the beginning of September, there were none of these birds to be found within at least one mile of the place, though I frequently made search for them. On the 4th of September, the whip-poor-will was again heard for two evenings successively in the same part of the woods. I also heard several of them passing, within the same week, between dusk and nine o'clock at night, it being then clear moonlight. These repeated their notes three or four times, and were heard no more. It is highly probable that they migrate during the evening and night.

The whip-poor-will is nine inches and a half long, and nineteen inches in extent; the bill is blackish, a full quarter of an inch long, much stronger than that of the night hawk, and bent a little at the point, the under mandible arched a little upwards, following the curvature of the upper; the nostrils are prominent and tubular, their openings directed forward; the mouth is extravagantly large, of a pale flesh colour within, and beset along the sides with a number of long, thick, elastic bristles, the longest of which extends more

than half an inch beyond the point of the bill, end in fine hair,
and curve inwards; these seem to serve as feelers; and prevent the escape of winged insects: the eyes are very large,
full, and bluish black; the plumage above is so variegated
with black, pale cream, brown, and rust colour, sprinkled and
powdered in such minute streaks and spots as to defy description; the upper part of the head is of a light brownish gray,
marked with a longitudinal streak of black, with others radiating from it; the back is darker, finely streaked with a less
deep black; the scapulars are very light whitish ochre, beautifully variegated with two or three oblique streaks of very deep
black; the tail is rounded, consisting of ten feathers, the exterior
one an inch and a quarter shorter than the middle ones, the
three outer feathers on each side are blackish brown for half
their length, thence pure white to the tips; the exterior one
is edged with deep brown nearly to the tip; the deep brown
of these feathers is regularly studded with light brown spots;
the four middle ones are without the white at the ends, but
beautifully marked with herring-bone figures of black and
light ochre finely powdered; cheeks and sides of the head, of
a brown orange or burnt colour; the wings, when shut, reach
scarcely to the middle of the tail, and are elegantly spotted
with very light and dark brown, but are entirely without the
large spot of white which distinguishes those of the night
hawk; chin, black streaked with brown; a narrow semicircle of white passes across the throat; breast and belly,
irregularly mottled and streaked with black and yellow
ochre; the legs and feet are of a light purplish flesh colour,
seamed with white; the former feathered before, nearly to
the feet; the two exterior toes are joined to the middle one,
as far as the first joint, by a broad membrane; the inner edge
of the middle claw is pectinated, and, from the circumstance
of its being frequently found with small portions of down
adhering to the teeth, is probably employed as a comb to rid
the plumage of its head of vermin; this being the principal
and almost only part so infested in all birds.

The female is about an inch less in length and in extent; the bill, mustaches, nostrils, &c., as in the male. She differs in being much lighter on the upper parts, seeming as if powdered with grains of meal; and, instead of the white on the three lateral tail-feathers, has them tipt for about three-quarters of an inch with a cream colour; the bar across the throat is also of a brownish ochre; the cheeks and region of the eyes are brighter brownish orange, which passes also to the neck, and is sprinkled with black and specks of white; the streak over the eye is also lighter.

The young was altogether covered with fine down, of a pale brown colour; the shafts, or rather sheaths, of the quills, bluish; the point of the bill, just perceptible.

Twenty species of this singular genus are now known to naturalists; of these, one only belongs to Europe, one to Africa, one to New Holland, two to India, and fifteen to America.

The present species, though it approaches nearer in its plumage to that of Europe than any other of the tribe, differs from it in being entirely without the large spot of white on the wing, and in being considerably less. Its voice and particular call are also entirely different.

Further to illustrate the history of this bird, the following notes are added, made at the time of dissection :—Body, when stript of the skin, less than that of the wood thrush; breast-bone, one inch in length; second stomach, strongly muscular, filled with fragments of pismires and grasshoppers; skin of the bird, loose, wrinkly, and scarcely attached to the flesh; flesh, also loose, extremely tender; bones, thin and slender; sinews and muscles of the wing, feeble; distance between the tips of both mandibles, when expanded, full two inches; length of the opening, one inch and a half, breadth, one inch and a quarter; tongue, very short, attached to the skin of the mouth; its internal parts, or *os hyoides,* pass up the hind head, and reach to the front, like those of the woodpecker, which enable the bird to revert the lower part of the mouth in the act of seizing insects and in calling; skull, extremely light and thin, being semi-transparent, its cavity nearly half occupied by the eyes ; aperture for the brain, very small, the quantity not exceeding that of a sparrow ; an owl of the same extent of wing has at least ten times as much.

Though this noted bird has been so frequently mentioned by name, and its manners taken notice of by almost every naturalist who has written on our birds, yet personally it has never yet been described by any writer with whose works I am acquainted. Extraordinary as this may seem, it is nevertheless true ; and in proof I offer the following facts :—

Three species only of this genus are found within the United States, the chuck-will's-widow, the night hawk, and the whip-poor-will. Catesby, in the eighth plate of his " Natural History of Carolina," has figured the first, and in the sixteenth of his Appendix the second ; to this he has added particulars of the whip-poor-will, believing it to be that bird, and has ornamented his figure of the night hawk with a large bearded appendage, of which in nature it is entirely destitute. After him, Mr Edwards in his sixty-third plate has in like manner figured the night hawk, also adding the bristles, and calling his figure the whip-poor-will, accompanying it with particulars of the notes, &c., of that bird, chiefly copied from Catesby. The next writer of eminence who has spoken of the whip-poor-will is Mr Pennant, justly considered as one of the most judicious and discriminating of English naturalists ; but, deceived by " the lights he had," he has, in his account of the short-winged goatsucker * (Arct. Zool., p. 434), given the size, markings of plumage, &c., of the chuck-will's-widow ; and, in the succeeding account of his long-winged goatsucker, describes pretty accurately the night hawk. Both of these birds he considers to be the whip-poor-will, and as having the same notes and manners.

After such authorities, it was less to be wondered at that many of our own citizens, and some of our naturalists and

* The figure is by mistake called the *long-winged* goatsucker. See " Arctic Zoology," vol. ii. pl. 18.

writers, should fall into the like mistake, as copies of the works of those English naturalists are to be found in several of our colleges, and in some of our public as well as private libraries. The means which the author of " American Ornithology " took to satisfy his own mind, and those of his friends, on this subject, were detailed at large in a paper published about two years ago in a periodical work of this city,* with which extract I shall close my account of the present species:—

" On the question, Is the whip-poor-will and the night hawk one and the same bird, or are they really two distinct species ? there has long been an opposition of sentiment, and many fruitless disputes. Numbers of sensible and observing people, whose intelligence and long residence in the country entitle their opinion to respect, positively assert that the night hawk and the whip-poor-will are very different birds, and do not even associate together. The naturalists of Europe, however, have generally considered the two names as applicable to one and the same species ; and this opinion has also been adopted by two of our most distinguished naturalists, Mr William Bartram of Kingsessing,† and Professor Barton of Philadelphia.‡ The writer of this, being determined to ascertain the truth by examining for himself, took the following effectual mode of settling this disputed point, the particulars of which he now submits to those interested in the question :—

" Thirteen of those birds usually called night hawks, which dart about in the air like swallows, and sometimes descend with rapidity from a great height, making a hollow sounding noise like that produced by blowing into the bunghole of an empty hogshead, were shot at different times and in different places, and accurately examined, both outwardly and by dissection. Nine of these were found to be males, and

* *The Portfolio.*

† *Caprimulgus Americanus*, night hawk or whip-poor-will (Travels, p. 292).

‡ *Caprimulgus Virginianus*, whip-poor-will or night hawk (Fragments of the Natural History of Pennsylvania, p. 3). See also American Phil. Trans., vol. iv. p. 208, 209, note.

four females. The former all corresponded in the markings and tints of their plumage; the latter also agreed in their marks, differing slightly from the males, though evidently of the same species. Two others were shot as they rose from the nests, or rather from the eggs, which, in both cases, were two in number, lying on the open ground. These also agreed in the markings of their plumage with the four preceding, and, on dissection, were found to be females. The eggs were also secured. A whip-poor-will was shot in the evening, while in the act of repeating his usual and well-known notes. This bird was found to be a male, differing in many remarkable particulars from all the former. Three others were shot at different times during the day, in solitary and dark shaded parts of the wood. Two of these were found to be females, one of which had been sitting on two eggs. The two females resembled each other almost exactly; the male also corresponded in its markings with the one first found, and all four were evidently of one species. The eggs differed from the former both in colour and markings.

"The differences between these two birds were as follows: —The sides of the mouth in both sexes of the whip-poor-will were beset with ranges of long and very strong bristles, extending more than half an inch beyond the point of the bill; both sexes of the night hawk were entirely destitute of bristles. The bill of the whip-poor-will was also more than twice the length of that of the night hawk. The long wing-quills of both sexes of the night hawk were of a deep brownish black, with a large spot of white nearly in their middle, and, when shut, the tips of the wings extended a little *beyond* the tail. The wing-quills of the whip-poor-will of both sexes were beautifully spotted with light brown—had no spot of white on them—and, when shut, the tips of the wings did not reach to the tip of the tail by at least *two inches*. The tail of the night hawk was handsomely *forked*, the exterior feathers being the longest, shortening gradually to the middle ones; the tail of the whip-poor-will was *rounded*, the exterior feathers being the shortest, lengthening gradually to the middle ones.

"After a careful examination of these and several other remarkable differences, it was impossible to withstand the conviction that these birds belonged to two distinct species of the same genus, differing in size, colour, and conformation of parts.

"A statement of the principal of these facts having been laid before Mr Bartram, together with a male and female of each of the above-mentioned species, and also a male of the great Virginian bat, or chuck-will's-widow, after a particular examination, that venerable naturalist was pleased to declare himself fully satisfied; adding, that he had now no doubt of the night hawk and the whip-poor-will being two very distinct species of *Caprimulgus*.

"It is not the intention of the writer of this to enter at present into a description of either the plumage, manners, migrations, or economy of these birds, the range of country they inhabit, or the superstitious notions entertained of them; his only object at present is the correction of an error, which, from the respectability of those by whom it was unwarily adopted, has been but too extensively disseminated, and received by too many as a truth."

RED OWL. (*Strix asio.*)

PLATE XLII.—Fig. 1, Female.

Little Owl, *Catesb.* i. 7.—*Lath.* i. 123.—*Linn. Syst.* 132.—*Arct. Zool.* ii. No. 117. —*Turton, Syst.* i. p. 166.—*Peale's Museum*, No. 428.

STRIX ASIO.—Linnæus.—Young.*

Strix asio, *Bonap. Synop.* p. 36.

This is another of our nocturnal wanderers, well known by its common name, the *little screech owl;* and noted for its melancholy quivering kind of wailing in the evenings, particularly towards the latter part of summer and autumn, near the farmhouse. On clear moonlight nights, they answer each

* See Vol. I. for description of the adult of this species, and Note.

1. Red Owl. 2. Warbling Flycatcher. 3. Purple Finch. 4. Brown Lark.

42.

other from various parts of the fields or orchard ; roost during the day in thick evergreens, such as cedar, pine, or juniper trees, and are rarely seen abroad in sunshine. In May, they construct their nest in the hollow of a tree, often in the orchard in an old apple tree ; the nest is composed of some hay and a few feathers ; the eggs are four, pure white, and nearly round. The young are at first covered with a whitish down.

The bird represented on the plate I kept for several weeks in the room beside me. It was caught in a barn, where it had taken up its lodging, probably for the greater convenience of mousing ; and being unhurt, I had an opportunity of remarking its manners. At first, it struck itself so forcibly against the window, as frequently to deprive it, seemingly, of all sensation for several minutes : this was done so repeatedly, that I began to fear that either the glass or the owl's skull must give way. In a few days, however, it either began to comprehend something of the matter, or to take disgust at the glass, for it never repeated its attempts ; and soon became quite tame and familiar. Those who have seen this bird only in the day can form but an imperfect idea of its activity, and even sprightliness, in its proper season of exercise. Throughout the day, it was all stillness and gravity,—its eyelids half shut, its neck contracted, and its head shrunk seemingly into its body ; but scarcely was the sun set, and twilight began to approach, when its eyes became full and sparkling, like two living globes of fire ; it crouched on its perch, reconnoitered every object around with looks of eager fierceness ; alighted and fed ; stood on the meat with clenched talons, while it tore it in morsels with its bill ; flew round the room with the silence of thought, and perching, moaned out its melancholy notes with many lively gesticulations, not at all accordant with the pitiful tone of its ditty, which reminded one of the shivering moanings of a half-frozen puppy.

This species is found generally over the United States, and is not migratory.

The red owl is eight inches and a half long, and twenty-one inches in extent ; general colour of the plumage above, a bright nut brown or tawny red ; the shafts, black ; exterior edges of the outer row of scapulars, white ; bastard wing, the five first primaries, and three or four of the first greater coverts, also spotted with white ; whole wing-quills, spotted with dusky on their exterior webs ; tail, rounded, transversely barred with dusky and pale brown ; chin, breast, and sides, bright reddish brown, streaked laterally with black, intermixed with white ; belly and vent, white, spotted with bright brown ; legs, covered to the claws with pale brown hairy down ; extremities of the toes and claws, pale bluish, ending in black ; bill, a pale bluish horn colour ; eyes, vivid yellow ; inner angles of the eyes, eyebrows, and space surrounding the bill, whitish ; rest of the face, nut brown ; head, horned or eared, each horn consisting of nine or ten feathers of a tawny red, shafted with black.

WARBLING FLYCATCHER. (*Muscicapa melodia.*)

PLATE XLII.—Fig. 2.

VIREO GILVUS.—Bonaparte.

Muscicapa gilva, *Vieill.* pl. 34. (auct. *Bonap.*)—Vireo gilvus, *Bonap. Synop.* p. 70.
Nomen. sp. 123.

This sweet little warbler is for the first time figured and described. In its general appearance it resembles the red-eyed flycatcher ; but, on a close comparison, differs from that bird in many particulars. It arrives in Pennsylvania about the middle of April, and inhabits the thick foliage of orchards and high trees ; its voice is soft, tender, and soothing, and its notes flow in an easy, continued strain, that is extremely pleasing. It is often heard among the weeping willows and Lombardy poplars of this city ; is rarely observed in the woods, but seems particularly attached to the society of man. It gleans among the leaves, occasionally darting after winged insects, and searching for caterpillars ; and seems by its manners to partake considerably of the nature of the genus *sylvia.* It is late in departing, and I have frequently heard its notes among the fading leaves of the poplar in October.

This little bird may be distinguished from all the rest of our songsters by the soft, tender, easy flow of its notes while hid among the foliage. In these there is nothing harsh, sudden, or emphatical; they glide along in a kind of meandering strain, that is peculiarly its own. In May and June it may be generally heard in the orchards, the borders of the city, and around the farmhouse.

This species is five inches and a half long, and eight inches and a half in extent; bill, dull lead colour above, and notched near the point, lower, a pale flesh colour; eye, dark hazel; line over the eye, and whole lower parts, white, the latter tinged with very pale greenish yellow near the breast; upper parts, a pale green olive; wings, brown, broadly edged with pale olive green; tail, slightly forked, edged with olive; the legs and feet, pale lead; the head inclines a little to ash; no white on the wings or tail. Male and female nearly alike.

PURPLE FINCH. *(Fringilla purpurea.)*

PLATE XLII.—Fig. 3.

ERYTHROSPIZA PURPUREA.—Bonaparte.*

This bird is represented as he appears previous to receiving his crimson plumage, and also when moulting. By recurring to the figure in Vol. I. pl. 7, fig. 4, of this work, which exhibits him in his full dress, the great difference of colour will be observed to which this species is annually subject.

It is matter of doubt with me whether this species ought not to be classed with the *Loxia;* the great thickness of the bill, and similarity that prevails between this and the pine grosbeak, almost induced me to adopt it into that class. But

---

* See description of adult male, Note and Synonyms, Vol. I. p. 119.

respect for other authorities has prevented me from making this alteration.

When these birds are taken in their crimson dress, and kept in a cage till they moult their feathers, they uniformly change to their present appearance, and sometimes never after receive their red colour. They are also subject, if well fed, to become so fat as literally to die of corpulency, of which I have seen several instances; being at these times subject to something resembling apoplexy, from which they sometimes recover in a few minutes, but oftener expire in the same space of time.

The female is entirely without the red, and differs from the present only in having less yellow about her.

These birds regularly arrive from the north, where they breed, in September, and visit us from the south again early in April, feeding on the cherry blossoms as soon as they appear. Of the particulars relative to this species, the reader is referred to the account in Vol. I., already mentioned.

The individual figured in the plate measured six inches and a quarter in length, and ten inches in extent; the bill was horn coloured; upper parts of the plumage, brown olive, strongly tinged with yellow, particularly on the rump, where it was brownish yellow; from above the eye, backwards, passed a streak of white, and another more irregular one from the lower mandible; feathers of the crown, narrow, rather long, and generally erected, but not so as to form a crest; nostrils and base of the bill, covered with reflected brownish hairs; eye, dark hazel; wings and tail, dark blackish brown, edged with olive; first and second row of coverts, tipt with pale yellow; chin, white; breast, pale cream, marked with pointed spots of deep olive brown; belly and vent, white; legs, brown. This bird, with several others marked nearly in the same manner, ·was shot 25th April, while engaged in eating the buds from the beech tree.

BROWN LARK. (*Alauda rufa.*)

PLATE XLII.—Fig. 4.

Red Lark, *Edw.* 297.—*Arct. Zool.* No. 279.—*Lath.* ii. 376.—L'Alouette aux joues brunes de Pennsylvanie, *Buff.* v. 58.—*Peale's Museum*, No. 5138.

ANTHUS LUDOVICIANUS.—Bonaparte.[*]

Synonyms of Anthus Ludovicianus, *Bonap.* (*from his Nomenclature*) : — " Alauda rubra, *Gmel. Lath.*—Alauda Ludoviciana, *Gmel. Lath.*—Alauda Pennsylvanica, *Briss.*—Farlouzanne, *Buff. Ois.*—Alouette aux joues brunes de Pennsylvanie, *Buff. Ois.*—Lark from Pennsylvania, *Ed. Glean.* p. 297.—Red Lark, *Penn. Brit. and Arct. Zool. Lath. Syn.*—Louisiana Lark, *Lath. Syn.*"— Anthus spinoletta, *Bonap. Synop.* p. 90.

In what particular district of the northern regions this bird breeds, I am unable to say. In Pennsylvania, it first arrives from the north about the middle of October ; flies in loose scattered flocks ; is strongly attached to flat, newly-ploughed

---

[*] *Anthus* is a genus of Bechstein's, formed to contain birds which have been generally called larks, but which have a nearer resemblance to the *Motacillæ*, or wagtails, and the accentors. They are also allied to *Seiurus* of Swainson.

The Prince of Musignano made this identical with the European rock lark, *Anthus aquaticus*, Bechst., *Alauda spinoletta*, Linn. ; but in his observations on Wilson's nomenclature, saw reason to change his opinion, and it will now stand as *A. Ludovicianus* of that gentleman. Audubon has, on the other hand, placed it in his "Biography" as the European bird, but I fear, with too slender comparison ; and the same name is mentioned in the "Northern Zoology," without comparing the arctic specimens with those of Britain or Europe. On these accounts, I rather trust to the observations of Bonaparte, which have been made from actual comparison. It must also be recollected, that the summer and winter dress of the *Anthi* differ very considerably in their shades.

Audubon has introduced in his "Biography" another *Anthus*, which he considers new, under the title of *pipiens*. It was only met with once, in the extensive prairies of the north-western States, where two were killed ; and though allied to the common brown titlark, were distinguished by the difference of their notes. If these specimens were not preserved, the species must rest on the authority of Mr Audubon's plate, and, of course, admitted with doubt.—Ed.

fields, commons, and such like situations ; has a feeble note, characteristic of its tribe ; runs rapidly along the ground ; and, when the flock takes to wing, they fly high, and generally to a considerable distance before they alight. Many of them continue in the neighbourhood of Philadelphia all winter, if the season be moderate. In the southern States, particularly in the lower parts of North and South Carolina, I found these larks in great abundance in the middle of February. Loose flocks of many hundreds were driving about from one corn-field to another ; and, in the low rice-grounds, they were in great abundance. On opening numbers of these, they appeared to have been feeding on various small seeds, with a large quantity of gravel. On the 8th of April, I shot several of these birds in the neighbourhood of Lexington, Kentucky. In Pennsylvania, they generally disappear, on their way to the north, about the beginning of May, or earlier. At Portland, in the district of Maine, I met with a flock of these birds in October. I do not know that they breed within the United States. Of their song, nest, eggs, &c., we have no account.

The brown lark is six inches long, and ten inches and a half in extent ; the upper parts, brown olive, touched with dusky ; greater coverts and next superior row, lighter ; bill, black, slender ; nostril, prominent ; chin and line over the eye, pale rufous ; breast and belly, brownish ochre, the former spotted with black ; tertials, black, the secondaries brown, edged with lighter ; tail, slightly forked, black ; the two exterior feathers, marked largely with white ; legs, dark purplish brown ; hind heel, long, and nearly straight ; eye, dark hazel. Male and female nearly alike. Mr Pennant says that one of these birds was shot near London.

Drawn from Nature by A. Wilson.　　　　　　　　　　*Engraved by W. H. Lizars.*

1. Turtle Dove.　2. Hermit Thrush.　3. Tawney Thrush.　4. Pine-swamp Warbler.

43.

CAROLINA PIGEON OR TURTLE DOVE. (*Columba Carolinensis.*)

PLATE XLIII.—Fig. 1.

Linn. Syst. 286.—*Catesb. Car.* i. 24.—*Buff.* ii. 557, *Pl. enl.* 175.—La Tourterelle de la Caroline, *Brisson*, i. 110.—*Peale's Museum*, No. 5088.—*Turton*, 479.—*Arct. Zool.* ii. No. 188.

ECTOPISTES CAROLINENSIS.—Swainson.

Genus Ectopistes, *Swain. N. Groups. Zool. Journ.* No. xi. p. 362.—Columba Carolinensis, *Bonap. Synop.* p. 119.—The Carolina Turtle Dove, *Aud. Orn. Biog.* i. 91, pl. 17, male and female.

This is a favourite bird with all those who love to wander among our woods in spring, and listen to their varied harmony. They will there hear many a singular and sprightly performer, but none so mournful as this. The hopeless woe of settled sorrow, swelling the heart of female innocence itself, could not assume tones more sad, more tender and affecting. Its notes are four; the first is somewhat the highest, and preparatory, seeming to be uttered with an inspiration of the breath, as if the afflicted creature were just recovering its voice from the last convulsive sobs of distress; this is followed by three long, deep, and mournful moanings, that no person of sensibility can listen to without sympathy. A pause of a few minutes ensues, and again the solemn voice of sorrow is renewed as before. This is generally heard in the deepest shaded parts of the woods, frequently about noon and towards the evening.

There is, however, nothing of real distress in all this; quite the reverse. The bird who utters it wantons by the side of his beloved partner, or invites her by his call to some favourite retired and shady retreat. It is the voice of love, of faithful connubial affection, for which the whole family of doves are so celebrated; and, among them all, none more deservingly so than the species now before us.

The turtle dove is a general inhabitant in summer of the United States, from Canada to Florida, and from the sea-coast to the Mississippi, and far to the westward. They are, however, partially migratory in the northern and middle States; and collect together in North and South Carolina, and their corresponding parallels, in great numbers, during the winter. On the 2d of February, in the neighbourhood of Newbern, North Carolina, I saw a flock of turtle doves of many hundreds; in other places, as I advanced farther south, particularly near the Savannah river, in Georgia, the woods were swarming with them, and the whistling of their wings was heard in every direction.

On their return to the north in March and early in April, they disperse so generally over the country, that there are rarely more than three or four seen together—most frequently only two. Here they commonly fly in pairs, resort constantly to the public roads to dust themselves and procure gravel; are often seen in the farmer's yard before the door, the stable, barn, and other outhouses, in search of food, seeming little inferior in familiarity, at such times, to the domestic pigeon. They often mix with the poultry while they are fed in the morning, visit the yard and adjoining road many times a day, and the pump, creek, horse-trough, and rills for water.

Their flight is quick, vigorous, and always accompanied by a peculiar whistling of the wings, by which they can easily be distinguished from the wild pigeon. They fly with great swiftness, alight on trees, fences, or on the ground indiscriminately; are exceedingly fond of buckwheat, hempseed, and Indian-corn; feed on the berries of the holly, the dogwood, and poke, huckleberries, partridge-berries, and the small acorns of the live oak and shrub oak. They devour large quantities of gravel, and sometimes pay a visit to the kitchen garden for peas, for which they have a particular regard.

In this part of Pennsylvania, they commence building about the beginning of May. The nest is very rudely constructed, generally in an evergreen, among the thick foliage of the vine, in an orchard, on the horizontal branches of an apple tree, and, in some cases, on the ground. It is composed of a handful of small twigs, laid with little art, on which are scattered dry fibrous roots of plants; and in this almost flat bed are deposited two eggs of a snowy whiteness. The male and female unite in feeding the young, and they have rarely more than two broods in the same season.

The flesh of this bird is considered much superior to that of the wild pigeon; but its seeming confidence in man, the tenderness of its notes, and the innocency attached to its character, are with many its security and protection; with others, however, the tenderness of its flesh, and the sport of shooting, overcome all other considerations. About the commencement of frost, they begin to move off to the south; numbers, however, remain in Pennsylvania during the whole winter.

The turtle dove is twelve inches long, and seventeen inches in extent; bill, black; eye, of a glossy blackness, surrounded with a pale greenish blue skin; crown, upper part of the neck and wings, a fine silky slate blue; back, scapulars, and lesser wing-coverts, ashy brown; tertials spotted with black; primaries, edged and tipt with white; forehead, sides of the neck, and breast, a pale brown vinous orange; under the ear-feathers, a spot or drop of deep black, immediately below which the plumage reflects the most vivid tints of green, gold, and crimson; chin, pale yellow ochre; belly and vent, whitish; legs and feet, coral red, seamed with white; the tail is long and cuneiform, consisting of fourteen feathers; the four exterior ones, on each side, are marked with black, about an inch from the tips, and white thence to the extremity; the next has less of the white at the tip; these gradually lengthen to the four middle ones, which are wholly dark slate; all of them taper towards the points, the two middle ones most so.

The female is an inch shorter, and is otherwise only distinguished by the less brilliancy of her colour; she also wants

the rich silky blue on the crown, and much of the splendour
of the neck; the tail is also somewhat shorter, and the white
with which it is marked less pure.*

HERMIT THRUSH. (*Turdus solitarius.*)

PLATE XLIII.—Fig. 2.

Little Thrush, *Catesby,* i. 31.—*Edwards,* 296.—Brown Thrush. *Arct. Zool.* 337,
No. 199.—*Peale's Museum,* No. 3542.

TURDUS SOLITARIUS.—Wilson.†

Turdus minor, *Bonap. Synop.* p. 75.—The Hermit Thrush, *Aud. Orn. Biog.*
i. p. 303, pl. 58, male and female.

The dark solitary cane and myrtle swamps of the southern
States are the favourite native haunts of this silent and recluse
species; and the more deep and gloomy these are, the more
certain we are to meet with this bird flitting among them.
This is the species mentioned in the first volume of this work,
while treating of the wood thrush, as having been figured and
described, more than fifty years ago, by Edwards, from a
dried specimen sent him by my friend Mr William Bartram,
under the supposition that it was the wood thrush (*Turdus*

---

* In addition to their history by Wilson, Audubon mentions, that
though regularly migrating in numbers, they are never in such vast
extent as the passenger pigeon, from two hundred and fifty to three
hundred being considered a large flock. He also mentions them differing in another more important particular—the manner of roosting.
They prefer sitting among the long grass of abandoned fields, at the
foot of the dry stalks of maize, and only occasionally resort to the dead
foliage of trees, or the different species of evergreens. They do not sit
near each other, but are dispersed over the field, whereas the passenger
pigeon roosts in compact masses on limbs of trees. In every respect
they run more into the ground doves, or bronze-winged pigeons, which
similarity some parts of the plumage will strengthen.—Ed.

† Bonaparte has wished to restore Gmelin's old name of *minor* to this
bird, which Wilson had thought in some manner erroneous, on account
of *solitarius* being preoccupied by another species. That, however,
will rank in the genus *Petrocincla;* and Mr Swainson has since described a small species under the name of *minor.*—Ed.

melodus). It is, however, considerably less, very differently marked, and altogether destitute of the clear voice and musical powers of that charming minstrel. It also differs in remaining in the southern States during the whole year; whereas the wood thrush does not winter even in Georgia, nor arrive within the southern boundary of that State until some time in April.

The hermit thrush is rarely seen in Pennsylvania, unless for a few weeks in spring, and late in the fall, long after the wood thrush has left us, and when scarcely a summer bird remains in the woods. In both seasons it is mute, having only in spring an occasional squeak, like that of a young stray chicken. Along the Atlantic coast, in New Jersey, they remain longer and later, as I have observed them there late in November. In the cane swamps of the Choctaw nation, they were frequent in the month of May, on the 12th of which I examined one of their nests on a horizontal branch, immediately over the path. The female was sitting, and left it with great reluctance, so that I had nearly laid my hand on her before she flew. The nest was fixed on the upper part of the body of the branch, and constructed with great neatness, but without mud or plaster, contrary to the custom of the wood thrush. The outside was composed of a considerable quantity of coarse rooty grass, intermixed with horse hair, and lined with a fine, green-coloured, thread-like grass, perfectly dry, laid circularly, with particular neatness. The eggs were four, of a pale greenish blue, marked with specks and blotches of olive, particularly at the great end. I also observed this bird on the banks of the Cumberland river in April. Its food consists chiefly of berries, of which these low swamps furnish a perpetual abundance, such as those of the holly, myrtle, gall bush (a species of *vaccinium*), yapon shrub, and many others.

A superficial observer would instantly pronounce this to be only a variety of the wood thrush; but taking into consideration its difference of size, colour, manners, want of song, secluded habits, differently formed nest, and spotted eggs, all unlike those of the former, with which it never associates, it is impossible not to conclude it to be a distinct and separate species, however near it may approach to that of the former. Its food, and the country it inhabits for half the year, being the same, neither could have produced those differences ; and we must believe it to be now, what it ever has been, and ever will be, a distinct connecting link in the great chain of this part of animated nature ; all the sublime reasoning of certain theoretical closet philosophers to the contrary notwithstanding.

Length of the hermit thrush, seven inches ; extent, ten inches and a 'half ; upper parts, plain deep olive brown ; lower, dull white ; upper part of the breast and throat, dull cream colour, deepest where the plumage falls over the shoulders of the wing, and marked with large dark brown pointed spots ; ear-feathers, and line over the eye, cream, the former mottled with olive ; edges of the wings, lighter ; tips, dusky ; tail-coverts and tail, inclining to a reddish fox colour. In the wood thrush, these parts incline to greenish olive. Tail, slightly forked ; legs, dusky ; bill, black above and at the tip, whitish below ; iris, black and very full ; chin, whitish.

The female differs very little,—chiefly in being generally darker in the tints, and having the spots on the breast larger and more dusky.

TAWNY THRUSH.   (*Turdus mustelinus.*)

PLATE XLIII.—Fig. 3.

Peale's Museum, No. 5570.

TURDUS WILSONII.—Bonaparte.*

Turdus Wilsonii, *Bonap. Synop.* p. 76.—Merula Wilsonii, *North. Zool.* ii. p. 183.

This species makes its appearance in Pennsylvania from the south regularly about the beginning of May, stays with us a

* The wood thrush of Vol. I., the hermit thrush, and our present species, have so much similarity to each other, that they have been confused together, and their synonyms often misquoted by different authors.

week or two, and passes on to the north and to the high mountainous districts to breed. It has no song, but a sharp chuck. About the 20th of May I met with numbers of them in the Great Pine Swamp, near Pocano; and on the 25th of September in the same year, I shot several of them in the neighbourhood of Mr Bartram's place. I have examined many of these birds in spring, and also on their return in fall, and found very little difference among them between the male and female. In some specimens the wing-coverts were brownish yellow; these appeared to be young birds. I have no doubt but they breed in the northern high districts of the United States; but I have not yet been able to discover their nests.

The tawny thrush is ten inches long, and twelve inches in extent; the whole upper parts are a uniform tawny brown; the lower parts, white; sides of the breast and under the wings, slightly tinged with ash; chin, white; throat, and upper parts of the breast, cream coloured, and marked with pointed spots of brown; lores, pale ash or bluish white; cheeks, dusky brown; tail, nearly even at the end, the shafts of all, as well as those of the wing-quills, continued a little beyond their webs; bill, black above and at the point, below

From these circumstances, the name of *mustelinus*, given by Wilson to this species, is incorrect; and Bonaparte has deservedly dedicated it to its first describer, a name which ought now to be used in our systems. Another bird has been also lost sight of in the alliance which exists among those, and which will now rank as an addition to the northern fauna, the *Turdus parvus* of Edwards, and confounded by Bonaparte with the *T. solitaria*. From the observations of Dr Richardson and Mr Swainson, in the second volume of the " Northern Zoology," there can be little doubt of its being distinct from any of the others just mentioned, and will be distinguished by the more rufous tinge of the upper parts. It was met by the Overland Expedition on the banks of the Saskatchewan, where it is migratory in summer, and appears as nearly allied to the others in its habits as it is in its external appearance. It spreads, no doubt, over the other parts of North America, getting more abundant, perhaps, towards the south. Mr Swainson has received it from Georgia, and remarks that the rufous tinge of the plumage is much clearer and more intense in the southern specimens.—ED.

at the base, flesh coloured ; corners of the mouth, yellow ; eye, large and dark, surrounded with a white ring ; legs, long, slender, and pale brown.

Though I have given this bird the same name that Mr Pennant has applied to one of our thrushes, it must not be considered as the same ; the bird which he has denominated the tawny thrush being evidently, from its size, markings, &c., the wood thrush, already described.

No description of the bird here figured has, to my knowledge, appeared in any former publication.

PINE-SWAMP WARBLER. (*Sylvia pusilla.*)

PLATE XLIII.—Fig. 4.

VIREO SPHAGNOSA.—Jardine.*
Sylvia sphagnosa, *Bonap. Synop.* p. 85.

This little bird is for the first time figured or described. Its favourite haunts are in the deepest and gloomiest pine and hemlock swamps of our mountainous regions, where every tree, trunk, and fallen log is covered with a luxuriant coat of moss, that even mantles over the surface of the ground, and prevents the sportsman from avoiding a thousand holes, springs, and swamps, into which he is incessantly plunged. Of the nest of this bird I am unable to speak. I found it associated with the Blackburnian warbler, the golden-crested wren, ruby-crowned wren, yellow rump, and others of that description, in such places as I have described, about the middle of May. It seemed as active in flycatching as in searching for other insects, darting nimbly about among the branches, and flirting its

* This species seems evidently a *Vireo.* Bonaparte thus observes, in his "Nomenclature," and we have used his name :—" A new species, called by a preoccupied name, but altered in the index to that of *leucoptera,* which is used for one of Vieillot's species, and was, therefore, changed to that of *palustris* by Stephens ; but as this also is preoccupied, I propose for it the name of *S. sphagnosa.*"—Ed.

Drawn from Nature by A.Wilson.

1. Passenger Pigeon. 2. Blue-mountain Warbler. 3. Hemlock W.

Engraved by W.H.Lizars.

44.

wings ; but I could not perceive that it had either note or song. I shot three, one male and two females. I have no doubt that they breed in those solitary swamps, as well as many other of their associates.

The pine-swamp warbler is four inches and a quarter long and seven inches and a quarter in extent; bill, black, not notched, but furnished with bristles ; upper parts, a deep green olive, with slight bluish reflections, particularly on the edges of the tail and on the head ; wings, dusky, but so broadly edged with olive green as to appear wholly of that tint ; immediately below the primary coverts, there is a single triangular spot of yellowish white ; no other part of the wings is white ; the three exterior tail-feathers with a spot of white on their inner vanes; the tail is slightly forked ; from the nostrils over the eye extends a fine line of white, and the lower eyelid is touched with the same tint; lores, blackish ; sides of the neck and auriculars, green olive ; whole lower parts, pale yellow ochre, with a tinge of greenish ; duskiest on the throat; legs, long, and flesh coloured.

The plumage of the female differs in nothing from that of the male.

PASSENGER PIGEON. (*Columba migratoria.*)

PLATE XLIV.—Fig. 1.

Catesby, i. 23.—*Linn. Syst.* 285.—*Turton*, 479.—*Arct. Zool.* p. 322, No. 187.— *Briss.* i. 100.—*Buff.* ii. 527.—*Peale's Museum*, No. 5084.

ECTOPISTES MIGRATORIA.—Swainson.[*]

Ectopistes, *Swain. N. Groups, Zool. Journ.* No. xi. p. 362.—Columba migratoria, *Bonap. Synop.* p. 120.—The Passenger Pigeon, *Aud. Orn. Biog.* i. p. 319, male and female.—Columba (Ectopistes) migratoria, *North. Zool.* ii. p. 363.

This remarkable bird merits a distinguished place in the annals of our feathered tribes,—a claim to which I shall endeavour to do justice ; and though it would be impossible,

[*] In all the large natural groups which have already come under our notice, we have seen a great variation of form, though the essential parts of it were always beautifully kept up. In the present immense

in the bounds allotted to this account, to relate all I have seen and heard of this species, yet no circumstance shall be omitted with which I am acquainted (however extraordinary some of these may appear) that may tend to illustrate its history.

family, Mr Swainson has characterised the passenger pigeons under the name of *Ectopistes*, at once distinguished by their graceful and lengthened make, and well represented by the common *Columba migratoria* and the Carolina pigeon of our author. The nicer distinctions will be found in the slender bill, and the relative proportions of the feet and wings. As far as our knowledge extends, the group is confined to both the continents of America. A single individual of this species was shot, while perched on a wall, in the neighbourhood of a pigeon-house at Westhall, in the parish of Monymail, Fifeshire, in December 1825. It came into the possession of Dr Fleming of Flisk, who has recorded its occurrence in his " British Zoology." He remarks that the feathers were quite fresh and entire, like a wild bird ; but we can only rank it as a very rare straggler.

Mr Audubon mentions having brought over 350 of these birds, when he last visited this country, and distributed them among different country gentlemen. Lord Stanley received fifty of them, which he intended to turn out in his park, in the neighbourhood of Liverpool.

We have the following additional account from Audubon of their flights, roosting, and destruction, in everything corroborating the history of Wilson, but too interesting to pass by :—

" Their great power of flight enables them to survey and pass over an astonishing extent of country in a very short time. Thus pigeons have been killed in the neighbourhood of New York, with their crops full of rice, which they must have collected in the fields of Georgia and Carolina, these districts being the nearest in which they could possibly have procured a supply of food. As their power of digestion is so great that they will decompose food entirely in twelve hours, they must, in this case, have travelled between three and four hundred miles in six hours, which shows their speed to be, at an average, about one mile in a minute. A velocity such as this would enable one of these birds, were it so inclined, to visit the European continent in less than three days.

" In the autumn of 1813, I left my house at Henderson, on the banks of the Ohio, on my way to Louisville. In passing over the Barrens, a few miles beyond Hardensburg, I observed the pigeons flying from north-east to south-west, in greater numbers than I thought I had ever seen them before. I travelled on, and still met more, the farther I proceeded. The air was literally filled with pigeons. The light of the noonday was obscured as by an eclipse. The dung fell in spots not unlike melting flakes of snow ; and the continued buzz of wings had a tendency to lull my senses to repose.

The wild pigeon of the United States inhabits a wide and extensive region of North America, on this side of the great Stony Mountains, beyond which, to the westward, I have not

" Before sunset I reached Louisville, distant from Hardensburg fifty-five miles. The pigeons were still passing in undiminished numbers, and continued to do so for three days in succession. The people were all in arms. The banks of the Ohio were crowded with men and boys, incessantly shooting at the pilgrims, which there flew lower as they passed the river. Multitudes were thus destroyed. For a week or more, the population fed on no other flesh than that of pigeons. The atmosphere, during this time, was strongly impregnated with the peculiar odour which emanates from the species." In estimating the number of these mighty flocks, and the food consumed by them daily he adds—" Let us take a column of one mile in breadth, which is far below the average size, and suppose it passing over us at the rate of one mile per minute. This will give us a parallelogram of 180 miles by 1, covering 180 square miles ; and allowing two pigeons to the square yard, we have one billion one hundred and fifteen millions one hundred and thirty-six thousand pigeons in one flock : and as every pigeon consumes fully half a pint per day, the quantity required to feed such a flock, must be eight millions seven hundred and twelve thousand bushels per day."

The accounts of their roosting places are as remarkable :—

" Let us now, kind reader, inspect their place of nightly rendezvous : —It was, as is always the case, in a portion of the forest where the trees were of great magnitude, and where there was little underwood. I rode through it upwards of forty miles, and, crossing it at different parts, found its average breadth to be rather more than three miles. Few pigeons were to be seen before sunset ; but a great number of persons, with horses and waggons, guns and ammunition, had already established encampments on the borders. Two farmers from the vicinity of Russelsville, distant more than a hundred miles, had driven upwards of three hundred hogs, to be fattened on the pigeons which were to be slaughtered. Here and there, the people employed in plucking and salting what had already been procured were seen sitting in the midst of large piles of these birds. The dung lay several inches deep, covering the whole extent of the roosting place like a bed of snow. Many trees, two feet in diameter, I observed, were broken off at no great distance from the ground ; and the branches of many of the largest and tallest had given way, as if the forest had been swept by a tornado. Everything proved to me that the number of birds resorting to this part of the forest must be immense beyond conception. As the period of their arrival approached, their foes anxiously prepared ;to

heard of their being seen. According to Mr Hutchins, they abound in the country round Hudson's Bay, where they seize them. Some were furnished with iron pots, containing sulphur, others with torches of pine-knots, many with poles, and the rest with guns. The sun was lost to our view ; yet not a pigeon had arrived. Everything was ready, and all eyes were gazing on the clear sky, which appeared in glimpses amidst the tall trees. Suddenly, there burst forth a general cry of, ' *Here they come !*' The noise which they made, though yet distant, reminded me of a hard gale at sea passing through the rigging of a close-reefed vessel. As the birds arrived, and passed over me, I felt a current of air that surprised me. Thousands were soon knocked down by polemen. The current of birds, however, still kept increasing. The fires were lighted, and a most magnificent, as well as a wonderful and terrifying sight, presented itself. The pigeons, coming in by thousands, alighted everywhere, one above another, until solid masses, as large as hogsheads, were formed on every tree, in all directions. Here and there the perches gave way under the weight with a crash, and, falling to the ground, destroyed hundreds of the birds beneath, forcing down the dense groups with which every stick was loaded. It was a scene of uproar and confusion. I found it quite useless to speak, or even to shout, to those persons who were nearest me. The reports, even, of the nearest guns, were seldom heard ; and I knew of the firing only by seeing the shooters reloading. No one dared venture within the line of devastation ; the hogs had been penned up in due time, the picking up of the dead and wounded being left for the next morning's employment. The pigeons were constantly coming ; and it was past midnight before I perceived a decrease in the number of those that arrived. The uproar continued, however, the whole night ; and, as I was anxious to know to what distance the sound reached, I sent off a man, accustomed to perambulate the forest, who, returning two hours afterwards, informed me that he had heard it distinctly when three miles from the spot. Towards the approach of day, the noise rather subsided ; but, long ere objects were at all distinguishable, the pigeons began to move off in a direction quite different from that in which they had arrived the evening before ; and at sunrise, all that were able to fly had disappeared. The howlings of the wolves now reached our ears ; and the foxes, lynxes, cougars, bears, racoons, opossums, and pole-cats, were seen sneaking off from the spot, whilst eagles and hawks, of different species, accompanied by a crowd of vultures, came to supplant them, and enjoy their share of the spoil. It was then that the authors of all this devastation began their entry amongst the dead, the dying, and the mangled. The pigeons were picked up, and piled in heaps, until each had as many as he could possibly dispose of, when the hogs were let loose to feed on the remainder."—E<small>D</small>.

usually remain as late as December, feeding, when the ground
is covered with snow, on the buds of juniper. They spread
over the whole of Canada; were seen by Captain Lewis and
his party near the Great Falls of the Missouri, upwards
of 2500 miles from its mouth, reckoning the meanderings
of the river; were also met with in the interior of Louisiana
by Colonel Pike; and extend their range as far south as the
Gulf of Mexico; occasionally visiting or breeding in almost
every quarter of the United States.

But the most remarkable characteristic of these birds is
their associating together, both in their migrations, and also
during the period of incubation, in such prodigious numbers
as almost to surpass belief, and which has no parallel among
any other of the feathered tribes on the face of the earth
with which naturalists are acquainted.

These migrations appear to be undertaken rather in quest
of food, than merely to avoid the cold of the climate, since
we find them lingering in the northern regions, around
Hudson's Bay, so late as December; and since their appearance is so casual and irregular, sometimes not visiting certain
districts for several years in any considerable numbers, while
at other times they are innumerable. I have witnessed these
migrations in the Gennesee country, often in Pennsylvania,
and also in various parts of Virginia, with amazement; but
all that I had then seen of them were mere straggling parties
when compared with the congregated millions which I have
since beheld in our western forests, in the States of Ohio,
Kentucky, and the Indiana territory. These fertile and extensive regions abound with the nutritious beech-nut, which
constitutes the chief food of the wild pigeon. In seasons
when these nuts are abundant, corresponding multitudes of
pigeons may be confidently expected. It sometimes happens
that, having consumed the whole produce of the beech trees
in an extensive district, they discover another, at the distance
perhaps of sixty or eighty miles, to which they regularly
repair every morning, and return as regularly in the course of

the day, or in the evening, to their place of general rendezvous, or, as it is usually called, the roosting place. These
roosting places are always in the woods, and sometimes occupy
a large extent of forest. When they have frequented one of
these places for some time, the appearance it exhibits is surprising. The ground is covered to the depth of several inches
with their dung; all the tender grass and underwood destroyed ; the surface strewed with large limbs of trees, broken
down by the weight of the birds clustering one above another;
and the trees themselves, for thousands of acres, killed as
completely as if girdled with an axe. The marks of this
desolation remain for many years on the spot ; and numerous
places could be pointed out where, for several years after,
scarcely a single vegetable made its appearance.

When these roosts are first discovered, the inhabitants, from
considerable distances, visit them in the night, with guns,
clubs, long poles, pots of sulphur, and various other engines
of destruction. In a few hours, they fill many sacks, and
load their horses with them. By the Indians, a pigeon roost
or breeding place is considered an important source of
national profit and dependence for that season, and all their
active ingenuity is exercised on the occasion. The breeding
place differs from the former in its greater extent. In the
western countries above mentioned, these are generally in
beech woods, and often extend, in nearly a straight line,
across the country for a great way. Not far from Shelbyville,
in the State of Kentucky, about five years ago, there was one
of these breeding places, which stretched through the woods
in nearly a north and south direction ; was several miles in
breadth, and was said to be upwards of forty miles in extent!
In this tract, almost every tree was furnished with nests,
wherever the branches could accommodate them. The pigeons
made their first appearance there about the 10th of April, and
left it altogether, with their young, before the 25th of May.

As soon as the young were fully grown, and before they
left the nests, numerous parties of the inhabitants, from all

parts of the adjacent country, came with waggons, axes, beds, cooking utensils, many of them accompanied by the greater part of their families, and encamped for several days at this immense nursery. Several of them informed me that the noise in the woods was so great as to terrify their horses, and that it was difficult for one person to hear another speak without bawling in his ear. The ground was strewed with broken limbs of trees, eggs, and young squab pigeons, which had been precipitated from above, and on which herds of hogs were fattening. Hawks, buzzards, and eagles were sailing about in great numbers, and seizing the squabs from their nests at pleasure; while, from twenty feet upwards to the tops of the trees, the view through the woods presented a perpetual tumult of crowding and fluttering multitudes of pigeons, their wings roaring like thunder, mingled with the frequent crash of falling timber; for now the axe-men were at work, cutting down those trees that seemed to be most crowded with nests, and contrived to fell them in such a manner, that, in their descent, they might bring down several others; by which means the falling of one large tree sometimes produced two hundred squabs, little inferior in size to the old ones, and almost one mass of fat. On some single trees, upwards of one hundred nests were found, each containing *one* young only; a circumstance in the history of this bird not generally known to naturalists. It was dangerous to walk under these flying and fluttering millions from the frequent fall of large branches, broken down by the weight of the multitudes above, and which, in their descent, often destroyed numbers of the birds themselves; while the clothes of those engaged in traversing the woods were completely covered with the excrements of the pigeons.

These circumstances were related to me by many of the most respectable part of the community in that quarter, and were confirmed, in part, by what I myself witnessed. I passed for several miles through this same breeding place, where every tree was spotted with nests, the remains of those above described. In many instances, I counted upwards of ninety nests on a single tree ; but the pigeons had abandoned this place for another, sixty or eighty miles off towards Green River, where they were said at that time to be equally numerous. From the great numbers that were constantly passing overhead to or from that quarter, I had no doubt of the truth of this statement. The mast had been chiefly consumed in Kentucky, and the pigeons, every morning, a little before sunrise, set out for the Indiana territory, the nearest part of which was about sixty miles distant. Many of these returned before ten o'clock, and the great body generally appeared on their return a little after noon.

I had left the public road to visit the remains of the breeding place near Shelbyville, and was traversing the woods with my gun, on my way to Frankfort, when, about one o'clock, the pigeons, which I had observed flying the greater part of the morning northerly, began to return, in such immense numbers as I never before had witnessed. Coming to an opening, by the side of a creek called the Benson, where I had a more uninterrupted view, I was astonished at their appearance. They were flying, with great steadiness and rapidity, at a height beyond gunshot, in several strata deep, and so close together, that, could shot have reached them, one discharge could not have failed of bringing down several individuals. From right to left, far as the eye could reach, the breadth of this vast procession extended, seeming everywhere equally crowded. Curious to determine how long this appearance would continue, I took out my watch to note the time, and sat down to observe them. It was then half-past one. I sat for more than an hour, but instead of a diminution of this prodigious procession, it seemed rather to increase both in numbers and rapidity; and anxious to reach Frankfort before night, I rose and went on. About four o'clock in the afternoon I crossed the Kentucky River, at the town of Frankfort, at which time the living torrent above my head seemed as numerous and as extensive as ever. Long after this I observed them in large bodies, that continued to pass for six or eight minutes, and these again were followed by other detached bodies, all moving in the same south-east direction, till after six in the evening. The great breadth of front which this mighty multitude preserved would seem to intimate a corresponding breadth of their breeding place, which, by several gentlemen, who had lately passed through part of it, was stated to me at several miles. It was said to be in Green county, and that the young began to fly about the middle of March. On the 17th of April, forty-nine miles beyond Danville, and not far from Green River, I crossed this same breeding place, where the nests, for more than three miles, spotted every tree: the leaves not being yet out, I had a fair prospect of them, and was really astonished at their numbers. A few bodies of pigeons lingered yet in different parts of the woods, the roaring of whose wings was heard in various quarters around me.

All accounts agree in stating that each nest contains only one young squab. These are so extremely fat, that the Indians, and many of the whites, are accustomed to melt down the fat for domestic purposes as a substitute for butter and lard. At the time they leave the nest, they are nearly as heavy as the old ones, but become much leaner after they are turned out to shift for themselves.

It is universally asserted in the western countries, that the pigeons, though they have only one young at a time, breed thrice, and sometimes four times, in the same season: the circumstances already mentioned render this highly probable. It is also worthy of observation, that this takes place during that period when acorns, beech nuts, &c., are scattered about in the greatest abundance, and mellowed by the frost. But they are not confined to these alone,—buckwheat, hempseed, Indian-corn, hollyberries, hackberries, huckleberries, and many others, furnish them with abundance at almost all seasons. The acorns of the live oak are also eagerly sought after by these birds, and rice has been frequently found in individuals

killed many hundred miles to the northward of the nearest
rice plantation.   The vast quantity of mast which these multitudes consume is a serious loss to the bears, pigs, squirrels,
and other dependants on the fruits of the forest.   I have taken
from the crop of a single wild pigeon a good handful of the
kernels of beech-nuts, intermixed with acorns and chestnuts.
To form a rough estimate of the daily consumption of one of
these immense flocks, let us first attempt to calculate the
numbers of that above mentioned, as seen in passing between
Frankfort and the Indiana territory.   If we suppose this column
to have been one mile in breadth (and I believe it to have
been much more), and that it moved at the rate of one mile
in a minute, four hours, the time it continued passing, would
make its whole length two hundred and forty miles.   Again,
supposing that each square yard of this moving body comprehended three pigeons, the square yards in the whole space,
multiplied by three, would give two thousand two hundred
and thirty millions, two hundred and seventy-two thousand
pigeons !—an almost inconceivable multitude, and yet probably
far below the actual amount.   Computing each of these to
consume half a pint of mast daily, the whole quantity at this
rate would equal seventeen millions, four hundred and twenty-
four thousand bushels per day !   Heaven has wisely and
graciously given to these birds rapidity of flight and a disposition to range over vast uncultivated tracts of the earth,
otherwise they must have perished in the districts where they
resided, or devoured up the whole productions of agriculture,
as well as those of the forests.

A few observations on the mode of flight of these birds must
not be omitted.   The appearance of large detached bodies of
them in the air, and the various evolutions they display, are
strikingly picturesque and interesting.   In descending the
Ohio by myself, in the month of February, I often rested on
my oars to contemplate their aerial manœuvres.   A column,
eight or ten miles in length, would appear from Kentucky,
high in air, steering across to Indiana.   The leaders of this

great body would sometimes gradually vary their course, until it formed a large bend, of more than a mile in diameter, those behind tracing the exact route of their predecessors. This would continue sometimes long after both extremities were beyond the reach of sight; so that the whole, with its glittery undulations, marked a space on the face of the heavens resembling the windings of a vast and majestic river. When this bend became very great, the birds, as if sensible of the unnecessary circuitous course they were taking, suddenly changed their direction, so that what was in column before became an immense front, straightening all its indentures, until it swept the heavens in one vast and infinitely extended line. Other lesser bodies also united with each other as they happened to approach, with such ease and elegance of evolution, forming new figures, and varying these as they united or separated, that I never was tired of contemplating them. Sometimes a hawk would make a sweep on a particular part of the column, from a great height, when, almost as quick as lightning, that part shot downwards out of the common track; but, soon rising again, continued advancing at the same height as before. This reflection was continued by those behind, who, on arriving at this point, dived down, almost perpendicularly, to a great depth, and rising, followed the exact path of those that went before. As these vast bodies passed over the river near me, the surface of the water, which was before smooth as glass, appeared marked with innumerable dimples, occasioned by the dropping of their dung, resembling the commencement of a shower of large drops of rain or hail.

Happening to go ashore one charming afternoon, to purchase some milk at a house that stood near the river, and while talking with the people within doors, I was suddenly struck with astonishment at a loud rushing roar, succeeded by instant darkness, which on the first moment I took for a tornado about to overwhelm the house and everything around in destruction. The people, observing my surprise, coolly said, " It is only the pigeons;" and, on running out, I beheld a flock, thirty or forty yards in width, sweeping along, very low, between the house and the mountain, or height, that formed the second bank of the river.   These continued passing for more than a quarter of an hour, and at length varied their bearing, so as to pass over the mountain, behind which they disappeared before the rear came up.

In the Atlantic States, though they never appear in such unparalleled multitudes, they are sometimes very numerous, and great havoc is then made amongst them with the gun, the clap-net, and various other implements of destruction.   As soon as it is ascertained in a town that the pigeons are flying numerously in the neighbourhood, the gunners rise *en masse ;* the clap-nets are spread out on suitable situations, commonly on an open height in an old buckwheat field ; four or five live pigeons, with their eyelids sewed up, are fastened on a movable stick—a small hut of branches is fitted up for the fowler, at the distance of forty or fifty yards—by the pulling of a string, the stick on which the pigeons rest is alternately elevated and depressed, which produces a fluttering of their wings similar to that of birds just alighting ; this being perceived by the passing flocks, they descend with great rapidity, and finding corn, buckwheat, &c., strewed about, begin to feed, and are instantly, by the pulling of a cord, covered by the net.   In this manner, ten, twenty, and even thirty dozen, have been caught at one sweep.   Meantime, the air is darkened with large bodies of them, moving in various directions ; the woods also swarm with them in search of acorns ; and the thundering of musketry is perpetual on all sides from morning to night.   Waggon loads of them are poured into market, where they sell from fifty to twenty-five, and even twelve cents, per dozen ; and pigeons become the order of the day at dinner, breakfast, and supper, until the very name becomes sickening.   When they have been kept alive, and fed for some time on corn and buckwheat, their flesh acquires great superiority ; but, in their common state, they are dry and blackish, and far inferior to the full-grown young ones, or squabs.

The nest of the wild pigeon is formed of a few dry slender twigs, carelessly put together, and with so little concavity, that the young one, when half grown, can easily be seen from below. The eggs are pure white. Great numbers of hawks, and sometimes the bald eagle himself, hover about those breeding places, and seize the old or the young from the nest amidst the rising multitudes, and with the most daring effrontery. The young, when beginning to fly, confine them-selves to the under part of the tall woods where there is no brush, and where nuts and acorns are abundant, searching among the leaves for mast, and appear like a prodigious torrent rolling along through the woods, every one striving to be in the front. Vast numbers of them are shot while in this situation. A person told me that he once rode furiously into one of these rolling multitudes, and picked up thirteen pigeons, which had been trampled to death by his horse's feet. In a few minutes they will beat the whole nuts from a tree with their wings, while all is a scramble, both above and below, for the same. They have the same cooing notes common to domestic pigeons, but much less of their gesticulations. In some flocks you will find nothing but young ones, which are easily distinguishable by their motley dress. In others, they will be mostly females; and again, great multitudes of males, with few or no females. I cannot account for this in any other way than that, during the time of incubation, the males are exclusively engaged in procuring food, both for themselves and their mates; and the young, being unable yet to undertake these extensive excursions, associate together accordingly. But, even in winter, I know of several species of birds who separate in this manner, particularly the red-winged starling, among whom thousands of old males may be found with few or no young or females along with them.

Stragglers from these immense armies settle in almost every part of the country, particularly among the beech woods, and in the pine and hemlock woods of the eastern and northern parts of the continent. Mr Pennant informs us that they breed near Moose Fort at Hudson's Bay, in N. lat. 51°, and I myself have seen the remains of a large breeding place as far south as the country of Choctaws, in lat. 32°. In the former of these places they are said to remain until December; from which circumstance it is evident that they are not regular in their migrations, like many other species, but rove about, as scarcity of food urges them. Every spring, however, as well as fall, more or less of them are seen in the neighbourhood of Philadelphia; but it is only once in several years that they appear in such formidable bodies, and this commonly when the snows are heavy to the north, the winter here more than usually mild, and acorns, &c., abundant.

The passenger pigeon is sixteen inches long, and twenty-four inches in extent; bill, black; nostril, covered by a high rounding protuberance; eye, brilliant fiery orange; orbit, or space surrounding it, purplish flesh-coloured skin; head, upper part of the neck, and chin, a fine slate blue, lightest on the chin; throat, breast, and sides, as far as the thighs, a reddish hazel; lower part of the neck, and sides of the same, resplendent changeable gold, green, and purplish crimson, the latter most predominant; the ground colour, slate; the plumage of this part is of a peculiar structure, ragged at the ends; belly and vent, white; lower part of the breast, fading into a pale vinaceous red; thighs, the same; legs and feet, lake, seamed with white; back, rump, and tail-coverts, dark slate, spotted on the shoulders with a few scattered marks of black; the scapulars, tinged with brown; greater coverts, light slate; primaries and secondaries, dull black, the former tipt and edged with brownish white; tail, long, and greatly cuneiform, all the feathers tapering towards the point, the two middle ones plain deep black, the other five, on each side, hoary white, lightest near the tips, deepening into bluish near the bases, where each is crossed on the inner vane with a broad spot of black, and nearer the root with another of ferruginous; primaries, edged with white; bastard wing, black.

The female is about half an inch shorter, and an inch less in extent; breast, cinereous brown; upper part of the neck, inclining to ash; the spot of changeable gold, green, and carmine, much less, and not so brilliant; tail-coverts, brownish slate; naked orbits, slate coloured; in all other respects like the male in colour, but less vivid, and more tinged with brown; the eye not so brilliant an orange. In both, the tail has only twelve feathers.

BLUE MOUNTAIN WARBLER. (*Sylvia montana.*)

PLATE XLIV.—Fig. 2.

SYLVICOLA MONTANA.—Jardine.*

Sylvia tigrina, *Bonap. Synop.* p. 82.

This new species was first discovered near that celebrated ridge or range of mountains with whose name I have honoured it. Several of these solitary warblers remain yet to be gleaned up from the airy heights of our alpine scenery, as well as from the recesses of our swamps and morasses, whither it is my design to pursue them by every opportunity. Some of these, I believe, rarely or never visit the lower cultivated parts of the country, but seem only at home among the glooms and silence of those dreary solitudes. The present species seems of that family or subdivision of the warblers that approach the flycatcher, darting after flies wherever they see them, and also searching with great activity among the leaves. Its song was a feeble screep, three or four times repeated.

This species is four inches and three-quarters in length; the upper parts, a rich yellow olive; front, cheeks, and chin, yellow, also the sides of the neck; breast and belly, pale

* Bonaparte is inclined to think that this is the *Sylvia tigrina* of Latham. He acknowledges, however, not having seen the bird, and, as we have no means at present of deciding the question, have retained Wilson's name. Both this and the following will range in *Sylvicola.*—Ed.

yellow, streaked with black or dusky; vent, plain pale yellow; wings, black; first and second row of coverts, broadly tipt with pale yellowish white; tertials, the same; the rest of the quills edged with whitish; tail, black, handsomely rounded, edged with pale olive; the two exterior feathers on each side, white on the inner vanes from the middle to the tips, and edged on the outer side with white; bill, dark brown; legs and feet, purple brown; soles, yellow; eye, dark hazel.

This was a male. The female I have never seen.

HEMLOCK WARBLER. (*Sylvia parus.*)

PLATE XLIV.—Fig. 3.

SYLVICOLA PARUS.—Jardine.
Sylvia parus, *Bonap. Synop.* p. 82.

This is another nondescript, first met with in the Great Pine Swamp, Pennsylvania. From observing it almost always among the branches of the hemlock trees, I have designated it by that appellation, the markings of its plumage not affording me a peculiarity sufficient for a specific name. It is a most lively and active little bird, climbing among the twigs, and hanging like a titmouse on the branches, but possessing all the external characters of the warblers. It has a few low and very sweet notes, at which times it stops and repeats them for a short time, then darts about as before. It shoots after flies to a considerable distance; often begins at the lower branches, and hunts with great regularity and admirable dexterity upwards to the top, then flies off to the next tree, at the lower branches of which it commences hunting upwards as before.

This species is five inches and a half long, and eight inches in extent; bill, black above, pale below; upper parts of the plumage, black, thinly streaked with yellow olive; head above, yellow, dotted with black; line from the nostril over the eye, sides of the neck, and whole breast, rich yellow; belly,

1.Sharp-shinn'd Hawk. 2.Redstart. 3.Yellow-rump.

paler, streaked with dusky; round the breast, some small streaks of blackish; wing, black, the greater coverts and next superior row, broadly tipt with white, forming two broad bars across the wing; primaries edged with olive, tertials with white; tail-coverts, black, tipt with olive; tail, slightly forked, black, and edged with olive; the three exterior feathers altogether white on their inner vanes; legs and feet, dirty yellow; eye, dark hazel; a few bristles at the mouth; bill, not notched.

This was a male. Of the female I can at present give no account.

SHARP-SHINNED HAWK. (*Falco velox.*)

PLATE XLV.—FIG. 1.

ACCIPITER PENNSYLVANICUS.—SWAINSON.—YOUNG FEMALE.

Autour à bec sineuse, *Temm. Pl. Col.* 67.

THIS is a bold and daring species, hitherto unknown to naturalists. The only hawk we have which approaches near it in colour is the pigeon hawk, already figured in this work, Plate XV.; but there are such striking differences in the present, not only in colour, but in other respects, as to point out decisively its claims to rank as a distinct species. Its long and slender legs and toes—its red fiery eye, feathered to the eyelids—its triangular grooved nostril, and length of tail,— are all different from the pigeon hawk, whose legs are short, its eyes dark hazel, surrounded with a broad bare yellow skin, and its nostrils small and circular, centered with a slender point that rises in it like the pistil of a flower. There is no hawk mentioned by Mr Pennant, either as inhabiting Europe or America, agreeing with this. I may, therefore, with confidence, pronounce it a nondescript, and have chosen a very singular peculiarity which it possesses for its specific appellation.

This hawk was shot on the banks of the Schuylkill, near Mr Bartram's. Its singularity of flight surprised me long before I succeeded in procuring it. It seemed to throw itself from one quarter of the heavens to the other with prodigious velocity, inclining to the earth, swept suddenly down into a thicket, and instantly reappeared with a small bird in its talons. This feat I saw it twice perform, so that it was not merely an accidental manœuvre. The rapidity and seeming violence of these zigzag excursions were really remarkable, and appeared to me to be for the purpose of seizing his prey by sudden surprise and main force of flight. I kept this hawk alive for several days, and was hopeful I might be able to cure him; but he died of his wound.

On the 15th of September, two young men whom I had despatched on a shooting expedition met with this species on one of the ranges of the Alleghany. It was driving around in the same furious headlong manner, and had made a sweep at a red squirrel, which eluded its grasp, and itself became the victim. These are the only individuals of this bird I have been able to procure, and fortunately they were male and female.

The female of this species (represented in the plate) is thirteen inches long, and twenty-five inches in extent; the bill is black towards the point on both mandibles, but light blue at its base; cere, a fine pea green; sides of the mouth, the same; lores, pale whitish blue, beset with hairs; crown and whole upper parts, very dark brown, every feather narrowly skirted with a bright rust colour; over the eye a stripe of yellowish white, streaked with deep brown; primaries, spotted on their inner vanes with black; secondaries, crossed on both vanes with three bars of dusky, below the coverts; inner vanes of both primaries and secondaries, brownish white; all the scapulars marked with large round spots of white, not seen unless the plumage be parted with the hand; tail long, nearly even, crossed with four bars of black and as many of brown ash, and tipt with white; throat and whole lower parts, pale yellowish white; the former marked with fine long pointed spots of dark brown, the latter with large oblong spots of reddish brown; femorals, thickly marked with spade-formed spots on a pale rufous ground; legs, long, and feathered a little below the knee, of a greenish yellow colour, most yellow at the joints; edges of the inside of the shins, below the knee, pro-jecting like the edge of a knife, hard and sharp, as if in-tended to enable the bird to hold its prey with more security between them; eye, brilliant yellow, sunk below a projecting cartilage.

The male was nearly two inches shorter; the upper parts, dark brown; the feathers skirted with pale reddish, the front streaked with the same; cere, greenish yellow; lores, bluish; bill, black, as in the female; streak over the eye, lighter than in the former; chin, white; breast the same, streaked with brown; bars on the tail, rather narrower, but in tint and number the same; belly and vent, white; feet and shins, exactly as in the female; the toes have the same pendulous lobes which mark those of the female, and of which the representation in the plate will give a correct idea; the wings barred with black, very noticeable on the lower side.

Since writing the above, I have shot another specimen of this hawk, corresponding in almost every particular with the male last mentioned, and which, on dissection, also proves to be a male. This last had within the grasp of its sharp talons a small lizard, just killed, on which he was about to feed. How he contrived to get possession of it appeared to me matter of surprise, as lightning itself seems scarcely more fleet than this little reptile. So rapid are its motions, that, in passing from one place to another, it vanishes, and actually eludes the eye in running a distance of twelve or fifteen feet. It is frequently seen on fences that are covered with grey moss and lichen, which in colour it very much resembles; it seeks shelter in hollow trees, and also in the ground about their decayed roots. They are most numerous in hilly parts of the country, particularly on the declivities of the Blue Mountain, among the crevices of rocks and stones. When they are disposed to run, it is almost impossible to shoot them, as they disappear at the first touch of the trigger. For the satisfaction of the curious, I have introduced a full-sized figure of this lizard, which is known in many parts of the country by the name of the Swift.

REDSTART. *(Muscicapa ruticilla.)*

PLATE XLV. – FIG. 2.

Edw. 257.—Yellow Tail, *Arct. Zool.* ii. p. 466, No. 301.

SETOPHAGA RUTICILLA.—SWAINSON.

By recurring to Vol. I. Plate VI. fig. 6, the male of this species may be seen in his perfect dress. The present figure represents the young bird as he appears for the first two seasons ; the female differs very little from this, chiefly in the green olive being more inclined to ash.

This is one of our summer birds, and, from the circumstance of being found off Hispaniola in November, is supposed to winter in the islands. They leave Pennsylvania about the 20th of September ; are dexterous flycatchers, though ranked by European naturalists among the warblers, having the bill notched and beset with long bristles.

In its present dress the redstart makes its appearance in Pennsylvania about the middle or 20th of April ; and, from being heard chanting its few sprightly notes, has been supposed by some of our own naturalists to be a different species. I have, however, found both parents of the same nest in the same dress nearly ; the female, eggs, and nest, as well as the notes of the male, agreeing exactly with those of the redstart— evidence sufficiently satisfactory to me.

Head above, dull slate ; throat, pale buff ; sides of the breast and four exterior tail-feathers, fine yellow, tipt with dark brown ; wings and back, greenish olive ; tail-coverts, blackish, tipt with ash ; belly, dull white ; no white or yellow on the wings ; legs, dirty purplish brown ; bill, black.

The redstart extends very generally over the United States, having myself seen it on the borders of Canada, and also on the Mississippi territory.

This species has the constant habit of flirting its expanded tail from side to side, as it runs along the branches, with its head levelled almost in a line with its body, occasionally shooting off after winged insects in a downward zigzag direction, and, with admirable dexterity, snapping its bill as it descends. Its notes are few and feeble, repeated at short intervals, as it darts among the foliage ; having at some times a resemblance to the sounds, *sic sic sàic ;* at others, *weesy weesy weesy ;* which last seems to be its call for the female, while the former appears to be its most common note.

YELLOW-RUMP WARBLER. (*Sylvia coronata.*)

PLATE XLV.—FIG. 3.

Edw. 255.—*Arct. Zool.* ii. p. 400, No. 288.

SYLVICOLA CORONATA.—SWAINSON.—WINTER PLUMAGE.

Sylvia coronata, *Bonap. Synop.* p. 78.—Sylvicola coronata, *North. Zool.* ii. p. 210.

I MUST again refer the reader to the first volume, Plate XVII. fig. 4, for this bird in his perfect colours ; the present figure exhibits him in his winter dress, as he arrives to us from the north early in September ; the former shows him in his spring and summer dress, as he visits us from the south about the 20th of March. These birds remain with us in Pennsylvania from September until the season becomes severely cold, feeding on the berries of the red cedar ; and, as December's snows come on, they retreat to the lower countries of the southern States, where, in February, I found them in great numbers among the myrtles, feeding on the berries of that shrub ; from which circumstance they are usually called, in that quarter, myrtle birds. Their breeding place I suspect to be in our northern districts, among the swamps and evergreens so abundant there, having myself shot them in the Great Pine Swamp about the middle of May.

They range along our whole Atlantic coast in winter, seeming particularly fond of the red cedar and the myrtle ; and I have found them numerous in October, on the low islands along the coast of New Jersey, in the same pursuit. They also dart after flies, wherever they can see them, generally skipping about with the wings loose.

Length, five inches and a quarter; extent, eight inches; upper parts and sides of the neck, a dark mouse brown, obscurely streaked on the back with dusky black; lower parts, pale dull yellowish white; breast, marked with faint streaks of brown; chin and vent, white; rump, vivid yellow; at each side of the breast, and also on the crown, a spot of fainter yellow; this last not observable without separating the plumage; bill, legs, and wings, black; lesser coverts, tipt with brownish white; tail-coverts, slate; the three exterior tail-feathers marked on their inner vanes with white; a touch of the same on the upper and lower eyelid. Male and female at this season nearly alike. They begin to change about the middle of February, and in four or five weeks are in their slate-coloured dress, as represented in the figure referred to.

SLATE-COLOURED HAWK. (*Falco Pennsylvanicus.*)

PLATE XLVI.—Fig. 1.

ACCIPITER PENNSYLVANICUS.—Swainson.*

Falco velox, *Bonap. Synop.* p. 29.—Autour à bec sineuse, *Temm. Pl. Col.* 67 (young).—Accipiter Pennsylvanicus, *North. Zool.* ii. p. 44.

This elegant and spirited little hawk is a native of Pennsylvania, and of the Altantic States generally, and is now for the first time introduced to the notice of the public. It frequents the more settled parts of the country, chiefly in winter; is at all times a scarce species; flies wide, very

---

* It is now satisfactorily ascertained that this and the *Falco velox* of the last plate are the same species, the latter representing the plumage of the young female. The changes and differences are the same with those of the common European sparrow hawk, *Accipiter nisus.*

This bird most probably extends to the intertropical parts of South America. Its occurrence far to the northward is not so common. It was not met with by Dr Richardson, and the authority of its existence in the Fur Countries rests on a specimen in the Hudson's Bay Company museum, killed at Moose Factory. It very nearly resembles two small species from Mexico, the *A. fringilloides* of Mr Vigors, and one newly characterised by Mr Swainson as *A. Mexicanus.*—Ed.

1. Slate coloured Hawk. 2.Ground Dove. 3.Female.

46.

irregular and swiftly ; preys on lizards, mice, and small birds, and is an active and daring little hunter. It is drawn of full size, from a very beautiful specimen shot in the neighbourhood of Philadelphia. The bird within his grasp is the *Tanagra rubra,* or black-winged red bird, in its green or first year's dress. In the spring of the succeeding year the green and yellow plumage of this bird becomes of a most splendid scarlet, and the wings and tail deepen into a glossy black. For a particular account of this tanager, see Vol. I. p. 192, of the present work.

The great difficulty of accurately discriminating between different species of the hawk tribe, on account of the various appearances they assume at different periods of their long lives, at first excited a suspicion that this might be one of those with which I was already acquainted, in a different dress, namely, the sharp-shinned hawk just described; for such are the changes of colour to which many individuals of this genus are subject, that unless the naturalist has recourse to those parts that are subject to little or no alteration in the full-grown bird, viz., the particular conformation of the legs, nostril, tail, and the relative length of the latter to that of the wings, also the peculiar character of the countenance, he will frequently be deceived. By comparing these, the same species may often be detected under a very different garb. Were all these changes accurately known, there is no doubt but the number of species of this tribe at present enumerated would be greatly diminished, the same bird having been described by certain writers three, four, and even five different times, as so many distinct species. Testing, however, the present hawk by the rules above mentioned, I have no hesitation in considering it as a species different from any hitherto described, and I have classed it accordingly.

The slate-coloured hawk is eleven inches long, and twenty-one inches in extent ; bill, blue black ; cere and sides of the mouth, dull green ; eyelid, yellow ; eye, deep sunk under the projecting eyebrow, and of a fiery orange colour ; upper parts of a fine slate ; primaries, brown black, and, as well as the secondaries, barred with dusky ; scapulars, spotted with white and brown, which is not seen unless the plumage be separated by the hand ; all the feathers above are shafted with black ; tail, very slightly forked, of an ash colour, faintly tinged with brown, crossed with four broad bands of black, and tipt with white ; tail, three inches longer than the wings ; over the eye extends a streak of dull white ; chin, white, mixed with fine black hairs ; breast and belly beautifully variegated with ferruginous and transverse spots of white ; femorals, the same ; vent, pure white ; legs, long, very slender, and of a rich orange yellow ; claws, black, large, and remarkably sharp ; lining of the wing, thickly marked with heart-shaped spots of black. This bird, on dissection, was found to be a male. In the month of February, I shot another individual of this species, near Hampton, in Virginia, which agreed almost exactly with the present.

GROUND DOVE.   (*Columba passerina.*)

PLATE XLVI.—Fig. 2, Male ; Fig. 3, Female.

Linn. Syst. 285.—*Sloan. Jam.* ii. 305.—Le Cocotzin, *Fernandez,* 24.—*Buff.* ii. 559, *Pl. enl.* 243.—*Turt. Syst.* 478.—Columba minuta, *Ibid.* p. 479.—*Arct. Zool.* p. 328, No. 191.—*Catesb.* i. 26.—La Petite Tourterelle d'Amerique, *Briss.* i. 113, pl. 9, fig. 1.

CHÆMEPELIA PASSERINA.—Swainson.

Chæmepelia, *Swain. N. Groups, Zool. Journ.* No. XI. p. 361.—Columba passerina (sub-genus Goura), *Bonap. Synop.* p. 120.

This is one of the least of the pigeon tribe, whose timid and innocent appearance forms a very striking contrast to the ferocity of the bird-killer of the same plate. Such as they are in nature, such I have endeavoured faithfully to represent them. I have been the more particular with this minute species, as no correct figure of it exists in any former work with which I am acquainted.

The ground dove is a native of North and South Carolina,

Georgia, the new State of Louisiana, Florida, and the islands
of the West Indies. In the latter, it is frequently kept in
cages; is esteemed excellent for the table, and honoured by
the French planters with the name of ortolan. They are
numerous in the sea islands on the coast of Carolina and
Georgia; fly in flocks or coveys of fifteen or twenty; seldom
visit the woods, preferring open fields and plantations; are
almost constantly on the ground, and, when disturbed, fly to
a short distance, and again alight. They have a frequent
jetting motion with the tail; feed on rice, various seeds and
berries, particularly those of the toothache tree,* under or near
which, in the proper season, they are almost sure to be found.
Of their nest or manner of breeding, I am unable at present
to give any account.

These birds seem to be confined to the districts lying south
of Virginia. They are plenty on the upper parts of Cape
Fear river, and in the interior of Carolina and Georgia; but
I never have met with them either in Maryland, Delaware, or
Pennsylvania. They never congregate in such multitudes as
the common wild pigeon, or even as the Carolina pigeon
or turtle dove; but, like the partridge, or quail, frequent the
open fields in small coveys. They are easily tamed, have a
low, tender, cooing note, accompanied with the usual gesticulations of their tribe.

The ground dove is a bird of passage, retiring to the islands,
and to the more southerly parts of the continent, on the approach of winter, and returning to its former haunts early in
April. It is of a more slender and delicate form, and less
able to bear the rigours of cold, than either of the other two
species common in the United States, both of which are found
in the northern regions of Canada, as well as in the genial
climate of Florida.

The dove, generally speaking, has long been considered as the
favourite emblem of peace and innocence, probably from the
respectful manner in which its name is mentioned in various
parts of Scripture; its being selected from among all the

* *Xanthoxylum clava Herculis.*

birds by Noah to ascertain the state of the deluge, and returning to the ark bearing the olive leaf, as a messenger of
peace and good tidings; the Holy Ghost, it is also said, was
seen to descend like a dove from heaven, &c. In addition to
these, there is in the dove an appearance of meekness and
innocency very interesting, and well calculated to secure our
partiality in its favour. These remarks are applicable to the
whole genus, but are more particularly so to the species now
before us, as being among the least, the most delicate, and
inoffensive of the whole.

The ground dove is six inches and a quarter long; bill,
yellow, black at the point; nostril, covered with a prominent
membrane, as is usual with the genus; iris of the eye, orange
red; front, throat, breast, and sides of the neck, pale vinaceous
purple; the feathers strongly defined by semicircular outlines,
those on the throat centered with dusky blue; crown and
hind head, a fine pale blue, intermixed with purple, the
plumage, like that on the throat, strongly defined; back,
cinerous brown, the scapulars deeply tinged with pale purple,
and marked with detached drops of glossy blue, reflecting
tints of purple; belly, pale vinaceous brown, becoming dark
cinerous towards the vent, where the feathers are bordered
with white; wing-quills, dusky outwardly, and at the tips;
lower sides, and whole interior vanes, a fine red chestnut,
which shows itself a little below their coverts; tail, rounded,
consisting of twelve feathers, the two middle ones cinereous
brown, the rest black, tipt and edged with white; legs and
feet, yellow.

The female has the back and tail-coverts of a mouse colour,
with little or none of the vinaceous tint on the breast and
throat, nor any of the light blue on the hind head; the throat
is speckled with dull white, pale clay colour, and dusky;
sides of the neck, the same, the plumage strongly defined;
breast, cinerous brown, slighly tinctured with purple; scapulars,
marked with large drops of a dark purplish blood colour,
reflecting tints of blue; rest of the plumage, nearly the same
as that of the male.

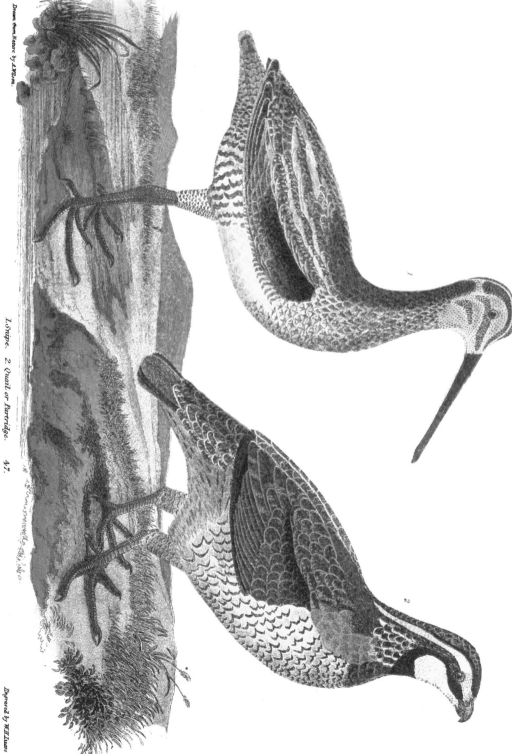

1 Snipe.   2. Quail or Partridge.   N.7.

SNIPE. *(Scolopax gallinago ?)*

PLATE XLVII.—Fig. 1.

La Beccassine, *Briss.* v. 298, pl. 26, fig. 1.—*Lath. Syn.* iii. 134.

SCOLOPAX WILSONII.—Temminck.*

Scolopax Wilsonii, *Temm. Pl. Col.*, Note to description of S. gigantea—*Bonap. Synop.* p. 330.—*Monog. del Gen. Scolopax Osserv. Sulla*, 2d edit., *Del. Reg. Anim.* p. 120.—Scolopax Brehmii, *Bonap. Observ. on Nomencl.*

THIS bird is well known to our sportsmen ; and, if not the same, has a very near resemblance to the common snipe of Europe. It is usually known by the name of the English

* Five or six species of snipes are so much allied in the colours and general marking of the plumage, that a very narrow examination is often necessary for their determination ; from this reason, the birds from America, Asia, and the Indian continent were considered as identical, and a much wider geographical range allotted to the European snipe than it was generally entitled to. Wilson had some doubts of this bird being the same with the European snipe, as he marks his name with a query, and observed the difference in the number of tail-feathers. Bonaparte observed the difference as soon as his attention was turned to the ornithology of America ; and, about the same time, a new snipe was described by Mr Kaup, in the Isis, as found occasionally in cold winters in the north of Germany. The Prince of Musignano, on comparing this description with the American species, from their very close alliance, judged them identical ; while, in the meantime, Temminck, comparing both together, perceived distinctions, and dedicated that of America to her own ornithologist, an opinion which Bonaparte afterwards confirmed and adopted in his monograph of that genus.

Mr Swainson has introduced a snipe, which he thinks is distinct, killed on the Rocky Mountains, and named by him *S. Drummondii ;* and another, killed on the Columbia, which he calls *S. Douglasii.* The first " is common in the Fur Countries up to lat. 65°, and is also found in the recesses of the Rocky Mountains. It is intermediate in size, between the *S. major* and *gallinago ;* it has a much longer bill than the latter, and two more tail-feathers. Its head is divided by a pale central stripe, as in *S. gallinula* and *major ;* its dorsal plumage more distinctly striped than that of the latter ; and the outer tail-feather is a quarter of an inch shorter than that of *S. Douglasii.*" The latter, in Mr Swainson's collection, has the tail of sixteen feathers, not narrowed, all banded with ferruginous except the outer pair, which are paler ; total length, eleven and a half inches.

snipe, to distinguish it from the woodcock, and from several others of the same genus. It arrives in Pennsylvania about the 10th of March, and remains in the low grounds for several weeks ; the greater part then move off to the north, and to the higher inland districts, to breed. A few are occasionally found, and consequently breed, in our low marshes during the summer. When they first arrive, they are usually lean ; but, when in good order, are accounted excellent eating. They are perhaps the most difficult to shoot of all our birds, as they fly in sudden zigzag lines, and very rapidly. Great numbers of these birds winter on the rice grounds of the southern States, where, in the month of February, they appeared to be much tamer than they are usually here, as I frequently observed them running about among the springs and watery thickets. I was told by the inhabitants that they generally disappeared early in the spring. On the 20th of March, I found these birds extremely numerous on the borders of the ponds near Louisville, Kentucky, and also in the neighbourhood of Lexington, in the same State, as late as the 10th of

Most of the snipes partially migrate in their native countries, and some perform a regular distant migration. Such is the case with the *S. gallinula* of Europe. The American species is a winter visitant in the northern States, and will most probably breed farther to the south, without leaving the country. In India, the snipes move according to the supply of water in the tanks, and at the season when they are comparatively dry, leave that district entirely. In this country, although many breed in the mosses, we have a large accession of numbers about the middle of September, both from the wilder high grounds, and from the continent of Europe ; and these, according to the weather, change their stations during the whole winter. Their movements are commenced generally about twilight, when they fly high, surveying the country as they pass, and one day may be found in abundance on the highest moorland ranges, while the next they have removed to some low and sheltered glade or marsh. In this we have a curious instance of that instinctive knowledge which causes so simultaneous a change of station in a single night. By close observation, during the winter months it may be regularly perceived, sometimes even daily, and some change certainly takes place before and after any sudden variation of weather.—ED.

April. I was told by several people that they are abundant in the Illinois country, up as far as Lake Michigan. They are but seldom seen in Pennsylvania during the summer, but are occasionally met with in considerable numbers on their return in autumn, along the whole eastern side of the Alleghany, from the sea to the mountains. They have the same soaring irregular flight in the air in gloomy weather as the snipe of Europe; the same bleating note and occasional rapid descent; spring from the marshes with the like feeble *squeak;* and in every respect resemble the common snipe of Britain, except in being about an inch less, and in having sixteen feathers in the tail, instead of fourteen, the number said by Bewick to be in that of Europe. From these circumstances, we must either conclude this to be a different species, or partially changed by difference of climate : the former appears to me the most probable opinion of the two.

These birds abound in the meadows and low grounds along our large rivers, particularly those that border the Schuylkill and Delaware, from the 10th of March to the middle of April, and sometimes later, and are eagerly sought after by many of our gunners. The nature of the grounds, however, which these birds frequent, the coldness of the season, and peculiar shyness and agility of the game, render this amusement attractive only to the most dexterous, active, and eager of our sportsmen.

The snipe is eleven inches long, and seventeen inches in extent ; the bill is more than two inches and a half long, fluted lengthwise, of a brown colour, and black towards the tip, where it is very smooth while the bird is alive, but, soon after it is killed, becomes dimpled, like the end of a thimble ; crown, black, divided by an irregular line of pale brown ; another broader one of the same tint passes over each eye ; from the bill to the eye, there is a narrow dusky line ; neck and upper part of the breast, pale brown, variegated with touches of white and dusky ; chin, pale ; back and scapulars, deep velvety black, the latter elegantly marbled with waving lines of ferruginous, and broadly edged exteriorly with white ;

wings, plain dusky, all the feathers, as well as those of the coverts, tipt with white ; shoulder of the wing, deep dusky brown, exterior quill edged with white ; tail-coverts, long, reaching within three-quarters of an inch of the tip, and of a pale rust colour, spotted with black ; tail, rounded, deep black, ending in a bar of bright ferruginous, crossed with a narrow waving line of black, and tipt with whitish ; belly, pure white ; sides, barred with dusky ; legs and feet, a very pale ashy green ; sometimes the whole thighs and sides of the vent are barred with dusky and white, as in the figure on the plate.

The female differs in being more obscure in her colours ; the white on the back being less pure, and the black not so deep.

QUAIL, OR PARTRIDGE. (*Perdix Virginianus.*)

PLATE XLVII.—Fig. 2.

Arct. Zool. 318, No. 185.—*Catesb. App.* p. 12.—Virginian Quail, *Turt. Syst.* p. 460.—Maryland Quail, *Ibid.*—La Perdrix d'Amerique, *Briss.* i. 230.—*Buff.* ii. 447.

ORTYX VIRGINIANUS.—Bonaparte.*

Perdix Virginiana, *Lath. Ind. Orn.* ii. p. 650.—Colin Colgnicui, *Temm. Pig. et Gall.* iii. p. 436.—Perdix Borealis, *Temm. Pig. et Gall. Ind.* p. 735.—Ortyx Borealis, *Steph. Cont. Shaw's Zool.* xi. p. 377.—Perdix (Ortyx) Virginiana, *Bonap. Synop.* p. 124.—The Virginian Partridge, *Aud.* i. p. 388, pl. 76.

THIS well-known bird is a general inhabitant of North America, from the northern parts of Canada and Nova Scotia, in which latter place it is said to be migratory, to the extremity

* The genus *Ortyx* was formed by Mr Stephens, the continuator of Shaw's Zoology, for the reception of the thick and strong-billed partridges peculiar to both continents of the New World, and holding the place there with the partridges, francolins, and quails of other countries. They live on the borders of woods, among brushwood, or on the thick grassy plains, and since the cultivation of the country, frequent cultivated fields. During the night they roost on trees, and occasionally perch during the day ; when alarmed, or chased by dogs, they fly to the middle branches ; and Mr Audubon remarks, " they walk with ease on the branches." In all these habits they show their alliance to the perching *Gallinæ*, and a variation from the true partridge. The same naturalist also remarks, that they occasionally perform partial migra-

of the peninsula of Florida; and was seen in the neighbourhood of the Great Osage village, in the interior of Louisiana.
They are numerous in Kentucky and Ohio. Mr Pennant

tions, from north-west to south-east, in the beginning of October, and
that for a few weeks the north-western shores of the Ohio are covered
with partridges.

Their general form is robust, the bill very strong, and apparently
fitted for a mode of feeding requiring considerable exertion, such as the
digging up of bulbous and tuberous roots. The head is crested in all
the known species, the feathers sometimes of a peculiar structure, the
shafts bare, and the extremity of the webs folding on each other. The
tail also exhibits different forms; in the more typical species short, as
in the partridges, and in others becoming broad and long, as seen in the
Indian genus *Crex*, or the more extensively distributed genus *Penelope*.
Considerable additions to the number of species have been lately made.
Those belonging to the northern continent, and consequently coming under
our notice, are two, discovered by Mr Douglas,—*Ortyx picta*, described
in the last volume of the " Linnean Transactions," and *O. Douglasii*, so
named by Mr Vigors, in honour of its discoverer, and also described
with the former. To these may be added the lovely *O. Californica*,
which, previous to this expedition, and the voyage of Captain Beechey
to the coast of California, was held in the light of a dubious species. I
have added the descriptions of these new species from Mr Douglas's
account in the " Transactions of the Linnean Society."

Ortyx picta.—DOUGLAS.

Male.—Bill, small, black; crown of the head and breast, lead colour;
crest, three linear black feathers, two inches long; irides, bright hazel
red; throat, purple red, bounded by a narrow white line, forming a
gorget above the breast, and extending round the eye and root of the
beak; back, scapulars, and outer coverts of the wings, fuscous brown;
belly, bright tawny or rusty colour, waved with black; the points of
the feathers white; quills, thirteen feathers, the fourth the longest;
under coverts, light brown, mixed with a rusty colour; tail, twelve
feathers, of unequal length, rounded, lead colour, but less bright than
the breast or crown of the head; tarsi, one inch and a quarter long,
reddish; toes, webbed nearly to the first joint.

Female.—Head and breast, light fuscous brown; the middle of the
feathers, black; crest, half an inch long; throat, whitish or light gray;
belly, light gray, waved with black, less bright than the male; under
coverts of the tail, foxy red; length, ten inches; girth, sixteen inches;
weight, about twelve ounces; flesh, brown, well-flavoured.

From October until March, these birds congregate in vast flocks, and
seem to live in a state of almost perpetual warfare; dreadful conflicts

remarks, that they have been lately introduced into the island of Jamaica, where they appear to thrive greatly, breeding in that warm climate twice in the year. Captain Henderson mentions them as being plenty near the Balize, at the Bay of

ensue between the males, which not unfrequently end in the destruction of one or both combatants, if we may judge from the number of dead birds daily seen plucked, mutilated, and covered with blood. When feeding, they move in compact bodies, each individual endeavouring to outdo his neighbour in obtaining the prize. The voice is *quick-quick-quick*, pronounced slowly, with a gentle suspension between each syllable. At such times, or when surprised, the crest is usually thrown forward over the back ; and the reverse when retreating, being brought backwards, and laid quite close. Their favourite haunts are dry upland, or undulating, gravelly, or sandy soils, in open woods or coppice thickets of the interior ; but during the severity of winter, when the ground is covered with snow, they migrate in large flocks to the more temperate places in the immediate vicinity of the ocean. Seeds of *Bromus altissimus, Madia sativa*, and a tribe of plants allied to *Wadelia*, catkins of *Corylus*, leaves of *Fragaria*, and various insects, are their common food. Nest on the ground, in thickets of *Pteris, Aspidium, Rubus, Rhamnus*, and *Ceanothus ;* neatly built with grass and dry leaves ; secreted with so much caution, that, without the help of a dog, they can hardly be found. Eggs, eleven to fifteen, yellowish white, with minute brown spots ; large in proportion to the bird. Pair in March. Common in the interior of California ; and, during the summer months, extending as far northward as 45° north latitude, that is, within a few miles of the Columbian Valley.

Ortyx Douglasii.—VIGORS.

Male.—Bill, brown ; crest, linear, black, one inch long; irides, hazel red ; body, fuscous brown, with a mixture of lead colour, and rusty or yellow streaks ; throat, whitish, with brown spots ; belly, foxy red or tawny, white spotted; scapulars and outer coverts, bright brown; under coverts, light reddish brown ; tail, twelve unequal rounded feathers ; legs, reddish ; length, nine inches ; girth, twelve inches ; weight, ten ounces; flesh, pleasant, dark coloured.

Female.—Crest, scarcely perceptible, dark.

This species appears to be an inhabitant of a more temperate climate than the preceding one, as it is never seen higher than 42° N. latitude, and even that very sparingly in comparison to *O. Picta* and *Californica*. The species do not associate together. In manner they are similar, at least as far as the opportunity I had of observing them went. I have never seen them but in winter dress, and know nothing of their nesting. —ED.

Honduras. They rarely frequent the forest, and are most numerous in the vicinity of well-cultivated plantations, where grain is in plenty. They, however, occasionally seek shelter in the woods, perching on the branches or secreting themselves among the brushwood; but are found most usually in open fields, or along fences sheltered by thickets of briers. Where they are not too much persecuted by the sportsmen, they become almost half domesticated; approach the barn, particularly in winter, and sometimes, in that severe season, mix with the poultry to glean up a subsistence. They remain with us the whole year, and often suffer extremely by long, hard winters, and deep snows. At such times, the arts of man combine with the inclemency of the season for their destruction. To the ravages of the gun are added others of a more insidious kind; traps are placed on almost every plantation, in such places as they are known to frequent. These are formed of lath, or thinly-split sticks, somewhat in the shape of an obtuse cone, laced together with cord, having a small hole at top, with a sliding lid to take out the game by. This is supported by the common figure 4 trigger; and grain is scattered below and leading to the place. By this contrivance, ten or fifteen have sometimes been taken at a time.*

* In addition to the common traps now described, Mr Audubon mentions that they are also netted, or *driven*, as it is called. He thus describes the method of driving :—

"A number of persons on horseback, provided with a net, set out in search of partridges, riding along the fences or brier thickets which the birds are known to frequent. One or two of the party whistle in imitation of the call-note, and, as partridges are plentiful, the call is soon answered by a covey, when the sportsmen immediately proceed to ascertain their position and number, seldom considering it worth while to set the net when there are only a few birds. They approach in a careless manner, talking and laughing as if merely passing by. When the birds are discovered, one of the party gallops off in a circuitous manner, gets in advance of the rest by a hundred yards or more, according to the situation of the birds, and their disposition to run, while the rest of the sportsmen move about on their horses, talking to each other, but, at the same time, watching every motion of the partridges. The person in advance being provided with the net, dismounts, and at once falls to

These are sometimes brought alive to market, and occasionally bought up by sportsmen, who, if the season be very severe, sometimes preserve and feed them till spring, when they are humanely turned out to their native fields again, to be put to death at some future time *secundum artem.* Between the months of August and March, great numbers of these birds are brought to the market of Philadelphia, where they are sold at from twelve to eighteen cents apiece.

The quail begins to build early in May. The nest is made on the ground, usually at the bottom of a thick tuft of grass, that shelters and conceals it. The materials are leaves and fine dry grass in considerable quantity. It is well covered above, and an opening left on one side for entrance. The female lays from fifteen to twenty-four eggs, of a pure white, without any spots. The time of incubation has been stated to me, by various persons, at four weeks, when the eggs were placed under the domestic hen. The young leave the nest as soon as they are freed from the shell, and are conducted about in search of food by the female; are guided by her voice, which at that time resembles the twittering of young chickens, and sheltered by her wings in the same manner as those of the domestic fowl, but with all that secrecy and precaution for their safety which their helplessness and greater danger require. In this situation, should the little timid family be unexpectedly surprised, the utmost alarm and consternation instantly prevail. The mother throws herself in the path,

---

placing it, so that his companions can easily drive the partridges into it. No sooner is the machine ready, than the net-bearer remounts and rejoins the party. The sportsmen separate to a short distance, and follow the partridges, talking, whistling, clapping their hands, or knocking the fence-rails. The birds move with great gentleness, following each other, and are kept in the right direction by the sportsmen. The leading bird approaches and enters the mouth of the net—the others follow in succession, when the net-bearer leaps from his horse, runs up and secures the entrance, and soon despatches the birds. In this manner fifteen or twenty partridges are caught at one driving, and sometimes many hundreds in the course of the day."—Ed.

fluttering along, and beating the ground with her wings, as if sorely wounded ; using every artifice she is master of to entice the passenger in pursuit of herself, uttering at the same time certain peculiar notes of alarm, well understood by the young, who dive separately amongst the grass, and secrete themselves till the danger is over ; and the parent, having decoyed the pursuer to a safe distance, returns by a circuitous route to collect and lead them off. This well-known manœuvre, which nine times in ten is successful, is honourable to the feelings and judgment of the bird, but a severe satire on man. The affectionate mother, as if sensible of the avaricious cruelty of his nature, tempts him with a larger prize, to save her more helpless offspring ; and pays him, as avarice and cruelty ought always to be paid, with mortification and disappointment.

The eggs of the quail have been frequently placed under the domestic hen, and hatched and reared with equal success as her own ; though, generally speaking, the young partridges, being more restless and vagrant, often lose themselves, and disappear. The hen ought to be a particular good nurse, not at all disposed to ramble, in which case they are very easily raised. Those that survive acquire all the familiarity of common chickens ; and there is little doubt that, if proper measures were taken, and persevered in for a few years, they might be completely domesticated. They have been often kept during the first season, and through the whole of the winter, but have uniformly deserted in the spring. Two young partridges that were brought up by a hen, when abandoned by her, associated with the cows, which they regularly followed to the fields, returned with them when they came home in the evening, stood by them while they were milked, and again accompanied them to the pasture. These remained during the winter, lodging in the stable, but as soon as spring came, they disappeared. Of this fact I was informed by a very respectable lady, by whom they were particularly observed.

It has been frequently asserted to me, that the quails lay occasionally in each other's nests. Though I have never

myself seen a case of this kind, I do not think it altogether
improbable, from the fact that they have often been known
to drop their eggs in the nest of the common hen, when that
happened to be in the fields, or at a small distance from the
house. The two partridges above mentioned were raised in
this manner; and it was particularly remarked by the lady
who gave me the information, that the hen sat for several days
after her own eggs were hatched, until the young quails made
their appearance.

The partridge, on her part, has sometimes been employed
to hatch the eggs of the common domestic hen. A friend of
mine, who himself made the experiment, informs me, that, of
several hen's eggs which he substituted in place of those of
the partridge, she brought out the whole; and that, for several
weeks, he occasionally surprised her in various parts of the
plantation with her brood of chickens; on which occasions
she exhibited all that distressful alarm, and practised her
usual manœuvres for their preservation. Even after they
were considerably grown, and larger than the partridge herself, she continued to lead them about; but, though their
notes or call were those of common chickens, their manners
had all the shyness, timidity, and alarm of young partridges;
running with great rapidity, and squatting in the grass exactly
in the manner of the partridge. Soon after this, they disappeared, having probably been destroyed by dogs, by the gun,
or by birds of prey. Whether the domestic fowl might not
by this method be very soon brought back to its original savage
state, and thereby supply another additional subject for the
amusement of the sportsman, will scarcely admit of a doubt.
But the experiment, in order to secure its success, would require
to be made in a quarter of the country less exposed than ours
to the ravages of guns, traps, dogs, and the deep snows of
winter, that the new tribe might have full time to become
completely naturalised, and well fixed in all their native habits.

About the beginning of September, the quails being now
nearly fully grown, and associated in flocks or coveys of from

four or five to thirty, afford considerable sport to the gunner. At this time the notes of the male are most frequent, clear, and loud. His common call consists of two notes, with sometimes an introductory one, and is similar to the sound produced by pronouncing the words "Bob White." This call may be easily imitated by whistling, so as to deceive the bird itself, and bring it near. While uttering this, he is usually perched on a rail of the fence, or on a low limb of an apple tree, where he will sometimes sit, repeating, at short intervals, " Bob White," for half an hour at a time. When a covey are assembled in a thicket or corner of a field, and about to take wing, they make a low twittering sound, not unlike that of young chickens; and when the covey is dispersed, they are called together again by a loud and frequently repeated note, peculiarly expressive of tenderness and anxiety.

The food of the partridge consists of grain, seeds, insects, and berries of various kinds. Buckwheat and Indian-corn are particular favourites. In September and October the buckwheat fields afford them an abundant supply, as well as a secure shelter. They usually roost at night in the middle of a field on high ground; and from the circumstance of their dung being often found in such places in one round heap, it is generally conjectured that they roost in a circle, with their heads outwards, each individual in this position forming a kind of guard to prevent surprise. They also continue to lodge for several nights in the same spot.

The partridge, like all the rest of the gallinaceous order, flies with a loud whirring sound, occasioned by the shortness, concavity, and rapid motion of its wings, and the comparative weight of its body. The steadiness of its horizontal flight, however, renders it no difficult mark to the sportsman, particularly when assisted by his sagacious pointer. The flesh of this bird is peculiarly white, tender, and delicate, unequalled in these qualities by that of any other of its genus in the United States.

The quail, as it is called in New England, or the partridge, as in Pennsylvania, is nine inches long, and fourteen inches in extent; the bill is black; line over the eye, down the neck, and whole chin, pure white, bounded by a band of black, which descends and spreads broadly over the throat; the eye is dark hazel; crown, neck, and upper part of the breast, red brown; sides of the neck, spotted with white and black on a reddish brown ground; back, scapulars, and lesser coverts, red brown, intermixed with ash, and sprinkled with black; tertials, edged with yellowish white; wings, plain dusky; lower part of the breast and belly, pale yellowish white, beautifully marked with numerous curving spots, or arrow-heads of black; tail, ash, sprinkled with reddish brown; legs, very pale ash.

The female differs in having the chin and sides of the head yellowish brown, in which dress it has been described as a different kind. There is, however, only one species of quail at present known within the United States.

RAIL. (*Rallus Carolinus.*)

PLATE XLVIII.—Fig. 1.

Soree, *Catesb.* i. 70.—*Arct. Zool.* p. 491, No. 409.—Little American Water-hen, *Edw.* 144.—Le Râle de Virginie, *Buff.* viii. 165.

CREX CAROLINUS.—Bonaparte.*

Rallus (Crex) Carolinus, *Bonap. Synop.* p. 335.

Of all our land or water fowl, perhaps none afford the sportsmen more agreeable amusement, or a more delicious repast, than the little bird now before us. This amusement is indeed temporary, lasting only two or three hours in the

* Almost every ornithologist has been at variance with regard to the propriety and limitation of the genera *Rallus, Crex,* and *Gallinula.* They appear to be sufficiently distinct, and not to run more into each other than many other groups, and, in the present state of ornithology, their separation is indispensable. *Crex* may be characterised by the bill shorter than the head, strong at the base, and tapering, the forehead feathered; the common land rail or corncrake of Europe, and our

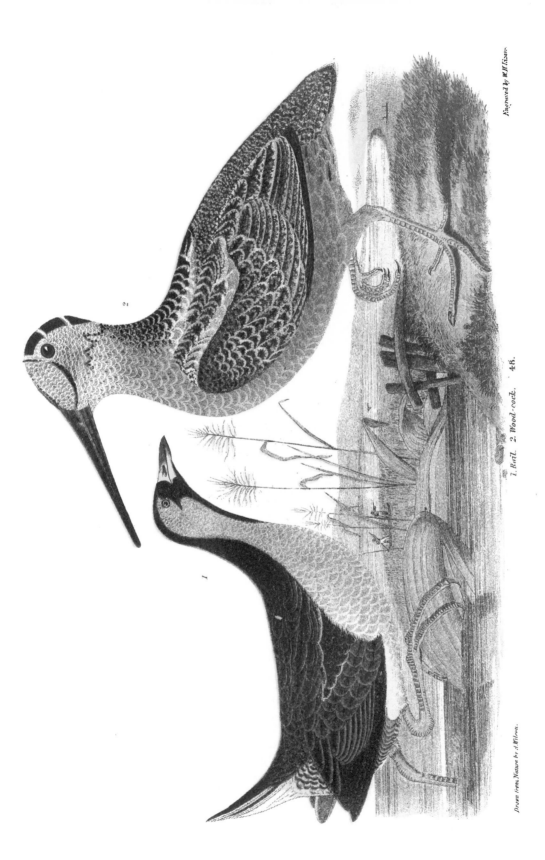

Drawn from Nature by A. Wilson.

1. Rail.　2. Wood-cock.　48.

Engraved by W.H.Lizars.

day, for four or five weeks in each year ; but as it occurs in
the most agreeable and temperate of our seasons, is attended
with little or no fatigue to the gunner, and is frequently
successful, it attracts numerous followers, and is pursued
in such places as the birds frequent with great eagerness and
enthusiasm.

present species, may be taken as very good typical examples. In
Gallinula, the forehead is defended with a flat cartilaginous shield, and
the habits are more open. In *Rallus,* the bill is longer than the head,
and comparatively slender.

In habit they nearly agree ; timid, and fond of concealment during
the day, they frequent low meadows or marshy grounds, and run
swiftly : the common land rail will beat a good runner for a short way, as
I have sometimes experienced. They run with the body near the ground,
and make their turns with astonishing celerity. When raised or surprised during the day, they fly clumsily ; but in the evening, and when
that faculty is exerted with their will, it is much more actively performed ; their time for exertion is evening and morning, often during
the night : then they feed, and, during breeding season, utter the incessant and inharmonious cry which almost all possess. The cry is
remarkable in all that I have heard, appearing to be uttered sometimes
within a few yards, and, in a second or two, as if at an opposite part of
the ground. The land rail possesses this ventriloquism to a great extent,
and, knowing their swift running powers, I at first thought that the
bird was actually traversing the field, and it was not until I had observed
one perched upon a stone utter its cry for some time, and give full
evidence of its powers, that I became convinced of the contrary. The
corncrake, and, indeed, I rather think most of the others, and also the
rails, seem to remain stationary when uttering the cry. A stone, clod
of earth, or old sod wall, is the common calling place of our own bird ;
and they may be easily watched, in the beginning of summer, if
approached with caution, before the herbage begins to thicken. They seem
to feed on larger prey than what are assigned to them : large water
insects and the smaller reptiles may assist in sustaining the aquatic
species, while slugs and larger snails will furnish subsistence to the
others. I have found the common short-tailed field mouse in the
stomach of our land rail.

Their flesh is generally delicate, some as much esteemed as the
American bird, and the young, before commencing their migrations,
become extremely fat.

Crex Carolinus is the only species of the genus yet discovered in
North America, and is peculiar to that continent.—ED.

The natural history of the rail, or, as it is called in Virginia, the sora, and in South Carolina, the coot, is, to the most of our sportsmen, involved in profound and inexplicable mystery. It comes they know not whence, and goes they know not where. No one can detect their first moment of arrival; yet all at once the reedy shores and grassy marshes of our large rivers swarm with them, thousands being sometimes found within the space of a few acres. These, when they do venture on wing, seem to fly so feebly, and in such short fluttering flights among the reeds, as to render it highly improbable to most people that they could possibly make their way over an extensive tract of country. Yet, on the first smart frost that occurs, the whole suddenly disappear, as if they had never been.

To account for these extraordinary phenomena, it has been supposed by some that they bury themselves in the mud; but as this is every year dug into by ditchers, and people employed in repairing the banks, without any of those sleepers being found, where but a few weeks before these birds were in- numerable, this theory has been generally abandoned. And here their researches into this mysterious matter generally end in the common exclamation of "What can become of them!" Some profound inquirers, however, not discouraged with these difficulties, have prosecuted their researches with more success; and one of those, living a few years ago near the mouth of James River, in Virginia, where the rail, or sora, are extremely numerous, has (as I was informed on the spot) lately discovered that they change into frogs! having himself found in his meadows an animal of an extraordinary kind, that appeared to be neither a sora nor a frog, but, as he expressed it, "something between the two." He carried it to his negroes, and afterwards took it home, where it lived three days; and, in his own and his negroes' opinion, it looked like nothing in this world but a real sora changing into a frog! What further confirms this grand discovery is the well-known circumstance of the frogs ceasing to hollow as soon as the sora comes in the fall.

This sagacious discoverer, however, like many others re-nowned in history, has found but few supporters, and, except his own negroes, has not, as far as I can learn, made a single convert to his opinion. Matters being so circumstanced, and some explanation necessary, I shall endeavour to throw a little more light on the subject by a simple detail of facts, leaving the reader to form his own theory as he pleases.

The rail, or sora, belongs to a genus of birds of which about thirty different species are enumerated by naturalists; and those are distributed over almost every region of the habitable parts of the earth. The general character of these is everywhere the same. They run swiftly, fly slowly, and usually with the legs hanging down; become extremely fat; are fond of concealment; and, wherever it is practicable, prefer running to flying. Most of them are migratory, and abound during the summer in certain countries, the inhabitants of which have very rarely an opportunity of seeing them. Of this last the land rail of Britain is a striking example. This bird, which during the summer months may be heard in almost every grass and clover field in the kingdom, uttering its common note *crek, crek*, from sunset to a late hour in the night, is yet unknown by sight to more than nine-tenths of the inhabitants. "Its well-known cry," says Bewick, "is first heard as soon as the grass becomes long enough to shelter it, and continues till the grass is cut; but the bird is seldom seen, for it constantly skulks among the thickest part of the herbage, and runs so nimbly through it, winding and doubling in every direction, that it is difficult to come near it; when hard pushed by the dog, it sometimes stops short, and squats down, by which means its too eager pursuer overshoots the spot, and loses the trace. It seldom springs but when driven to extremity, and generally flies with its legs hanging down, but never to a great distance; as soon as it alights, it runs off, and, before the fowler has reached the spot, the bird is at a considerable distance."* The water crake, or spotted rail,

* Bewick's British Birds, vol. i. p. 308.

of the same country, which in its plumage approaches nearer to our rail, is another notable example of the same general habit of the genus. "Its common abode," says the same writer, "is in low swampy grounds, in which are pools or streamlets overgrown with willows, reeds, and rushes, where it lurks and hides itself with great circumspection; it is wild, solitary, and shy, and will swim, dive, or skulk under any cover, and sometimes suffer itself to be knocked on the head rather than rise before the sportsman and his dog." The water rail of the same country is equally noted for the like habits. In short, the whole genus possess this strong family character in a very remarkable degree.

These three species are well known to migrate into Britain early in spring, and to leave it for the more southern parts of Europe in autumn. Yet they are rarely or never seen on their passage to or from the countries where they are regularly found at different seasons of the year, and this for the very same reasons that they are so rarely seen even in the places where they inhabit.

It is not, therefore, at all surprising, that the regular migrations of the American rail, or sora, should in like manner have escaped notice in a country like this, whose population bears so small a proportion to its extent, and where the study of natural history is so little attended to. But that these migrations do actually take place, from north to south, and *vice versa*, may be fairly inferred from the common practice of thousands of other species of birds less solicitous of concealment, and also from the following facts.

On the 22d day of February, I killed two of these birds in the neighbourhood of Savannah, in Georgia, where they have never been observed during the summer. On the 2d of May following, I shot another in a watery thicket below Philadelphia, between the rivers Schuylkill and Delaware, in what is usually called the Neck. This last was a male, in full plumage. We are also informed that they arrive at Hudson's Bay early in June, and again leave that settlement for the south early in autumn. That many of them also remain here to breed is proven by the testimony of persons of credit and intelligence with whom I have conversed, both here and on James River, in Virginia, who have seen their nests, eggs, and young. In the extensive meadows that border the Schuylkill and Delaware it was formerly common, before the country was so thickly settled there, to find young rail in the first mowing time among the grass. Mr James Bartram, brother to the botanist, a venerable and still active man of eighty-three, and well acquainted with this bird, says that he has often seen and caught young rail in his own meadows in the month of June ; he has also seen their nest, which he says is usually in a tussock of grass, is formed of a little dry grass, and has four or five eggs, of a dirty whitish colour, with brown or blackish spots : the young run off as soon as they break the shell, are then quite black, and run about among the grass like mice. The old ones he has very rarely observed at that time, but the young often. Almost every old settler along these meadows with whom I have conversed has occasionally seen young rail in mowing time ; and all agree in describing them as covered with blackish down. There can, therefore, be no reasonable doubt as to the residence of many of these birds, both here and to the northward, during the summer. That there can be as little doubt relative to their winter retreat will appear more particularly towards the sequel of the present account. During their residence here, in summer, their manners exactly correspond with those of the water crake of Britain, already quoted, so that, though actually a different species, their particular habits common places of resort, and eagerness for concealment, are as nearly the same as the nature of the climates will admit.

Early in August, when the reeds along the shores of the Delaware have attained their full growth, the rail resort to them in great numbers to feed on the seeds of this plant, of which they, as well as the rice birds, and several others, are immoderately fond. These reeds, which appear to be the

Zizania panicula effusa of Linnæus, and the *Zizania clavulosa*
of Willdenow, grow up from the soft muddy shores of the tide
water, which are alternately dry, and covered with four or five
feet of water. They rise with an erect, tapering stem, to
the height of eight or ten feet, being nearly as thick below as
a man's wrist, and cover tracts along the river of many acres.
The cattle feed on their long green leaves with avidity, and
wade in after them as far as they dare safely venture. They
grow up so close together, that, except at or near high water,
a boat can with difficulty make its way through among them.
The seeds are produced at the top of the plant, the blossoms,
or male parts, occupying the lower branches of the panicle,
and the seeds the higher. These seeds are nearly as long as
a common-sized pin, somewhat more slender, white, sweet to
the taste, and very nutritive, as appears by their effects on the
various birds that at this season feed on them.

When the reeds are in this state, and even while in blossom,
the rail are found to have taken possession of them in great
numbers. These are generally numerous in proportion to the
full and promising crop of the former. As you walk along
the embankment of the river at this season, you hear them
squeaking in every direction like young puppies. If a stone
be thrown among the reeds, there is a general outcry, and a
reiterated *kuk*, *kuk*, *kuk*, something like that of a guineafowl.
Any sudden noise, or the discharge of a gun, produces the
same effect. In the meantime none are to be seen, unless it
be at or near high water; for when the tide is low, they
universally secrete themselves among the interstices of
the reeds, and you may walk past, and even over, where
there are hundreds, without seeing a single individual. On
their first arrival, they are generally lean, and unfit for the
table; but, as the reeds ripen, they rapidly fatten, and from
the 20th of September to the middle of October are excellent,
and eagerly sought after. The usual method of shooting
them, in this quarter of the country, is as follows:—The
sportsman furnishes himself with a light batteau, and a stout

experienced boatman, with a pole of twelve or fifteen feet
long, thickened at the lower end to prevent it from sinking
too deep into the mud.   About two hours or so before high
water, they enter the reeds, and each takes his post, the
sportsman standing in the bow ready for action, the boatman,
on the stern seat, pushing her steadily through the reeds.
The rail generally spring singly, as the boat advances, and
at a short distance ahead, and are instantly shot down, while
the boatman, keeping his eye on the spot where the bird fell,
directs the boat forward, and picks it up as the gunner is
loading.   It is also the boatman's business to keep a sharp
look-out, and give the word "Mark!" when a rail springs
on either side without being observed by the sportsman, and
to note the exact spot where it falls until he has picked it
up; for this once lost sight of, owing to the sameness in the
appearance of the reeds, is seldom found again.   In this
manner the boat moves steadily through and over the reeds,
the birds flushing and falling, the gunner loading and firing,
while the boatman is pushing and picking up.   The sport
continues till an hour or two after high water, when the
shallowness of the water, and the strength and weight of the
floating reeds, as also the backwardness of the game to spring
as the tide decreases, oblige them to return.   Several boats
are sometimes within a short distance of each other, and a
perpetual cracking of musketry prevails along the whole reedy
shores of the river.   In these excursions it is not uncommon
for an active and expert marksman to kill ten or twelve dozen
in a tide.   They are usually shot singly, though I have
known five killed at one discharge of a double-barrelled piece.
These instances, however, are rare.

The flight of these birds among the reeds is usually low,
and, shelter being abundant, is rarely extended to more than
fifty or one hundred yards.   When winged, and uninjured in
their legs, they swim and dive with great rapidity, and are
seldom seen to rise again.   I have several times, on such
occasions, discovered them clinging with their feet to the

reeds under the water, and at other times skulking under the floating reeds, with their bill just above the surface. Sometimes, when wounded, they dive, and rising under the gunwale of the boat, secrete themselves there, moving round as the boat moves, until they have an opportunity of escaping unnoticed. They are feeble and delicate in everything but the legs, which seem to possess great vigour and energy; and their bodies being so remarkably thin or compressed as to be less than an inch and a quarter through transversely, they are enabled to pass between the reeds like rats. When seen, they are almost constantly jetting up the tail. Yet, though their flight among the reeds seems feeble and fluttering, every sportsman who is acquainted with them here must have seen them occasionally rising to a considerable height, stretching out their legs behind them, and flying rapidly across the river where it is more than a mile in width.

Such is the mode of rail-shooting in the neighbourhood of Philadelphia. In Virginia, particularly along the shores of James River, within the tide water, where the rail, or sora, are in prodigious numbers, they are also shot on the wing, but more usually taken at night in the following manner:—A kind of iron grate is fixed on the top of a stout pole, which is placed like a mast in a light canoe, and filled with fire. The darker the night the more successful is the sport. The person who manages the canoe is provided with a light paddle ten or twelve feet in length, and, about an hour before high water, proceeds through among the reeds, which lie broken and floating on the surface. The whole space, for a considerable way round the canoe, is completely enlightened; the birds stare with astonishment, and, as they appear, are knocked on the head with the paddle, and thrown into the canoe. In this manner, from twenty to eighty dozen have been killed by three negroes in the short space of three hours!

At the same season, or a little earlier, they are very numerous in the lagoons near Detroit, on our northern frontiers, where another species of reed (of which they are equally fond) grows in shallows in great abundance. Gentlemen who have shot them there, and on whose judgment I can rely, assure me that they differ in nothing from those they have usually killed on the shores of the Delaware and Schuylkill: they are equally fat, and exquisite eating. On the sea-coast of New Jersey, where these reeds are not to be found, this bird is altogether unknown; though along the marshes of Maurice River, and other tributary streams of the Delaware, and wherever the reeds abound, the rail are sure to be found also. Most of them leave Pennsylvania before the end of October, and the southern States early in November, though numbers linger in the warm southern marshes the whole winter. A very worthy gentleman, Mr Harrison, who lives in Kittiwan, near a creek of that name, on the borders of James River, informed me, that, in burning his meadows early in March, they generally raise and destroy several of these birds. That the great body of these rail winter in countries beyond the United States is rendered highly probable from their being so frequently met with at sea, between our shores and the West India islands. A Captain Douglas informed me, that on his voyage from St Domingo to Philadelphia, and more than a hundred miles from the capes of the Delaware, one night the man at the helm was alarmed by a sudden crash on deck that broke the glass in the binnacle, and put out the light. On examining into the cause, three rail were found on deck, two of which were killed on the spot, and the other died soon after. The late Bishop Madison, president of William and Mary College, Virginia, assured me that a Mr Skipwith, for some time our consul in Europe, on his return to the United States, when upwards of three hundred miles from the capes of the Chesapeake, several rail, or soras, I think five or six, came on board, and were caught by the people. Mr Skipwith, being well acquainted with the bird, assured him that they were the very same with those usually killed on James River. I have received like assurances from several other gentlemen and captains of vessels who have met

with these birds between the mainland and the islands, so as
to leave no doubt on my mind of the fact. For why should
it be considered incredible that a bird which can both swim
and dive well, and at pleasure fly with great rapidity, as I
have myself frequently witnessed, should be incapable of
migrating, like so many others, over extensive tracts of land
or sea? Inhabiting, as they do, the remote regions of
Hudson's Bay, where it is impossible they could subsist during the rigours of their winter, they must either emigrate
from thence or perish ; and as the same places in Pennsylvania which abound with them in October are often laid
under ice and snow during the winter, it is as impossible that
they could exist here in that inclement season. Heaven has,
therefore, given them, in common with many others, certain
prescience of these circumstances, and judgment, as well as
strength of flight, sufficient to seek more genial climates
abounding with their suitable food.

The rail is nine inches long, and fourteen inches in extent ;
bill, yellow, blackish towards the point ; lores, front, crown,
chin, and stripe down the throat, black ; line over the eye,
cheeks, and breast, fine light ash ; sides of the crown, neck,
and upper parts generally, olive brown, streaked with black,
and also with long lines of pure white, the feathers being
centred with black on a brown olive ground, and edged with
white ; these touches of white are shorter near the shoulder
of the wing, lengthening as they descend ; wing, plain olive
brown ; tertials, streaked with black, and long lines of white ;
tail, pointed, dusky olive brown, centred with black ; the four
middle feathers bordered for half their length with lines of
white ; lower part of the breast marked with semicircular
lines of white on a light ash ground ; belly, white ; sides under
the wings, deep olive, barred with black, white, and reddish
buff ; vent, brownish buff ; legs, feet, and naked part of the
thighs, yellowish green ; exterior edge of the wing, white ;
eyes, reddish hazel.

The females and young of the first season have the throat

white, the breast pale brown, and little or no black on the head. The males may always be distinguished by their ashy blue breasts and black throats.

During the greater part of the months of September and October, the market of Philadelphia is abundantly supplied with rail, which are sold from half a dollar to a dollar a dozen. Soon after the 20th of October, at which time our first smart frosts generally take place, these birds move off to the south. In Virginia, they usually remain until the first week in November.

Since the above was written, I have received from Mr George Ord of Philadelphia some curious particulars relative to this bird, which, as they are new, and come from a gentleman of respectability, are worthy of being recorded, and merit further investigation.

"My personal experience," says Mr Ord, "has made me acquainted with a fact in the history of the rail which perhaps is not generally known, and I shall, as briefly as possible, communicate it to you. Some time in the autumn of the year 1809, as I was walking in a yard, after a severe shower of rain, I perceived the feet of a bird projecting from a spout. I pulled it out, and discovered it to be a rail, very vigorous, and in perfect health. The bird was placed in a small room, on a gin-case, and I was amusing myself with it, when, in the act of pointing my finger at it, it suddenly sprang forward, apparently much irritated, fell to the floor, and, stretching out its feet, and bending its neck until the head nearly touched the back, became to all appearance lifeless. Thinking the fall had killed the bird, I took it up, and began to lament my rashness in provoking it. In a few minutes it again breathed, but it was some time before it perfectly recovered from the fit into which, it now appeared evident, it had fallen. I placed the rail in a room wherein canary birds were confined, and resolved that, on the succeeding day, I would endeavour to discover whether or not the passion of anger had produced the fit. I entered the room at the appointed time, and approached the bird, which had retired, on beholding me, in a sullen humour, to a corner. On pointing my finger at it, its feathers were immediately ruffled, and in an instant it sprang forward, as in the first instance, and fell into a similar fit. The following day, the experiment was repeated with the like effect. In the fall of 1811, as I was shooting amongst the reeds, I perceived a rail rise but a few feet before my batteau. The bird had risen about a yard, when it became entangled in the tops of a small bunch of reeds, and immediately fell. Its feet and neck were extended as in the instances above mentioned, and, before it had time to recover, I killed it. Some few days afterwards, as a friend and I were shooting in the same place, he killed a rail, and, as we approached the spot to pick it up, another was perceived, not a foot off, in a fit. I took up the latter, and placed it in the crown of my hat. In a few moments it revived, and was as vigorous as ever. These facts go to prove that the rail is subject to gusts of passion, which operate to so violent a degree as to produce a disease similar in its effects to epilepsy. I leave the explication of the phenomenon to those pathologists who are competent and willing to investigate it. It may be worthy of remark, that the birds affected as described were all females of the *Gallinula Carolina*, or common rail.

"The rail, though generally reputed a simple bird, will sometimes manifest symptoms of considerable intelligence. To those acquainted with rail-shooting, it is hardly necessary to mention that the tide, in its flux, is considered an almost indispensable auxiliary; for, when the water is off the marsh, the lubricity of the mud, the height and compactness of the reed, and the swiftness of foot of the game, tend to weary the sportsman and to frustrate his endeavours. Even should he succeed in a tolerable degree, the reward is not commensurate to the labour. I have entered the marsh in a batteau at a common tide, and in a well-known haunt have beheld but few birds. The next better tide, on resorting to the same spot, I have perceived abundance of game. The fact is, the rail dive, and conceal themselves beneath the fallen reed, merely projecting their heads above the surface of the water for air, and remain in that situation until the sportsman has passed them ; and it is well known that it is a common practice with wounded rail to dive to the bottom, and, holding on by some vegetable substance, support themselves in that situation until exhausted. During such times, the bird, in escaping from one enemy, has often to encounter another not less formidable. Eels and catfish swarm in every direction prowling for prey, and it is ten to one if a wounded rail escapes them. I myself have beheld a large eel make off with a bird that I had shot, before I had time to pick it up ; and one of my boys, in bobbing for eels, caught one with a whole rail in its belly.

" I have heard it observed, that on the increase of the moon the rail improves in fatness, and decreases in a considerable degree with that planet. Sometimes I have conceited that the remark was just. If it be a fact, I think it may be explained on the supposition that the bird is enabled to feed at night as well as by day while it has the benefit of the moon, and with less interruption than at other periods."

I have had my doubts as to the propriety of classing this bird under the genus *Rallus*. Both Latham and Pennant call it a Gallinule ; and when one considers the length and formation of its bill, the propriety of their nomenclature is obvious. As the article was commenced by our printers before I could make up my mind on the subject, the reader is requested to consider this species the *Gallinula Carolina* of Dr Latham.

WOODCOCK.   (*Scolopax minor.*)

PLATE XLVIII.—Fig. 2.

Arct. Zool. p. 463, No. 365.—*Turt. Syst.* 396.—*Lath. Syn.* iii. 131.

RUSTICOLA MINOR.—Vieillot.*

Rusticola minor, *Vieill. Gal. des Ois.* 242.—Great Red Woodcock, Scolopax Americana rufa, *Bart. Trav.* p. 292.—Scolopax rusticola minor, *Bonap. Synop.* p. 331.—*Monog. del Gen. Scolopax Osser. Sulla,* 2d ed. *del Reg. Anim. Cuv.*

This bird, like the preceding, is universally known to our sportsmen.   It arrives in Pennsylvania early in March, some-

---

* Among many natural groups, such as *Scolopax* of Linnæus, there are gradations of form which have not been thought of sufficient importance to constitute a genus, but have been mentioned as divisions only.   Such is the case with the present, which is generally classed under those with the tibiæ feathered and the tibiæ bare.   Vieillot, following this division, proposed *Rusticola* for the woodcocks, or those with plumed tibiæ ; and, as far as artificial systems are concerned, and facility of reference, we should prefer keeping them as a sub-genus.

The woodcocks, in addition to the plumed tibiæ, differ in other respects ; and an individual, technically unacquainted with ornithology, would at once pick them out from the snipes from a something in their *tourneur,* as Mr Audubon would call it.   The tarsi are much shorter, and show that the bird is not intended to wade, or to frequent very marshy situations, like the snipes.   They are all inhabitants of woods, and it is only during severe storms that they are constantly found near a rill or streamlet.   Their food is as much found by searching under the fallen leaves and decayed grasses as in wet places ; and in this country, where woodcocks are abundant, they may be traced through a wood by the newly scratched-up leaves.   There is a marked difference, also, in the plumage ; it is invariably of a more sombre shade, sometimes the under parts are closely barred with a darker colour ; while, in the snipes, the latter part is oftener pure white.   We have a beautiful connection between the divisions in the *Scolopax Sabini* of Vigors,* which, though of the lesser size of the snipes, has the entire plumage of the woodcock, and also the thighs feathered to a greater length downwards.

The species are few in number, amounting only to three or four. America, Europe, and India seem as yet their only countries.   The habits of most agree, and all partially migrate from north to south to breed.—Ed.

* Is this the *Scolopax Sakhalina* of Vieillot, *Nouv. Dict?*—Ed.

times sooner ; and I doubt not but in mild winters some few remain with us the whole of that season. During the day they keep to the woods and thickets, and at the approach of evening seek the springs and open watery places to feed in. They soon disperse themselves over the country to breed. About the beginning of July, particularly in long-continued hot weather, they descend to the marshy shores of our large rivers, their favourite springs and watery recesses inland being chiefly dried up. To the former of these retreats they are pursued by the merciless sportsman, flushed by dogs, and shot down in great numbers. This species of amusement, when eagerly followed, is still more laborious and fatiguing than that of snipe-shooting ; and, from the nature of the ground, or cripple, as it is usually called, viz., deep mire intersected with old logs, which are covered and hid from sight by high reeds, weeds, and alder bushes, the best dogs are soon tired out; and it is customary with sportsmen who regularly pursue this diversion to have two sets of dogs, to relieve each other alternately.

The woodcock usually begins to lay in April. The nest is placed on the ground, in a retired part of the woods, frequently at the root of an old stump. It is formed of a few withered leaves and stalks of grass, laid with very little art. The female lays four, sometimes five eggs, about an inch and a half long, and an inch or rather more in diameter, tapering suddenly to the small end. These are of a dun clay colour, thickly marked with spots of brown, particularly at the great end, and interspersed with others of a very pale purple. The nest of the woodcock has, in several instances that have come to my knowledge, been found with eggs in February ; but its usual time of beginning to lay is early in April. In July, August, and September, they are considered in good order for shooting.

The woodcock is properly a nocturnal bird, feeding chiefly at night, and seldom stirring about till after sunset. At such times, as well as in the early part of the morning, particularly in spring, he rises, by a kind of spiral course, to a considerable height in the air, uttering at times a sudden *quack*, till, having gained his utmost height, he hovers around in a wild irregular manner, making a sort of murmuring sound; then descends with rapidity as he rose. When uttering his common note on the ground, he seems to do it with difficulty, throwing his head towards the earth, and frequently jetting up his tail. These notes and manœuvres are most usual in spring, and are the call of the male to his favourite female. Their food consists of various larva, and other aquatic worms, for which, during the evening, they are almost continually turning over the leaves with their bill, or searching in the bogs. Their flesh is reckoned delicious, and prized highly. They remain with us till late in autumn, and, on the falling of the first snows, descend from the ranges of the Alleghany to the lower parts of the country in great numbers; soon after which, viz., in November, they move off to the south.

This bird, in its general figure and manners, greatly resembles the woodcock of Europe, but is considerably less, and very differently marked below, being an entirely distinct species. A few traits will clearly point out their differences. The lower parts of the European woodcock are thickly barred with dusky waved lines, on a yellowish white ground. The present species has those parts of a bright ferruginous. The male of the American species weighs from five to six ounces, the female, eight; the European, twelve. The European woodcock makes its first appearance in Britain in October and November, that country being in fact only its winter quarters; for early in March they move off to the northern parts of the Continent to breed. The American species, on the contrary, winters in countries south of the United States, arrives here early in March, extends its migrations as far, at least, as the river St Lawrence, breeds in all the intermediate places, and retires again to the south on the approach of winter. The one migrates from the torrid to the temperate regions, the other, from the temperate to the arctic. The two birds, therefore, notwithstanding their names are the same, differ not only in size and markings, but also in native climate. Hence the absurdity of those who would persuade us that the woodcock of America crosses the Atlantic to Europe, and *vice versa.* These observations have been thought necessary, from the respectability of some of our own writers, who seem to have adopted this opinion.

How far to the north our woodcock is found, I am unable to say. It is not mentioned as a bird of Hudson's Bay, and, being altogether unknown in the northern parts of Europe, it is very probable that its migrations do not extend to a very high latitude; for it may be laid down as a general rule, that those birds which migrate to the arctic regions, in either continent, are very often common to both. The head of the woodcock is of singular conformation, large, somewhat triangular, and the eye fixed at a remarkable distance from the bill, and high in the head. This construction was necessary to give a greater range of vision, and to secure the eye from injury while the owner is searching in the mire. The flight of the woodcock is slow. When flushed at any time in the woods, he rises to the height of the bushes or underwood, and almost instantly drops behind them again at a short distance, generally running off for several yards as soon as he touches the ground. The notion that there are two species of woodcock in this country probably originated from the great difference of size between the male and female, the latter being considerably the larger.

The male woodcock is ten inches and a half long, and sixteen inches in extent; bill, a brownish flesh colour, black towards the tip, the upper mandible ending in a slight knob, that projects about one-tenth of an inch beyond the lower,[*] each grooved, and in length somewhat more than two inches

[*] Mr Pennant (Arctic Zoology, p. 463), in describing the American woodcock, says that the lower mandible is much shorter than the upper. From the appearance of his figure, it is evident that the specimen from which that and his description were taken had lost nearly half an inch from the lower mandible, probably broken off by accident. Turton and others have repeated this mistake.

and a half; forehead, line over the eye, and whole lower
parts, reddish tawny; sides of the neck, inclining to ash;
between the eye and bill, a slight streak of dark brown;
crown, from the forepart of the eye backwards, black, crossed
by three narrow bands of brownish white; cheeks, marked
with a bar of black, variegated with light brown; edges of
the back and of the scapulars, pale bluish white; back and
scapulars, deep black, each feather tipt or marbled with light
brown and bright ferruginous, with numerous fine zigzag lines
of black crossing the lighter parts; quills, plain dusky brown;
tail, black, each feather marked along the outer edge with
small spots of pale brown, and ending in narrow tips, of a pale
drab colour above, and silvery white below; lining of the
wing, bright rust; legs and feet, a pale reddish flesh colour;
eye, very full and black, seated high and very far back in the
head; weight, five ounces and a half, sometimes six.

The female is twelve inches long, and eighteen in extent;
weighs eight ounces; and differs also in having the bill very
near three inches in length: the black on the back is not quite
so intense, and the sides under the wings are slightly barred
with dusky.

The young woodcocks of a week or ten days old are
covered with down of a brownish white colour, and are marked
from the bill along the crown to the hind head with a broad
stripe of deep brown; another line of the same passes through
the eyes to the hindhead, curving under the eye; from the
back to the rudiments of the tail, runs another of the same
tint, and also on the sides under the wings; the throat and
breast are considerably tinged with rufous; and the quills at
this age are just bursting from their light blue sheaths, and
appear marbled, as in the old birds; the legs and bill are of a
pale purplish ash colour, the latter about an inch long. When
taken, they utter a long, clear, but feeble *peep*, not louder
than that of a mouse. They are far inferior to young partridges in running and skulking; and, should the female
unfortunately be killed, may easily be taken on the spot.

Drawn from Nature by A Wilson

Ruffed Grous or Pheasant.

49.

Engraved by W.H.Lizars

RUFFED GROUSE. (*Tetrao umbellus.*)

PLATE XLIX.

Arct. Zool. p. 301, No. 179.—Ruffed Heathcock or Grouse, *Edw.* 248.—La Gelinote Huppée de Pennsylvanie, *Briss.* i. 214, *Pl. enl.* 104.—*Buff.* ii. 281.—*Phil. Trans.* 62, 393.—*Turt. Syst.* 454.—*Peale's Museum,* No. 4702.

BONASIA UMBELLUS.—Bonaparte.*

Tetrao umbellus, *Temm. Pig. et Gall. Ind.* p. 704.—Tetrao hurpecal, *Temm. Pig. et Gall.* iii. p. 161.—Bonasia umbellus, *Steph. Cont. Sh. Zool.* xi. p. 300.— Bonasia umbellus, *Bonap. Synop.* p. 126.—The Ruffed Grouse, *Aud. Orn. Biog.* i. p. 211, pl. 41, male and female.

THIS is the partridge of the eastern States, and the pheasant of Pennsylvania and the southern districts. It is represented in the plate of its full size, and was faithfully copied from a perfect and very beautiful specimen.

This elegant species is well known in almost every quarter of the United States, and appears to inhabit a very extensive range of country. It is common at Moose Fort, on Hudson's Bay, in lat. 51°; is frequent in the upper parts of Georgia; very abundant in Kentucky and the Indiana territory; and was found by Captains Lewis and Clarke in crossing the great range of mountains that divide the waters of the Columbia and Missouri, more than three thousand miles, by their measurement, from the mouth of the latter. Its favourite places of resort are high mountains, covered with the balsam pine, hemlock, and such like evergreens. Unlike the pinnated grouse, it always prefers the woods; is seldom or never found in open plains; but loves the pine-sheltered declivities of mountains near streams of water. This great difference of disposition in two species, whose food seems to be nearly the same, is very extraordinary. In those open plains called the Barrens of Kentucky, the pinnated grouse was seen in great numbers, but none of the ruffed; while in the high groves

---

* *Bonasia* is a sub-genus, formed by the Prince of Musignano for the reception of this bird. The distinctions are, the unplumed tarsi and toes, contrasted with *Tetrao,* where the former are thickly clothed.—ED.

with which that singular tract of country is interspersed, the latter, or pheasant, was frequently met with ; but not a single individual of the former.

The native haunts of the pheasant being a cold, high, mountainous, and woody country, it is natural to expect that, as we descend from thence to the sea-shores, and the low, flat, and warm climate of the southern States, these birds should become more rare; and such indeed is the case. In the lower parts of Carolina, Georgia, and Florida, they are very seldom observed ; but as we advance inland to the mountains, they again make their appearance. In the lower parts of New Jersey, we indeed occasionally meet with them ; but this is owing to the more northerly situation of the country ; for even here they are far less numerous than among the mountains.

Dr Turton, and several other English writers, have spoken of a long-tailed grouse, said to inhabit the back parts of Virginia, which can be no other than the present species ; there being, as far as I am acquainted, only these two, the ruffed and pinnated grouse, found native within the United States.

The manners of the pheasant are solitary ; they are seldom found in coveys of more than four or five together, and more usually in pairs, or singly. They leave their sequestered haunts in the woods early in the morning, and seek the path or road, to pick up gravel, and glean among the droppings of the horses. In travelling among the mountains that bound the Susquehanna, I was always able to furnish myself with an abundant supply of these birds every morning without leaving the path. If the weather be foggy or lowering, they are sure to be seen in such situations. They generally move along with great stateliness, their broad fanlike tail spread out in the manner exhibited in the drawing. The drumming, as it is usually called, of the pheasant, is another singularity of this species. This is performed by the male alone. In walking through solitary woods frequented by these birds, a stranger is surprised by suddenly hearing a kind of thumping very similar to that produced by striking two full-blown ox-bladders together, but much louder ; the strokes at first are slow and distinct ; but gradually increase in rapidity, till they run into each other, resembling the rumbling sound of very distant thunder, dying away gradually on the ear. After a few minutes' pause, this is again repeated, and, in a calm day, may be heard nearly half a mile off. This drumming is most common in spring, and is the call of the cock to his favourite female. It is produced in the following manner :— The bird, standing on an old prostrate log, generally in a retired and sheltered situation, lowers his wings, erects his expanded tail, contracts his throat, elevates the two tufts of feathers on the neck, and inflates his whole body, something in the manner of the turkey cock, strutting and wheeling about with great stateliness. After a few manœuvres of this kind, he begins to strike with his stiffened wings in short and quick strokes, which become more and more rapid until they run into each other, as has been already described. This is most common in the morning and evening, though I have heard them drumming at all hours of the day. By means of this, the gunner is led to the place of his retreat ; though, to those unacquainted with the sound, there is great deception in the supposed distance, it generally appearing to be much nearer than it really is.*

* Mr Audubon confirms the correctness of Wilson's comparison of the drumming noise produced by this bird. He mentions having often called them within shot by imitating the sound, which he accomplished "by beating a large inflated bullock's bladder with a stick, keeping up as much as possible the same *time* as that in which the bird beats. At the sound produced by the bladder and the stick, the male grouse, inflamed with jealousy, has flown directly towards me, when, being prepared, I have easily shot it. An equally successful stratagem is employed to decoy the males of our little partridge, by imitating the call-note of the female during spring and summer ; but in no instance, after repeated trials, have I been able to entice the pinnated grouse to come towards me whilst imitating the *booming* sounds of that bird."

Most game are very easily called by those expert at imitating sounds. Grouse are often called by poachers, and partridges may be brought near by a quill and horse-hair. Many of the *Tringæ* and *Totani* are easily whistled.—ED.

The pheasant begins to pair in April, and builds its nest early in May. This is placed on the ground, at the root of a bush, old log, or other sheltered and solitary situation, well surrounded with withered leaves. Unlike that of the quail, it is open above, and is usually composed of dry leaves and grass. The eggs are from nine to fifteen in number, of a brownish white, without any spots, and nearly as large as those of a pullet. The young leave the nest as soon as hatched, and are directed by the cluck of the mother, very much in the manner of the common hen. On being surprised, she exhibits all the distress and affectionate manœuvres of the quail, and of most other birds, to lead you away from the spot. I once started a hen pheasant with a single young one, seemingly only a few days old : there might have been more, but I observed only this one. The mother fluttered before me for a moment ; but, suddenly darting towards the young one, seized it in her bill, and flew off along the surface through the woods, with great steadiness and rapidity, till she was beyond my sight, leaving me in great surprise at the incident. I made a very close and active search around the spot for the rest, but without success. Here was a striking instance of something more than what is termed blind instinct, in this remarkable deviation from her usual manœuvres when she has a numerous brood. It would have been impossible for me to have injured this affectionate mother, who had exhibited such an example of presence of mind, reason, and sound judgment, as must have convinced the most bigoted advocates of mere instinct. To carry off a whole brood in this manner at once would have been impossible, and to attempt to save one at the expense of the rest would be unnatural. She therefore usually takes the only possible mode of saving them in that case, by decoying the person in pursuit of herself, by such a natural imitation of lameness as to impose on most people. But here, in the case of a single solitary young one, she instantly altered her plan, and adopted the most simple and effectual means for its preservation.

The pheasant generally springs within a few yards, with a loud whirring noise,* and flies with great vigour through the

' * Mr Audubon has the following observations on the flight and whirring noise produced during it :—" When this bird rises from the ground, at a time when pursued by an enemy or tracked by a dog, it produces a loud whirring sound, resembling that of the whole tribe, excepting the blackcock of Europe, which has less of it than any other species. This whirring sound is never heard when the grouse rises of its own accord for the purpose of removing from one place to another ; nor, in similar circumstances, is it commonly produced by our little partridge. In fact, I do not believe that it is emitted by any species of grouse, unless when surprised and forced to rise. I have often been lying on the ground in the woods or the fields for hours at a time, for the express purpose of observing the movements and habits of different birds, and have frequently seen a partridge or a grouse rise on wing from within a few yards of the spot in which I lay, unobserved by them, as gently and softly as any other bird, and without producing any whirring sound. Nor even when this grouse ascends to the top of a tree does it make any greater noise than other birds of the same size would do."

The structure of the wings among all the *Tetraonidœ* and *Phasianidœ* is such as to preclude the possibility of an entirely noiseless flight when the members are actively used ; but I have no doubt that it can be, and is sometimes, increased. When any kind of game is suddenly sprung or alarmed, the wings are made use of with more violence than when the flight is fairly commenced, or a rise to the branch of a tree is only contemplated. I have heard it produced by all our British game to a certain extent, when flying over me perfectly unalarmed. The noise is certainly produced by the rapid action of the wings, and I believe the birds cannot exert that with a totally noiseless flight. Sounds at variance from that occasioned by ordinary flight are produced by many birds, particularly during the breeding season, when different motions are employed ; and it appears to me to be rather a consequence depending on the peculiar flight, than the flight employed to produce the sound as a love or other call. Such is the booming noise produced by snipes in spring, always accompanied by the almost imperceptible motion of the wings in the very rapid descent of the bird. A somewhat similar sound is produced by the lapwing when flying near her nest or young, and is always heard during a rapid flight performed diagonally downwards. The cock pheasant produces a loud *whirr* by a violent motion of his wings after calling. A very peculiar rustling is heard when the peacock raises his train, and the cause, a rapid, trembling motion of the feathers, is easily perceived ; and the strut of the turkey

woods, beyond reach of view, before it alights. With a good
dog, however, they are easily found; and at some times
exhibit a singular degree of infatuation, by looking down
from the branches where they sit on the dog below, who, the
more noise he keeps up, seems the more to confuse and stupify
them, so that they may be shot down, one by one, till the
whole are killed, without attempting to fly off. In such cases,
those on the lower limbs must be taken first; for, should the
upper ones be first killed, in their fall they alarm those below,
who immediately fly off. In deep snows they are usually
taken in traps, commonly dead traps, supported by a figure
4 trigger. At this season, when suddenly alarmed, they
frequently dive into the snow, particularly when it has newly
fallen, and coming out at a considerable distance, again take
wing. They are pretty hard to kill, and will often carry off
a large load to the distance of two hundred yards, and drop
down dead. Sometimes, in the depth of winter, they approach
the farmhouse, and lurk near the barn or about the garden.
They have also been often taken young, and tamed, so as to
associate with the fowls; and their eggs have frequently been
hatched under the common hen; but these rarely survive
until full grown. They are exceedingly fond of the seeds of

cock is produced apparently by the rapid exertion of the muscles acting
on the roots of the quills.

Under this species may be mentioned the *T. Sabinii* of Douglas. It
is so very closely allied, that Dr Richardson remarks, " After a careful
comparison of Mr Douglas's *T. Sabinii*, deposited in the Edinburgh
Museum, they appeared to me to differ in no respect from the young of
T. umbellus."

The characters of *T. Sabinii*, given by Mr Douglas, are—Rufus, nigro
notatus; dorso maculis cordiformibus, nucha alisque lineis ferrugineo-
flovis; abdomine albo brunneo fasciato; rectricibus fasciatis, fascia
subapicali lata nigra.

Mr Douglas thinks that there is some difference between the specimens
of *T. umbellus* killed on the Rocky Mountains and more northern parts,
from those in the States of New York and Pennsylvania, and proposes,
if they should be hereafter found distinct, that it should stand as *T.
umbelloides.*—ED.

grapes ; occasionally eat ants, chestnuts, blackberries, and various vegetables. Formerly they were numerous in the immediate vicinity of Philadelphia ; but as the woods were cleared and population increased, they retreated to the interior.

At present there are very few to be found within several miles of the city, and those only singly, in the most solitary and retired woody recesses.

The pheasant is in best order for the table in September and October. At this season they feed chiefly on whortle-berries, and the little red aromatic partridge-berries ; the last of which give their flesh a peculiar delicate flavour. With the former our mountains are literally covered from August to November, and these constitute, at that season, the greater part of their food. During the deep snows of winter, they have recourse to the buds of alder and the tender buds of the laurel. I have frequently found their crops distended with a large handful of these latter alone ; and it has been confidently asserted, that, after having fed for some time on the laurel buds, their flesh becomes highly dangerous to eat of, partaking of the poisonous qualities of the plant. The same has been asserted of the flesh of the deer, when, in severe weather and deep snows, they subsist on the leaves and bark of the laurel. Though I have myself ate freely of the flesh of the pheasant, after emptying it of large quantities of laurel buds, without experiencing any bad consequences, yet, from the respec-tability of those, some of them eminent physicians, who have particularised cases in which it has proved deleterious, and even fatal, I am inclined to believe that, in certain cases, where this kind of food has been long continued, and the birds allowed to remain undrawn for several days, until the contents of the crop and stomach have had time to diffuse themselves through the flesh, as is too often the case, it may be unwhole-some, and even dangerous. Great numbers of these birds are brought to our markets at all times during fall and winter, some of which are brought from a distance of more than a hundred miles, and have been probably dead a week or two, unpicked and undrawn, before they are purchased for the table. Regulations prohibiting them from being brought to market unless picked and drawn would very probably be a sufficient security from all danger. At these inclement seasons, however, they are generally lean and dry; and, indeed, at all times their flesh is far inferior to that of the pinnated grouse. They are usually sold in Philadelphia market at from three-quarters of a dollar to a dollar and a quarter a pair, and sometimes higher.

The pheasant, or partridge, of New England, is eighteen inches long, and twenty-three inches in extent; bill, a horn colour, paler below; eye, reddish hazel, immediately above which is a small spot of bare skin of a scarlet colour; crested; head and neck, variegated with black, red brown, white, and pale brown; sides of the neck furnished with a tuft of large black feathers, twenty-nine or thirty in number, which it occasionally raises; this tuft covers a large space of the neck destitute of feathers; body above, a bright rust colour, marked with oval spots of yellowish white, and sprinkled with black; wings, plain olive brown, exteriorly edged with white, spotted with olive; the tail is rounded, extends five inches beyond the tips of the wings, is of a bright reddish brown, beautifully marked with numerous waving transverse bars of black, is also crossed by a broad band of black, within half an inch of the tip, which is bluish white, thickly sprinkled and specked with black; body below, white, marked with large blotches of pale brown; the legs are covered half way to the feet with hairy down of a brownish white colour; legs and feet, pale ash; toes, pectinated along the sides; the two exterior ones joined at the base, as far as the first joint, by a membrane; vent, yellowish rust colour.

The female and young birds differ in having the ruff or tufts of feathers on the neck of a dark brown colour, as well as the bar of black on the tail inclining much to the same tint.

Drawn from Nature by A. Wilson.

Engraved by W.H.Lizars.

1. Great Horned Owl. 2. Barn O. 3. Meadow Mouse. 4. Red Bat. 5. Small-headed Flycatcher. 6. Hawk Owl.

GREAT HORNED OWL. *(Strix Virginiana.)*

PLATE L.—Fig. 1.

Arct. Zool. p. 228, No. 114.—*Edw.* 60.—*Lath.* i. 119.—*Turt. Syst.* p. 166.—
Peale's Museum, No. 410.

BUBO VIRGINIANA.—Cuvier.*

Le Grand Hibou d'Amerique, *Cuv. Reg. Anim.* i. p. 329.—Strix Virginiana,
Bonap. Synop. p. 37.—The Great Horned Owl, *Aud. Orn. Biog.* i. p. 313,
pl. 61, male and female.—Strix (Bubo) Virginiana, *North. Zool.* ii. p. 82.

THE figure of this bird, as well as of those represented in the
same plate, is reduced to one half its natural dimensions.

* Cuvier uses the title *Bubo* to distinguish those species which, as in
the genus *Otus*, have the tarsi feathered, and are furnished with egrets,
but have the disk surrounding the face less distinctly marked, and have
a small external conch. He assumes as the type the eagle owl of
Europe, but places the Virginian species in his genus *Otus*, with
the small long-eared owl of Britain : the latter has the disk very distinct, and the ears large, the characters of *Otus ;* but the American bird
is in every way a true *Bubo*, as defined by the great French naturalist.
It is a genus of very extensive geographical distribution ; individuals
exist in almost every latitude, and in the four quarters of the world.
Their abodes are the deep and interminable forests, their habits
nocturnal, though they are not so much annoyed or stupified if disturbed in the day, and much more difficult to approach, earnestly
watching their pursuer.

An eagle owl in my possession remains quiet during the day, unless
he is shown some prey, when he becomes eager to possess it, and when
it is put within his reach, at once clutches it, and retires to a corner to
devour it at leisure. During night he is extremely active, and sometimes
keeps up an incessant bark. It is so similar to that of a cur or terrier
as to annoy a large Labrador house-dog, who expressed his dissatisfac-
tion by replying to him, and disturbing the inmates nightly. I at first
mistook the cry also for that of a dog, and, without any recollection of
the owl, sallied forth to destroy this disturber of our repose ; and it was
not until tracing the sound to the cage, that I became satisfied of the
author of the annoyance. I have remarked that he barks more incessantly during a clear winter night than at any other time, and the thin
air at that season makes the cry very distinctly heard to a considerable
distance. This bird also shows a great antipathy to dogs, and will perceive one at a considerable distance, nor is it possible to distract his

By the same scale the greater part of the hawks and owls of the present volume are drawn, their real magnitude rendering this unavoidable.

attention so long as the animal remains in sight. When first perceived, the feathers are raised and the wings lowered as when feeding, and the head moved round, following the object while in sight : if food is thrown, it will be struck with the foot and held, but no further attention paid to it.

The Virginian owl seems to be very extensively distributed over America, is tolerably common over every part of the continent, and Mr Swainson has seen specimens from the tableland of Mexico. The southern specimens present only a brighter colouring in the rufous parts of the plumage.

According to all authorities, owls have been regarded as objects of superstition ; and this has sometimes been taken advantage of by the well-informed for purposes far from what ought to be the duty of a better education to inculcate. None are more accessible to such superstitions than the primitive natives of Ireland and the north of Scotland. Dr Richardson thus relates an instance, which came to his own knowledge, of the consequences arising from a visit of this nocturnal wanderer.

" A party of Scottish Highlanders, in the service of the Hudson's Bay Company, happened, in a winter journey, to encamp after nightfall in a dense clump of trees, whose dark tops and lofty stems, the growth of more than one century, gave a solemnity to the scene that strongly tended to excite the superstitious feelings of the Highlanders. The effect was heightened by the discovery of a tomb, which, with a natural taste often exhibited by the Indians, had been placed in this secluded spot. Our travellers having finished their supper, were trimming their fire preparatory to retiring to rest, when the slow and dismal notes of the horned owl fell on the ear with a startling nearness. None of them being acquainted with the sound, they at once concluded that so unearthly a voice must be the moaning of the spirit of the departed, whose repose they supposed they had disturbed by inadvertently making a fire of some of the wood of which his tomb had been constructed. They passed a tedious night of fear, and, with the first dawn of day, hastily quitted the ill-omened spot."

In India there is a large owl, known by the native name of *Googoo*, or *Ooloo*, which, according to some interesting notices, accompanying a large box of birds sent to Mr Selby from the vicinity of Hyderabad, is held as an object of both fear and veneration. " If an *Ooloo* should alight on the house of a Hindoo, he would leave it immediately, take the thatch off, and put fresh on. The eyes and brain are considered an infallible cure for fits in children, and both are often given to women

This noted and formidable owl is found in almost every quarter of the United States. His favourite residence, how-

in labour. The flesh, bones, &c., boiled down to a jelly, are used to cure spasms or rheumatism. Some of the fat, given to a child newly born, averts misfortune from him for life." Independent of these, says our correspondent, " there are innumerable superstitions regarding this bird, and a native will always kill one when he has an opportunity.

We must mention here a very beautiful species, which is certainly *first accurately* described in the second volume of the "Northern Zoology," though Wilson appears to have had some information regarding a large white owl ; and Dr Richardson is of opinion that the *Strix Scandiaca* of Linnæus, if not actually the species, at least resembles it. It is characterised and figured by the northern travellers under the name of *Bubo Arctica,* arctic or white-horned owl ; and we add the greater part of their description.

" This very beautiful owl appears to be rare, only one specimen having been seen by the members of the expedition. It was observed flying, at mid-day, in the immediate vicinity of Carlton House, and was brought down with an arrow by an Indian boy. I obtained no information respecting its habits.

" The facial disk is very imperfect ; the ears, small, and without an operculum, as in *Strix Virginiana ;* the ear-feathers, ample ; but the disk even smaller than in the last-mentioned bird, and the tarsi somewhat longer. The toes are similarly connected. The tail is of moderate length, and considerably rounded. The bill is strong, and rather short.

" *Description.*—Colour of the bill and claws, bluish black. Irides, yellow. The face is white, bounded posteriorly by blackish brown, succeeded by white, which two latter colours are continued in a mixed band across the throat. Egrets, coloured at the base, like the adjoining plumage ; the longer feathers tipped with blackish brown, their inner webs, white, varied with wood brown. The whole dorsal aspect is marked with undulated lines, or fine bars, of umber brown, alternating with white ; the markings bearing some resemblance to those of the Virginian owl, but being much more lively and handsome. On the greater wing-coverts, on the inner half of the scapularies, and also partially on the neck and lesser wing-coverts, the white is tinged or replaced by pale wood brown. The primaries and secondaries are wood brown, with a considerable portion of white along the margins of their inner webs. They are crossed by from five to six distant umber brown bars on both webs, the intervening spaces being finely speckled with the same. Near the tips of the primaries, the fine sprinkling of the dark colour nearly obscures the wood brown. On the tertiaries, the

ever, is in the dark solitudes of deep swamps, covered with a
growth of gigantic timber ; and here, as soon as evening draws

wood brown is mostly replaced by white. The tail-feathers are white,
deeply tinged on their inner webs by wood brown, and crossed by six
bars of umber brown, about half as broad as the intervening spaces ;
their tips are white.

"*Under surface.*—Chin, white. Throat, crossed by the band above
mentioned, behind which there is a large space of pure snow white, that
is bounded on the breast by blotches of liver brown, situated on the
tips of the feathers. The belly and long plumage of the flanks are
white, crossed by narrow, regular bars of dark brown. The vent-
feathers, under tail-coverts, thighs, and feet, are pure white. The
linings of the wings are also white, with the exception of a brown spot
on the tips of the greater interior coverts."

Audubon has the following remarks on their incubation, which are
somewhat at variance with Wilson. It would also appear that this bird
makes love during the day :—

"Early in February, the great horned owls are seen to pair. The
curious evolutions of the male in the air, or his motions when he has
alighted near his beloved, it is impossible to describe. His bowings,
and the snappings of his bill, are extremely ludicrous ; and no sooner
is the female assured that the attentions paid her by the beau are the
result of a sincere affection, than she joins in the motions of her future
mate.

"The nest, which is very bulky, is usually fixed on a large horizontal
branch, not far from the trunk of the tree. It is composed externally
of crooked sticks, and is lined with coarse grasses and some feathers.
The whole measures nearly three feet in diameter. The eggs, which
are from three to six, are almost globular in form, and of a dull white
colour. The male assists the female in sitting on the eggs. Only one
brood is raised in the season. The young remain in the nest until fully
fledged, and afterwards follow the parents for a considerable time, uttering a mournful sound, to induce them to supply them with food.
They acquire the full plumage of the old birds in the first spring, and
until then are considerably lighter, with more dull buff in their tints.
I have found nests belonging to this species in large hollows of decayed
trees, and twice in the fissures of rocks. In all these cases, little preparation had been made previous to the laying of the eggs, as I found
only a few grasses and feathers placed under them.

"The great horned owl lives retired, and it is seldom that more than
one is found in the neighbourhood of a farm after the breeding season ;
but as almost every detached farm is visited by one of these dangerous
and powerful marauders, it may be said to be abundant. The havoc

on, and mankind retire to rest, he sends forth such sounds as seem scarcely to belong to this world, startling the solitary pilgrim as he slumbers by his forest fire—

Making night hideous.

Along the mountainous shores of the Ohio, and amidst the deep forests of Indiana, alone, and reposing in the woods, this ghostly watchman has frequently warned me of the approach of morning, and amused me with his singular exclamations, sometimes sweeping down and around my fire, uttering a loud and sudden *Waugh O! Waugh O!* sufficient to have alarmed a whole garrison. He has other nocturnal solos, no less melodious, one of which very strikingly resembles the half-suppressed screams of a person suffocating or throttled, and cannot fail of being exceedingly entertaining to a lonely benighted traveller, in the midst of an Indian wilderness!

This species inhabits the country round Hudson's Bay; and, according to Pennant, who considers it a mere variety of the eagle owl (*Strix bubo*) of Europe, is found in Kamtschatka; extends even to the arctic regions, where it is often found white, and occurs as low as Astrakan. It has also been seen white in the United States, but this has doubtless been owing to disease or natural defect, and not to climate. It preys on

which it commits is very great. I have known a plantation almost stripped of the whole of the poultry raised upon it during spring by one of these daring foes of the feathered race in the course of the ensuing winter.

" This species is very powerful, and equally spirited. It attacks wild turkeys when half grown, and often masters them. Mallards, guinea-fowls, and common barn fowls prove an easy prey; and on seizing them, it carries them off in its talons from the farmyards to the interior of the woods. When wounded, it exhibits a revengeful tenacity of spirit, scarcely surpassed by any of the noblest of the eagle tribe, disdaining to scramble away like the barred owl, but facing its enemy with undaunted courage, protruding its powerful talons and snapping its bill as long as he continues in its presence. On these occasions, its large goggle eyes are seen to open and close in quick succession, and the feathers of its body, being raised, swell out its apparent bulk to nearly double the natural size."—ED.

young rabbits, squirrels, rats, mice, partridges, and small birds of various kinds. It has been often known to prowl about the farmhouse, and carry off chickens from roost. A very large one, wing-broken while on a foraging excursion of this kind, was kept about the house for several days, and at length disappeared, no one knew how. Almost every day after this, hens and chickens also disappeared, one by one, in an unaccountable manner, till, in eight or ten days, very few were left remaining. The fox, the minx, and weasel were alternately the reputed authors of this mischief, until one morning, the old lady herself, rising before day to bake, in passing towards the oven, surprised her late prisoner, the owl, regaling himself on the body of a newly-killed hen ! The thief instantly made for his hole under the house, from whence the enraged matron soon dislodged him with the brush handle, and without mercy despatched him. In this snug retreat were found the greater part of the feathers, and many large.fragments, of her whole family of chickens.

There is something in the character of the owl so recluse, solitary, and mysterious, something so discordant in the tones of its voice, heard only amid the silence and gloom of night, and in the most lonely and sequestered situations, as to have strongly impressed the minds of mankind in general with sensations of awe and abhorrence of the whole tribe. The poets have indulged freely in this general prejudice ; and in their descriptions and delineations of midnight storms and gloomy scenes of nature, the owl is generally introduced to heighten the horror of the picture. Ignorance and superstition, in all ages and in all countries, listen to the voice of the owl, and even contemplate its physiognomy with feelings of disgust, and a kind of fearful awe. The priests or conjurors among some of our Indian nations have taken advantage of the reverential horror for this bird, and have adopted the *great horned owl*, the subject of the present account, as the symbol or emblem of their office. "Among the Creeks," says Mr Bartram, in his Travels, p. 504, "the junior priests or students constantly wear a white mantle, and have a great owl skin cased and stuffed very ingeniously, so well executed as almost to appear like the living bird, having large sparkling glass beads or buttons fixed in the head for eyes. This insignia of wisdom and divination they wear sometimes as a crest on the top of the head, at other times the image sits on the arm, or is borne on the hand. These bachelors are also distinguished from the other people by their taciturnity, grave and solemn countenance, dignified step, and singing to themselves songs or hymns in a low, sweet voice, as they stroll about the town.

Nothing is a more effectual cure for superstition than a knowledge of the general laws and productions of nature, nor more forcibly leads our reflections to the first, great, self-existent CAUSE of all, to whom our reverential awe is then humbly devoted, and not to any of His dependent creatures. With all the gloomy habits and ungracious tones of the owl, there is nothing in this bird supernatural or mysterious, or more than that of a simple bird of prey, formed for feeding by night, like many other animals, and of reposing by day. The harshness of its voice, occasioned by the width and capacity of its throat, may be intended by Heaven as an alarm and warning to the birds and animals on which it preys to secure themselves from danger. The voices of all carnivorous birds and animals are also observed to be harsh and hideous, probably for this very purpose.

The great horned owl is not migratory, but remains with us the whole year. During the day he slumbers in the thick evergreens of deep swamps, or seeks shelter in large hollow trees. He is very rarely seen abroad by day, and never but when disturbed. In the month of May they usually begin to build. The nest is generally placed in the fork of a tall tree, and is constructed of sticks piled in considerable quantities, lined with dry leaves and a few feathers. Sometimes they choose a hollow tree ; and in that case carry in but few materials. The female lays four eggs, nearly as large as those of a hen, almost globular, and of a pure white. In one of these nests, after the young had flown, were found the heads and bones of two chickens, the legs and head of the golden-winged woodpecker, and part of the wings and feathers of several other birds. It is generally conjectured that they hatch but once in the season.

The length of the male of this species is twenty inches ; the bill is large, black, and strong, covered at the base with a cere ; the eyes, golden yellow ; the horns are three inches in length, and very broad, consisting of twelve or fourteen feathers, their webs black, broadly edged with bright tawny ; face, rusty, bounded on each side by a band of black ; space between the eyes and bill, whitish ; whole lower parts elegantly marked with numerous transverse bars of dusky on a bright tawny ground, thinly interspersed with white ; vent, pale yellow ochre, barred with narrow lines of brown ; legs and feet large, and covered with feathers or hairy down of a pale brown colour ; claws, very large, blue black ; tail, rounded, extending about an inch beyond the tips of the wings, crossed with six or seven narrow bars of brown, and variegated or marbled with brown and tawny ; whole upper parts finely pencilled with dusky, on a tawny and whitish ground ; chin, pure white, under that a band of brown, succeeded by another narrow one of white ; eyes, very large.

The female is full two feet in length, and has not the white on the throat so pure. She has also less of the bright ferruginous or tawny tint below ; but is principally distinguished by her superior magnitude.

WHITE OR BARN OWL. (*Strix flammea.*)

PLATE L.—FIG. 2.

Lath. i. 138.—*Arct. Zool.* p. 235, No. 124.—*Phil. Trans.* iii. 138.—L'Effraie, ou la Fresaie, *Buff.* i. 366, pl. 26, *Pl. enl.* 440.—*Bewick's Brit. Birds*, i. p. 89. —Common Owl, *Turt. Syst.* p. 170.—*Peale's Museum*, No. 486.

ULULA FLAMMEA.—Cuvier.*

Strix flammea, *Bonap. Synop.* p. 38.

This owl, though so common in Europe, is much rarer in this part of the United States than the preceding, and is only

* From the authority of most writers, this owl is common to both continents. Temminck says those from America are *exactly* the same. I have not personally had an opportunity of comparing them.

In all true night-feeding birds, or those that require to steal upon their prey unobserved, the general plumage is formed for a light, smooth, and noiseless flight ; but the members are not adapted for great swiftness, or for seizing their prey by quick and sudden evolutions. The form is comparatively light, as far as the necessary requisites for sufficient strength can be combined with it ; and the plumage being ample and loose, assists by its buoyancy, and does not offer the same resistance to the air as one of a stiff and rigid texture. The wings, the great organs of locomotion, and which, in flight, produce the most noise, are rounded, having the webs of the feathers very broad, calculated for a powerful and *sustaining* flight ; and the mechanism of the feathers at once bespeaks an intention to destroy the sound produced by motion. In all those birds which perform very swift and rapid flights—the falcons, for instance, swifts, or swallows, many of the sea-fowl, the frigate bird —the wings are very pointed (a contrariety of form to the *Strigidæ*), with the plumules very closely united, and locked together, so as to form almost a thin or solid slip. These produce more resistance, and act as a strong propelling medium when vigorously used. In the owls, the wings present a larger surface, but are not so capable of swift motion ; and to prevent the noise which would necessarily be produced by the violent percussion of so great an expanse, the webs are entirely detached at the tips, and the plumules of the inner ones being drawn to a fine point, thus offer a free passage to the air, and a gradual diminution of resistance. As a further proof that this structure is so intended, we find it to a much less extent in those species that feed occasionally during the day, and we have also the narrowing and accumination of the wings, denoting superior flight ; while, in some, there is a still greater digression in the elongated tail.—Ed.

found here during very severe winters. This may possibly be owing to the want of those favourite recesses in this part of the world which it so much affects in the eastern continent. The multitudes of old ruined castles, towers, monasteries, and cathedrals, that everywhere rise to view in those countries, are the chosen haunts of this well-known species. Its savage cries at night give, with vulgar minds, a cast of supernatural horror to those venerable mouldering piles of antiquity. This species, being common to both continents, doubtless extends to the arctic regions. It also inhabits Tartary, where, according to Pennant, "The Monguls and natives almost pay it divine honours, because they attribute to this species the preservation of the founder of their empire, Ginghis Khan. That prince, with his small army, happened to be surprised and put to flight by his enemies, and forced to conceal himself in a little coppice ; an owl settled on the bush under which he was hid, and induced his pursuers not to search there, as they thought it impossible that any man could be concealed in a place where that bird would perch. From thenceforth they held it to be sacred, and every one wore a plume of the feathers of this species on his head. To this day the Kalmucs continue the custom on all great festivals ; and some tribes have an idol in form of an owl, to which they fasten the real legs of one." *

This species is rarely found in Pennsylvania in summer. Of its place and manner of building, I am unable, from my own observation, to speak. The bird itself has been several times found in the hollow of a tree, and was once caught in a barn in my neighbourhood. European writers inform us that it makes no nest, but deposits its eggs in the holes of walls, and lays five or six, of a whitish colour ; it is said to feed on mice and small birds, which, like the most of its tribe, it swallows whole, and afterwards emits the bones, feathers, and other indigestible parts, at its mouth, in the form of small round cakes, which are often found in the empty buildings it frequents. During its repose it is said to make a blowing noise resembling the snoring of a man.†

* Arctic Zoology, p. 235.   † Bewick, i. p. 90.

It is distinguished in England by various names, the barn owl, the church owl, gillihowlet and screech owl. In the Lowlands of Scotland it is universally called the hoolet.

The white or barn owl is fourteen inches long, and upwards of three feet six inches in extent; bill, a whitish horn colour, longer than is usual among its tribe; space surrounding each eye remarkably concave, the radiating feathers meeting in a high projecting ridge, arching from the bill upwards; between these lies a thick tuft of bright tawny feathers, that are scarcely seen unless the ridges be separated; face, white, surrounded by a border of narrow, thick-set, velvety feathers, of a reddish cream colour at the tip, pure silvery white below, and finely shafted with black; whole upper parts, a bright tawny yellow, thickly sprinkled with whitish and pale purple, and beautifully interspersed with larger drops of white, each feather of the back and wing-coverts ending in an oblong spot of white, bounded by black; head, large, tumid; sides of the neck, pale yellow ochre, thinly sprinkled with small touches of dusky; primaries and secondaries the same, thinly barred, and thickly sprinkled with dull purplish brown; tail, two inches shorter than the tips of the wings, even, or very slightly forked, pale yellowish, crossed with five bars of brown, and thickly dotted with the same; whole lower parts, pure white, thinly interspersed with small round spots of blackish; thighs, the same; legs, long, thinly covered with short white down nearly to the feet, which are of a dirty white, and thickly warted; toes, thinly clad with white hairs; legs and feet large and clumsy; the ridge, or shoulder of the wing, is tinged with bright orange brown. The aged bird is more white; in some, the spots of black on the breast are wanting, and the colour below a pale yellow; in others, a pure white,

The female measures fifteen inches and a half in length, and three feet eight inches in extent; is much darker above; the lower parts tinged with tawny, and marked also with round spots of black. One of these was lately sent me, which was shot on the border of the meadows below Philadelphia.

Its stomach contained the mangled carcasses of four large meadow mice, hair, bones, and all.   The common practice of most owls is, after breaking the bones, to swallow the mouse entire; the bones, hair, and other indigestible parts, are afterwards discharged from the mouth in large roundish dry balls, that are frequently met with in such places as these birds usually haunt.

As the meadow mouse is so eagerly sought after by those birds, and also by great numbers of hawks, which regularly, at the commencement of winter, resort to the meadows below Philadelphia, and to the marshes along the sea-shore, for the purpose of feeding on these little animals, some account of them may not be improper in this place.   Fig. 3 represents the meadow mouse drawn by the same scale, viz., reduced to one half its natural dimensions.   This species appears not to have been taken notice of by Turton in the latest edition of his translation of Linnæus.   From the nose to the insertion of the tail, it measures four inches; the tail is between three-quarters and an inch long, hairy, and usually curves upwards; the fore feet are short, five-toed, the inner toe very short, but furnished with a claw; hind feet also five-toed; the ears are shorter than the fur, through which, though large, they are scarcely noticeable; the nose is blunt; the colour of the back is dark brown, that of the belly, hoary; the fur is long, and extremely fine; the hind feet are placed very far back, and are also short; the eyes exceeding small.   This mischievous creature is a great pest to the meadows, burrowing in them in every direction; but is particularly injurious to the embankments raised along the river, perforating them in numerous directions, and admitting the water, which afterwards effects dangerous breaches, inundating large extents of these low grounds,—and thus they become the instruments of their own destruction.   In their general figure they bear great resemblance to the common musk rat, and, like them, swim and dive well.   They feed on the bulbous roots of plants, and also on garlic, of which they are remarkably fond.

Another favourite prey of most of our owls is the bat, one species of which is represented at fig. 4, as it hung during the day in the woods where I found it. This also appears to be a nondescript. The length of this bat, from the nose to the tip of the tail, is four inches ; the tail itself is as long as the body, but generally curls up inwards ; the general colour is a bright iron gray, the fur being of a reddish cream at bottom, then strongly tinged with lake, and minutely tipt with white ; the ears are scarcely half an inch long, with two slight valves ; the nostrils are somewhat tubular ; fore teeth, in the upper jaw none, in the lower four, not reckoning the tusks ; the eyes are very small black points ; the chin, upper part of the breast, and head, are of a pale reddish cream colour ; the wings have a single hook or claw each, and are so constructed that the animal may hang either with its head or tail downward. I have several times found two hanging fast locked together behind a leaf, the hook of one fixed in the mouth of the other ; the hind feet are furnished with five toes, sharp-clawed ; the membrane of the wings is dusky, shafts, light brown ; extent, twelve inches. In a cave, not far from Carlisle, in Pennsylvania, I found a number of these bats in the depth of winter, in very severe weather : they were lying on the projecting shelves of the rocks, and when the brand of fire was held near them, wrinkled up their mouths, showing their teeth ; when held in the hand for a short time ; they became active, and, after being carried into a stove room, flew about as lively as ever.

SMALL-HEADED FLYCATCHER. (*Muscicapa minuta.*)

PLATE L.—Fig. 5.

SYLVICOLA ? MINUTA.—Jardine.

Sylvia minuta, *Bonap. Synop*. p. 86.

This very rare species is the only one I have met with, and is reduced to half its size, to correspond with the rest of the figures on the same plate. It was shot on the 24th of April, in an orchard, and was remarkably active, running, climbing, and darting about among the opening buds and blossoms with extraordinary agility. From what quarter of the United States or of North America it is a wanderer, I am unable to determine, having never before met with an individual of the species. Its notes and manner of breeding are also alike unknown to me. This was a male : it measured five inches long, and eight and a quarter in extent ; the upper parts were dull yellow olive ; the wings, dusky brown, edged with lighter ; the greater and lesser coverts, tipt with white ; the lower parts, dirty white, stained with dull yellow, particularly on the upper parts of the breast ; the tail, dusky brown, the two exterior feathers, marked, like those of many others, with a spot of white on the inner vanes ; head, remarkably small ; bill, broad at the base, furnished with bristles, and notched near the tip ; legs, dark brown ; feet, yellowish ; eye, dark hazel.

Since writing the above, I have shot several individuals of this species in various quarters of New Jersey, particularly in swamps : they all appear to be nearly alike in plumage. Having found them there in June, there is no doubt of their breeding in that State, and probably in such situations far to the southward ; for many of the southern summer birds that rarely visit Pennsylvania are yet common to the swamps and pine woods of New Jersey. Similarity of soil and situation, of plants and trees, and consequently of fruits, seeds, and insects, &c., are, doubtless, their inducements. The summer red bird, great Carolina wren, pine-creeping warbler, and many others, are rarely seen in Pennsylvania, or to the northward, though they are common in many parts of West Jersey.

HAWK OWL. (*Strix Hudsonia.*)

PLATE L.—Fig. 6.

Little Hawk Owl, *Edw.* 62.—*Lath.* i. 142, No. 29.—*Phil. Trans.* 61, 385.—Le Chat-huant de Canada, *Briss.* i. 518.—*Buff.* i. 391.—Chouette à longue queue de Siberie, *Pl. enl.* 463.—*Arct. Zool.* p. 234, No. 123.—*Peale's Museum,* No. 500.

SURNIA FUNEREA. —Dumeril.*

Strix (sub-genus Surnia) funerea, *Bonap. Synop.* p. 35.—Strix funerea, *Temm. Man.* i. p. 86.—*North. Zool.* ii. p. 92.

THIS is another inhabitant of both continents, a kind of equivocal species, or rather a connecting link between the hawk and owl tribes, resembling the latter in the feet, and in the radiating feathers round the eye and bill; but approaching nearer to the former in the smallness of its head, narrowness of its face, and in its length of tail. In short, it seems just such a figure as one would expect to see generated between a hawk and an owl of the same size, were it possible for them to produce, and yet is as distinct, independent, and original a species as any other. The figure on the plate is reduced to one half the size of life. It has also another strong trait of the hawk tribe,—in flying and preying by day, contrary to the general habit of owls. It is characterised as a bold and active species, following the fowler, and carrying off his game as soon as it is shot. It is said to prey on partridges and other birds; and is very common at Hudson's Bay, where it is called by the Indians *coparacoch.*† We are also informed

---

* In this we have the true form of a diurnal owl. The head is comparatively small; facial disk, imperfect; the ears hardly larger than in birds of prey, and not operculated; the wings and tail more hawk-like, the former, as Wilson observes, with the webs scarcely divided at the tips. Flies by day, and, according to Dr Richardson, preys during winter on ptarmigan, which it constantly attends in their spring migrations northward, and is even so bold, on a bird being killed by the hunters, as to pounce down upon it, though it may be unable, from its size, to carry it off.—ED.

† Edwards.

that this same species inhabits Denmark and Sweden, is frequent in all Siberia, and on the west side of the Uralian chain as far as Casan and the Volga ; but not in Russia.* It was also seen by the navigators near Sandwich Sound, in lat. 61 deg. north.

This species is very rare in Pennsylvania and the more southern parts of the United States.   Its favourite range seems to be along the borders of the arctic regions, making occasional excursions southwardly when compelled by severity of weather, and consequent scarcity of food. I some time ago received a drawing of this bird from the district of Maine, where it was considered rare : that, and the specimen from which the drawing in the plate was taken, which was shot in the neighbourhood of Philadelphia, are the only two that have come under my notice. These having luckily happened to be male and female, have enabled me to give a description of both. Of their nest or manner of breeding we have no account.

The male of this species is fifteen inches long; the bill, orange yellow, and almost hid among the feathers ; plumage of the chin, curving up over the under mandible ; eyes, bright orange ; head, small ; face, narrow, and with very little concavity ; cheeks, white ; crown and hind head, dusky black, thickly marked with round spots of white ; sides of the neck, marked with a large curving streak of brown black, with another a little behind it of a triangular form ; back, scapulars, rump, and tail-coverts, brown olive, thickly speckled with broad spots of white ; the tail extends three inches beyond the tips of the wings, is of a brown olive colour, and crossed with six or seven narrow bars of white, rounded at the end, and also tipt with white ; the breast and chin are marked with a large spot of brown olive ; upper part of the breast, light ; lower, and all the parts below, elegantly barred with dark brown and white ; legs and feet, covered to and beyond the claws with long whitish plumage, slightly yellow, and barred

* Pennant.

with fine lines of olive; claws, horn colour. The weight of this bird was twelve ounces.

The female is much darker above; the quills are nearly black; and the upper part of the breast is blotched with deep blackish brown.

It is worthy of remark, that, in all owls that fly by night, the exterior edges and sides of the wing-quills are slightly recurved, and end in fine hairs or points; by which means the bird is enabled to pass through the air with the greatest silence, a provision necessary for enabling it the better to surprise its prey. In the hawk owl now before us, which flies by day, and to whom this contrivance would be of no consequence, it is accordingly omitted, or at least is scarcely observable. So judicious, so wise, and perfectly applicable are all the dispositions of the Creator!

MARSH HAWK. (*Falco uliginosus.*)

PLATE LI.—Fig. 2.

Edw. iv. 291.—*Lath.* i. 90.—*Arct. Zool.* p. 208, No. 105.—*Bartram*, p. 290.—
Peale's Museum, No. 318.

CIRCUS CYANEUS.—Bechstein.[*]

Falco (sub-genus Circus), *Bonap. Synop.* p. 33.—Buteo (Circus) cyaneus? var.
Americanus, *North. Zool.* ii. p. 55.

A DRAWING of this hawk was transmitted to Mr Edwards, more than fifty years ago, by Mr William Bartram, and engraved in plate 291 of Edwards' "Ornithology." At that time, and I believe till now, it has been considered as a species peculiar to this country.

I have examined various individuals of this hawk, both in summer and in the depth of winter, and find them to correspond so nearly with the ring-tail of Europe, that I have no doubt of their being the same species.

This hawk is most numerous where there are extensive

* See note in Vol. III. accompanying description of the male.—ED.

meadows and salt marshes, over which it sails very low, making frequent circuitous sweeps over the same ground, in search of a species of mouse, figured in plate 50, and very abundant in such situations. It occasionally flaps the wings, but is most commonly seen sailing about within a few feet of the surface. They are usually known by the name of the mouse hawk along the sea-coast of New Jersey, where they are very common. Several were also brought me last winter from the meadows below Philadelphia. Having never seen its nest, I am unable to describe it from my own observation. It is said by European writers to build on the ground, or on low limbs of trees. Mr Pennant observes that it sometimes changes to a rust-coloured variety, except on the rump and tail. It is found, as was to be expected, at Hudson's Bay, being native in both this latitude and that of Britain. We are also informed that it is common in the open and temperate parts of Russia and Siberia; and extends as far as Lake Baikal, though it is said not to be found in the north of Europe.*

The marsh hawk is twenty-one inches long, and three feet eleven inches in extent; cere and legs, yellow, the former tinged with green, the latter long and slender; nostril, large, triangular; this and the base of the bill thickly covered with strong curving hairs, that rise from the space between the eye and bill, arching over the base of the bill and cere; this is a particular characteristic; bill, blue, black at the end; eye, dark hazel; cartilage overhanging the eye, and also the eyelid, bluish green; spot under the eye, and line from the front over it, brownish white; head above and back, dark glossy chocolate brown, the former slightly seamed with bright ferruginous; scapulars, spotted with the same *under the surface;* lesser coverts and band of the wing, here and there edged with the same; greater coverts and primaries, tipt with whitish; quills, deep brown at the extreme half, some of the outer ones hoary on the exterior edge; all the primaries, yellowish white on the inner vanes and upper half, also barred on the inner vanes

* Pallas, as quoted by Pennant.

with black; tail, long, extending three inches beyond the wings, rounded at the end, and of a pale sorrel colour, crossed by four broad bars of very dark brown, the two middle feathers excepted, which are barred with deep and lighter shades of chocolate brown; chin, pale ferruginous; round the neck, a collar of bright rust colour; breast, belly, and vent, pale rust, shafted with brown; femorals, long, tapering, and of the same pale rust tint; legs, feathered near an inch below the knee. This was a female. The male differs chiefly in being rather lighter and somewhat less.

This hawk is particularly serviceable to the rice-fields of the southern States, by the havoc it makes among the clouds of rice buntings that spread such devastation among that grain in its early stage. As it sails low and swiftly over the surface of the field, it keeps the flocks in perpetual fluctuation, and greatly interrupts their depredations. The planters consider one marsh hawk to be equal to several negroes for alarming the rice-birds. Formerly the marsh hawk used to be numerous along the Schuylkill and Delaware, during the time the reeds were ripening, and the reed-birds abundant; but they have of late years become less numerous here.

Mr Pennant considers the " *strong, thick, and short legs* " of this species as specific distinctions from the ring-tailed hawk; the legs, however, are *long* and *slender ;* and a marsh hawk such as he has described, with strong, thick, and short legs, is nowhere to be found in the United States.

SWALLOW-TAILED HAWK. (*Falco furcatus.*)

PLATE LI.—Fɪɢ. 3.

Linn. Syst. 129.—*Lath.* i. 60.—Hirundo maxima Peruviana avis prædatoris calca-
ribus instructa, *Feuillee, Voy. Peru*, tom. ii. 33.—*Catesb.* i. 4.—Le Milan de
la Caroline, *Briss.* i. 418.—*Buff.* i. 221.—*Turt. Syst.* 149.—*Arct. Zool.* p. 210,
No. 108.—*Peale's Museum*, No. 142.

ELANUS FURCATUS.—Sᴀᴠɪɢɴʏ.*

Le Milan de Caroline, *Cuv. Reg. Anim.* i. p. 322.—Elanus furcatus, *Bonap. Synop.*
p. 31.—Nauclerus furcatus, *Vig. Zool. Journ.* No. vii. p. 387.—*Less. Man.
d'Ornith.* i. p. 101.—The Swallow-tailed Hawk, *Aud.* pl. 72; *Orn. Biog.* i.
p. 368.

Tʜɪꜱ very elegant species inhabits the southern districts of
the United States in summer; is seldom seen as far north as
Pennsylvania, but is very abundant in South Carolina and

* The characters of the birds composing this genus are,—general
form of less strength than most of the *Falconidæ*; bill, rather weak;
tooth, little seen; the tarsi, short, thick, reticulated, and partly
feathered in front; wings, greatly elongated; timorous, and, like the
kites, excel in flight, circling in the air. Mr Vigors has formed a genus,
Nauclerus, of this and a small African species, dividing them from
Elanus, where they were placed by most prior ornithologists. In these
two birds the tail is forked to a great extent, while in the others it
only commences to assume that form, and in one is altogether square.
The claws, also are not circular underneath, as in the others, to which
Mr Vigors would restrict *Elanus*. The wings of the two birds, however, show considerable difference; the quills, in the American, being
abruptly emarginated, the third longest; in the African, the second is
longest, and only a slight emargination on the two first. Altogether
we are not quite satisfied with the distinctions. I have for the present
retained *Elanus*, notwithstanding the differences that do exist between
some of its members.

According to Audubon, they feed chiefly on the wing; and having
pounced on any prey upon the ground, rise with it, and devour it
while flying. "In calm and warm weather," he remarks, "they soar
to an immense height, pursuing the large insects called *mosquito hawks*,
and performing the most singular evolutions that can be conceived,
using their tail with an elegance peculiar to themselves." They thus
show a manner of feeding entirely different from most birds of prey,
which generally retire to some distance, and devour in quiet on the
ground. There are some partly insectivorous hawks—*Penis*, for in-

Georgia, and still more so in West Florida, and the extensive prairies of Ohio and the Indiana territory. I met with these birds in the early part of May at a place called Duck Creek, in Tennessee, and found them sailing about in great numbers near Bayo Manchac on the Mississippi, twenty or thirty being within view at the same time. At that season a species of cicada, or locust, swarmed among the woods, making a deafening noise, and I could perceive these hawks frequently snatching them from the trees. A species of lizard, which is very numerous in that quarter of the country, and has the faculty of changing its colour at will, also furnishes the swallow-tailed hawk with a favourite morsel. These lizards are sometimes of the most brilliant light green, in a few minutes change to a dirty clay colour, and again become nearly black. The swallow-tailed hawk and Mississippi kite feed eagerly on this lizard, and, it is said, on a small green snake also, which is the mortal enemy of the lizard, and frequently pursues it to the very extremity of the branches, where both become the prey of the hawk.*

The swallow-tailed hawk retires to the south in October, at which season, Mr Bartram informs me, they are seen, in Florida, at a vast height in the air, sailing about with great

---

stance—which seize and devour the insect during flight ; but larger prey is treated at leisure. I am aware of none that feed so decidedly on the wing as that now described ; in everything it will appear more like a large swallow than an accipitrine bird.

Mr Audubon remarks another curious circumstance at variance with the wary manners of the *Falconidæ.* "When one is killed, and falls to the ground, the whole flock comes over the dead bird, as if intent upon carrying it off. I have killed several of these hawks in this manner, firing as fast as I could load my gun."

This bird occurred to the late Dr Walker, at Ballachulish, in Argyleshire, in 1792. Another specimen was taken near Howes, in Wensleydale, Yorkshire, by W. Fotheringill, Esq., and communicated to the London Society, November 1823.—ED.

* This animal, if I mistake not, is the *Lacerta bullaris,* or *bladder lizard,* of Turton, vol. i. p. 666. The facility with which it changes colour is surprising, and not generally known to naturalists.

steadiness ; and continue to be seen thus, passing to their winter quarters, for several days. They usually feed from their claws as they fly along. Their flight is easy and graceful, with sometimes occasional sweeps among the trees, the long feathers of their tail spread out, and each extremity of it used alternately to lower, elevate, or otherwise direct their course. I have never yet met with their nests.

These birds are particularly attached to the extensive prairies of the western countries, where their favourite snakes, lizards, grasshoppers, and locusts, are in abundance. They are sometimes, though rarely, seen in Pennsylvania and New Jersey, and that only in warm and very long summers. A specimen now in the Museum of Philadelphia was shot within a few miles of that city. We are informed that one was taken in the South Sea, off the coast which lies between Ylo and Arica, in about lat. 23 deg. south, on the 11th of September, by the Reverend the Father Louis Feuillee.* They are also common in Mexico, and extend their migrations as far as Peru.

The swallow-tailed hawk measures full two feet in length, and upwards of four feet six inches in extent ; the bill is black ; cere, yellow, covered at the base with bristles ; iris of the eye, silvery cream, surrounded with a blood-red ring ; whole head and neck, pure white, the shafts, fine black hairs ; the whole lower parts also pure white ; the throat and breast, shafted in the same manner ; upper parts, or back, black, glossed with green and purple ; whole lesser coverts, very dark purple ; wings long, reaching within two inches of the tip of the tail, and black ; tail also very long, and remarkably forked, consisting of twelve feathers, all black, glossed with green and purple ; several of the tertials, white, or edged with white, but generally covered by the scapulars ; inner vanes of the secondaries, white on their upper half, black towards their points ; lining of the wings, white ; legs, yellow, short, and thick, and feathered before half way below the knee ; claws,

* Jour. des Obs., tom. ii. 33.

Drawn from Nature by A. Wilson

Engraved by W.H. Lizars

1. Long-eared Owl. 2. Marsh Hawk. 3. Swallow-tailed H.

51.

much curved, whitish ; outer claw, very small. The greater part of the plumage is white at the base; and when the scapulars are a little displaced, they appear spotted with white.

This was a male in perfect plumage. The colour and markings of the male and female are nearly alike.

LONG-EARED OWL. (*Strix otus.*)

PLATE LI.—Fig. 1.

Turt. Syst. p. 167.—*Bewick,* i. p. 84.—*Peale's Museum,* No. 434.

OTUS VULGARIS.—Fleming.*

Strix otus, *Bonap. Synop.* p. 37.—*North. Zool.* ii. p. 72.

This owl is common to both continents, and is much more numerous in Pennsylvania than the white or barn owl : six or seven were found in a single tree, about fifteen miles from Philadelphia. There is little doubt but this species is found inhabiting America to a high latitude, though we have no certain accounts of the fact. Except in size, this species has more resemblance to the great horned owl than any other of

* Upon the authority of the Prince of Musignano, and the examinations of the various writers who have mentioned this bird, it appears very near indeed, if not identical with, the *O. vulgaris* of Europe ; and I have ventured to retain it as such, until I can decide from personal observation. The opinions of Vieillot, &c., have been confused by the existence of a second species in the United States, which will appear in the fifth volume of the elegant continuation of Wilson, now in progress by Bonaparte, under the title of *Otus Mexicanus.*

In the second volume of the " Northern Zoology," we have the long-eared owl referred to this species, and no mention is made of any difference arising even from climate, The habits described by Wilson and Dr Richardson are precisely similar to those exhibited by our European bird.

Otus has been formed by Cuvier for the reception of those species with aigrettes, where the facial disk is conspicuous and the head proportionally small, as in *Bubo ;* and where the ear-conch is large, extending, as in this species, from the posterior part of the orbit to behind the limb of the lower jaw. The plumage is loose and downy, the habits nocturnal.—Ed.

its tribe.   It resembles it also in breeding among the branches
of tall trees ; lays four eggs, of nearly a round form, and pure
white.*   The young are grayish white until nearly full grown,
and roost during the day close together on a limb, among the
thickest of the foliage.   This owl is frequently seen abroad
during the day, but is not remarkable for its voice or habits.

The long-eared owl is fourteen inches and a half long, and
three feet two inches in extent ; ears, large, composed of six
feathers, gradually lengthening from the front one backwards,
black, edged with rusty yellow; irides, vivid yellow ; inside
of the circle of the face, white, outside or cheeks, rusty ; at
the internal angle of the eye, a streak of black ; bill, blackish
horn colour ; forehead and crown, deep brown, speckled with
minute points of white and pale rusty ; outside circle of the
face, black, finely marked with small curving spots of white ;
back and wings, dark brown, sprinkled and spotted with
white, pale ferruginous, and dusky ; primaries, barred with
brownish yellow and dusky, darkening towards the tips ;
secondaries, more finely barred, and powdered with white and
dusky ; tail, rounded at the end, of the same length with the
wings, beautifully barred and marbled with dull white and
pale rusty, on a dark brown ground ; throat and breast,
clouded with rusty, cream, black, and white ; belly, beautifully
streaked with large arrow-heads of black ; legs and thighs,
plain pale rusty, feathered to the claws, which are blue black,
large, and sharp ; inside of the wing, brownish yellow, with a
large spot of black at the root of the primaries.   This was a
female.   Of the male I cannot speak precisely ; though, from
the number of these birds which I have examined in the fall,
when it is difficult to ascertain their sex, I conjecture that
they differ very little in colour.

About six or seven miles below Philadelphia, and not far
from the Delaware, is a low swamp, thickly covered with
trees, and inundated during great part of the year.   This

* Buffon remarks that it rarely constructs a nest of its own, but not
unfrequently occupies that of others, particularly the magpie.

place is the resort of great numbers of the qua-bird, or night raven (*Ardea nycticorax*), where they build in large companies. On the 25th of April, while wading among the dark recesses of this place, observing the habits of these birds, I discovered a *long-eared owl*, which had taken possession of one of their nests, and was sitting; on mounting to the nest, I found it contained four eggs, and, breaking one of these, the young appeared almost ready to leave the shell. There were numbers of the qua-birds' nests on the adjoining trees all around, and one of them actually on the same tree. Thus we see how unvarying are the manners of this species, however remote and different the countries may be where it has taken up its residence.

RED-TAILED HAWK. (*Falco borealis.*)

PLATE LII.—Fig. 2.

Arct. Zool. p. 205, No. 100.—American Buzzard, *Lath.* i. 50.—*Turt. Syst.* p. 151.—F. aquilinus cauda ferruga, Great Eagle Hawk, *Bartram*, p. 290.— *Peale's Museum*, No. 182.

BUTEO BOREALIS.—Swainson.[*]

Falco (sub-genus Buteo) borealis, *Bonap. Synop.* p. 32.—The Red-tailed Hawk, *Aud.* pl. 51, male and female; *Orn. Biog.* i. p. 265.—Buteo borealis, *North. Zool.* ii. p. 50.

The figure of this bird, and those of the other two hawks on the same plate, are reduced to exactly half the dimensions of the living subjects. These representations are offered to the public with confidence in their fidelity; but *these*, I am

[*] The red-tailed buzzard is a species peculiar to America, and, in its adult state, seems perfectly known to ornithologists. The figure on the same plate, and next described by our author, has been subject to more discussion, and has been variously named. From the testimonies of Bonaparte and Audubon, it may, however, be certainly considered as the young or immature bird—an idea which Wilson himself entertained, and showed by his mark of interrogation to the young, and the quotation of its synonyms. The figure at fig. 2 is the young in immature plumage, where the red tail has not yet appeared, and which is known to authors under the name of *F. Leverianus.*—Ed.

sorry to say, are almost all I have to give towards elucidating their history. Birds, naturally thinly dispersed over a vast extent of country ; retiring during summer to the depth of the forests to breed ; approaching the habitations of man, like other thieves and plunderers, with shy and cautious jealousy ; seldom permitting a near advance ; subject to great changes of plumage ; and, since the decline of falconry, seldom or never domesticated,—offer to those who wish eagerly to investigate their history, and to delineate their particular character and manners, great and insurmountable difficulties. Little more can be done in such cases than to identify the species, and trace it through the various quarters of the world where it has been certainly met with.

The red-tailed hawk is most frequently seen in the lower parts of Pennsylvania during the severity of winter. Among the extensive meadows that border the Schuylkill and Delaware, below Philadelphia, where flocks of larks (*Alauda magna*), and where mice and moles are in great abundance, many individuals of this hawk spend the greater part of the winter. Others prowl around the plantations, looking out for vagrant chickens ; their method of seizing which is by sweeping swiftly over the spot, and, grappling them with their talons, bear them away to the woods. The bird from which the figure in the plate was drawn was surprised in the act of feeding on a hen he had just killed, and which he was compelled to abandon. The remains of the chicken were immediately baited to a steel trap, and early the next morning the unfortunate red-tail was found a prisoner, securely fastened by the leg. The same hen which the day before he had massacred was, the very next, made the means of decoying him to his destruction,—in the eye of the farmer, a system of fair and just retribution.

This species inhabits the whole United States, and, I believe, is not migratory, as I found it in the month of May as far south as Fort Adams, in the Mississippi territory. The young were, at that time, nearly as large as their parents, and were very clamorous, making an incessant squealing noise. One, which I shot, contained in its stomach mingled fragments of frogs and lizards.

The red-tailed hawk is twenty inches long, and three feet nine inches in extent ; bill, blue black ; cere, and sides of the mouth, yellow, tinged with green ; lores, and spot on the under eyelid, white, the former marked with fine radiating hairs ; eyebrow, or cartilage, a dull eel-skin colour, prominent, pro-jecting over the eye ; a broad streak of dark brown extends from the sides of the mouth backwards ; crown and hind head, dark brown, seamed with white and ferruginous ; sides of the neck, dull ferruginous, streaked with brown ; eye, large ; iris, pale amber ; back and shoulders, deep brown ; wings, dusky, barred with blackish ; ends of the five first primaries, nearly black ; scapulars, barred broadly with white and brown ; sides of the tail-coverts, white, barred with ferruginous, middle ones dark, edged with rust ; tail, rounded, extending two inches beyond the wings, and of a bright red brown, with a single band of black near the end, and tipt with brownish white ; on some of the lateral feathers are slight indications of the remains of other narrow bars ; lower parts, brownish white ; the breast, ferruginous, streaked with dark brown ; across the belly, a band of interrupted spots of brown ; chin, white ; femorals and vent, pale brownish white, the former marked with a few minute heart-shaped spots of brown ; legs, yellow, feathered half way below the knees.

This was a male. Another specimen, shot within a few days after, agreed in almost every particular of its colour and markings with the present, and, on dissection, was found to be a female.

AMERICAN BUZZARD, OR WHITE-BREASTED HAWK. (*Falco borealis?*)

PLATE LII.—Fig. 1.

Lath. Syn. Sup. p. 31.—*Ind. Orn.* i. p. 18, No. 31.—*Peale's Museum*, No. 400.

BUTEO BOREALIS.—Young of the year.—Bonaparte.

Falco (sub-genus Buteo) borealis, *Bonap. Synop.* p. 32.

It is with some doubt and hesitation that I introduce the present as a distinct species from the preceding. In their size and general aspect they resemble each other considerably ; yet I have found both males and females among each ; and in the present species I have sometimes found the ground colour of the tail strongly tinged with ferruginous, and the bars of dusky but slight ; while in the preceding, the tail is sometimes wholly red brown, the single bar of black near the tip excepted ; in other specimens evident remains of numerous other bars are visible. In the meantime both are figured, and future observations may throw more light on the matter.

This bird is more numerous than the last, but frequents the same situations in winter. One, which was shot on the wing, lived with me several weeks, but refused to eat. It amused itself by frequently hopping from one end of the room to the other, and sitting for hours at the window looking down on the passengers below. At first, when approached by any person, he generally put himself in the position in which he is represented ; but after some time he became quite familiar, permitting himself to be handled, and shutting his eyes, as if quite passive. Though he lived so long without food, he was found on dissection to be exceedingly fat, his stomach being enveloped in a mass of solid fat of nearly an inch in thickness.

The white-breasted hawk is twenty-two inches long, and four feet in extent ; cere, pale green ; bill, pale blue, black at the point ; eye, bright straw colour ; eyebrow, projecting greatly ; head, broad, flat, and large ; upper part of the head, sides of the neck, and back, brown, streaked and seamed with

1.Red-tailed Hawk. 2.American Buzzard. 3.Ash-coloured Hawk.

52.

white and some pale rust ; scapulars and wing-coverts, spotted with white ; wing-quills much resembling the preceding species ; tail-coverts, white, handsomely barred with brown ; tail, slightly rounded, of a pale brown colour, varying in some to a sorrel, crossed by nine or ten bars of black, and tipt for half an inch with white ; wings, brown, barred with dusky ; inner vanes nearly all white ; chin, throat, and breast, pure white, with the exception of some slight touches of brown that enclose the chin ; femorals, yellowish white, thinly marked with minute touches of rust ; legs, bright yellow, feathered half way down ; belly, broadly spotted with black or very deep brown ; the tips of the wings reach to the middle of the tail.

My reason for inclining to consider this a distinct species from the last is the circumstance of having uniformly found the present two or three inches larger than the former, though this may possibly be owing to their greater age.

ASH-COLOURED OR BLACK-CAP HAWK.
(*Falco atricapillus.*)

PLATE LII.—Fig. 3.

Ash-coloured Buzzard? *Lath. Syn.* i. p. 55, No. 35.—*Peale's Museum*, No. 406.

ASTUR ATRICAPILLUS.—Bonaparte.*

Falco palumbarius, *Bonap. Synop.* p. 28.—Autour royal, Falco regalis, *Temm. Pl. Col.* tab. 495.—Accipiter (Astur) atricapillus, *North. Zool.* ii. p. 39.—Astur atricapillus, *Jard. and Selb. Illust. Orn.* pl. 121.

Of this beautiful species I can find no precise description. The ash-coloured buzzard of Edwards differs so much from

* The *Falco atricapillus* of Wilson has been confounded by all writers, except the Prince of Musignano in his review of Cuvier, and the authors of the "Northern Zoology," with the goshawk of Europe. Wilson expresses his doubt, from being unable to compare it with actual specimens. Sabine makes out the arctic specimens to be identical. Audubon is of opinion, also, that they were identical ; but from what I recollect of that gentleman's drawing, it must have been made from this

this, particularly in wanting the fine zigzag lines below, and
the black cap, that I cannot for a moment suppose them to be

bird.   While Temminck makes a new species altogether in his *Autour
royal*, without noticing Wilson.

The greatest difference between the two birds is the marking of the
breast and under parts, and it is so distinct as to be at once perceived.
In the American species, the under parts are of a uniform pale grayish
white, having the tail and centre of each feather black, forming a dark
streak.   This extends to those in the centre of the belly, after which it
is hardly visible ; every feather in addition is clouded transversely with
irregular bars of gray.   In the European bird, the markings are in the
shape of two decided transverse bars on each feather, with the shaft
dark, but not exceeding its own breadth,—each, as a whole, having a
very different appearance.   The upper parts of the American bird are
also of a blue shade, and the markings of the head and auriculars are
darker and more decided.   Wilson's figure is a most correct representation.

The genus *Astur*, of Bechstein, has now been used for this form, and
is generally synonymous with *Les autours* of the French.   Mr Swainson,
however, is inclined to make it rather a sub-genus of *Accipiter*, in which
the sparrow hawks and lesser species have been placed.   There is some
difference in the construction of the tarsi, but the habits and general
form are nearly similar.   In the ornithology of America, the *Astur
Pennsylvanicus* will show an example of the one ; the bird now in
question that of the other.

In general form, the birds of this group are strong, but do not show
the firm and compact structure of the true falcon.   The wings are short
and rounded, and present a considerable under surface, favourable to a
smooth and sailing flight, which power is rendered more perfect by the
lengthened and expanded tail.   The tarsi and feet bear a relative proportion of strength to their bodies, and the claws are more than usually
hooked and sharp ; that of the inner toe always equal to the hallux.
Their favourite abodes are woods, or well-clothed countries, where they
build and rear their young, hunting for prey about the skirts.   They
are extremely active and bold ; their flight is sailing in circles, or, when
in search of prey, skimming near to the ground, about fences and brush,
and darting at anything, either on the ground or on wing, with great
celerity.   I have seen some of our native species pick up a bird, when
flying near the ground, so rapidly, that the motion of stooping and
clutching was hardly perceptible, and the flight continued as if nothing
had happened.   During their higher flights, or when threading through
a thick wood, which they do with great dexterity, the motions of the
tail are perceived directing their movements, and, in the latter case, is

the same. The individual from which the drawing was made is faithfully represented in the plate, reduced to one half its natural dimensions. This bird was shot within a few miles of Philadelphia, and is now preserved, in good order, in Mr Peale's Museum.

Its general make and aspect denotes great strength and spirit ; its legs are strong, and its claws of more than proportionate size. Should any other specimen or variety of this hawk, differing from the present, occur during the publication of this work, it will enable me more accurately to designate the species.

The black-cap hawk is twenty-one inches in length ; the bill and cere are blue ; eye, reddish amber ; crown, black, bordered on each side by a line of white finely specked with black ; these lines of white meet on the hind head ; whole upper parts, slate, tinged with brown, slightest on the quills ; legs, feathered half way down, and, with the feet, of a yellow colour ; whole lower parts and femorals, white, most elegantly speckled with fine transverse pencilled zigzag lines of dusky, all the shafts being a long black line ; vent, pure white.

If this be not the celebrated *goshawk*, formerly so much esteemed in falconry, it is very closely allied to it. I have never myself seen a specimen of that bird in Europe, and the descriptions of their best naturalists vary considerably ; but, from a careful examination of the figure and account of the goshawk, given by the ingenious Mr Bewick (Brit. Birds, vol. i. p. 65), I have very little doubt that the present will be found to be the same.

The goshawk inhabits France and Germany ; is not very common in South Britain, but more frequent in the northern

most conspicuously necessary. When perched at rest, the position is unusually erect ; so much, that the line of the back and tail is almost perpendicular. The plumage in the adults is often of a dark leaden colour above, with bars and crosses on the under parts ; in the young, the upper surface assumes different shades of brown, while the markings beneath are longitudinal.—ED.

parts of the island, and is found in Russia and Siberia. Buffon, who reared two young birds of this kind, a male and female, observes, that " the goshawk, before it has shed its feathers, that is, in its first year, is marked on the breast and belly with longitudinal brown spots ; but after it has had two moultings they disappear, and their place is occupied by transverse waving bars, which continue during the rest of its life ; " he also takes notice, that though the male was much smaller than the female, it was fiercer and more vicious.

Mr Pennant informs us that the goshawk is used by the Emperor of China in his sporting excursions, when he is usually attended by his grand falconer and a thousand of inferior rank. Every bird has a silver plate fastened to its foot, with the name of the falconer who has charge of it, that, in case it should be lost, it may be restored to the proper person ; but, if he should not be found, the bird is delivered to another officer, called " the guardian of lost birds," who, to make his situation known, erects his standard in a conspicuous place among the army of hunters. The same writer informs us, that he examined, in the Leverian Museum, a specimen of the goshawk which came from America, and which was superior in size to the European. He adds, "they are the best of all hawks for falconry." *

BLACK HAWK. (*Falco Sancti Johannis ?*)

PLATE LIII.—Fig. 1.

Lath. Ind. Orn. p. 34, No. 74.—Chocolate-coloured Falcon, *Penn. Arct. Zool.* No. 94.

BUTEO SANCTI JOHANNIS ?—Bonaparte.

Falco (sub-genus Buteo) Sancti Johannis, *Bonap. Synop.* p. 32.

This, and the other two figures on the same plate, are reduced from the large drawings, which were taken of the exact size of nature, to one half their dimensions. I regret

* Arctic Zoology, p. 204.

the necessity which obliges me to contract the figures of these birds, by which much of the grandeur of the originals is lost ; particular attention, however, has been paid in the reduction to the accurate representation of all their parts.

This is a remarkably shy and wary bird, found most frequently along the marshy shores of our large rivers ; feeds on mice, frogs, and moles ; sails much, and sometimes at a great height ; has been seen to kill a duck on wing ; sits by the side of the marshes on a stake for an hour at a time, in an almost perpendicular position, as if dozing ; flies with great ease, and occasionally with great swiftness, seldom flapping the wings ; seems particularly fond of river shores, swamps, and marshes ; is most numerous with us in winter, and but rarely seen in summer ; is remarkable for the great size of its eye, length of its wings, and shortness of its toes. The breadth of its head is likewise uncommon.

The black hawk is twenty-one inches long, and four feet two inches in extent ; bill, bluish black ; cere, and sides of the mouth, orange yellow ; feet, the same ; eye, very large ; iris, bright hazel ; cartilage overhanging the eye, prominent, of a dull greenish colour ; general colour above, brown black, slightly dashed with dirty white ; nape of the neck, pure white under the surface ; front, white ; whole lower parts, black, with slight tinges of brown ; and a few circular touches of the same on the femorals ; legs, feathered to the toes, and black, touched with brownish ; the wings reach rather beyond the tip of the tail ; the five first primaries are white on their inner vanes ; tail, rounded at the end, deep black, crossed with five narrow bands of pure white, and broadly tipt with dull white ; vent, black, spotted with white ; inside vanes of the primaries, snowy ; claws, black, strong, and sharp ; toes, remarkably short.

I strongly suspect this bird to be of the very same species with the next, though both were found to be males. Although differing greatly in plumage, yet, in all their characteristic features, they strikingly resemble each other. The chocolate-coloured hawk of Pennant, and St John's falcon of the same author (Arct. Zool., No. 93 and 94), are doubtless varieties of this; and, very probably, his rough-legged falcon also. His figures, however, are bad, and ill calculated to exhibit the true form and appearance of the bird.

This species is a native of North America alone. We have no account of its ever having been seen in any part of Europe; nor have we any account of its place or manner of breeding.

BLACK HAWK.

PLATE LIII.—FIG. 2.—YOUNG.

Peale's Museum, No. 405.

BUTEO SANCTI JOHANNIS.—YOUNG.—BONAPARTE.

Falco (sub-genus Buteo) Sancti Johannis, young, *Bonap. Synop.* p. 32.

THIS is probably a younger bird of the preceding species, being, though a male, somewhat less than its companion. Both were killed in the same meadow, at the same place and time. In form, features, and habitudes, it exactly agreed with the former.

This bird measures twenty inches in length, and in extent, four feet; the eyes, bill, cere, toes, and claws, were as in the preceding; head above, white, streaked with black and light brown; along the eyebrows, a black line; cheeks, streaked like the head; neck, streaked with black and reddish brown, on a pale yellowish white ground; whole upper parts, brown black, dashed with brownish white and pale ferruginous; tail, white for half its length, ending in brown, marked with one or two bars of dusky and a larger bar of black, and tipt with dull white; wings as in the preceding, their lining variegated with black, white, and ferruginous; throat and breast, brownish yellow, dashed with black; belly, beautifully variegated with spots of white, black, and pale ferruginous; femorals and feathered legs, the same, but rather darker; vent, plain brownish white.

1 Black Hawk. 2 Variety of d? 3 Red shouldered H. 4 Female Baltimore Oriole. 5 Female Towheé Bunting.

The original colour of these birds in their young state may probably be pale brown, as the present individual seemed to be changing to a darker colour on the neck and sides of the head. This change, from pale brown to black, is not greater than some of the genus are actually known to undergo. One great advantage of examining living or newly killed specimens is, that whatever may be the difference of colour between any two, the eye, countenance, and form of the head instantly betray the common family to which they belong ; for this family likeness is never lost in the living bird, though in stuffed skins and preserved specimens it is frequently entirely obliterated. I have no hesitation, therefore, in giving it as my opinion that the present and preceding birds are of the same species, differing only in age, both being males. Of the female I am unable at present to speak.

Pennant, in his account of the chocolate-coloured hawk, which is, very probably, the same with the present and preceding species, observes that it preys much on ducks, sitting on a rock, and watching their rising, when it instantly strikes them.

While traversing our sea-coast and salt marshes, between Cape May and Egg Harbour, I was everywhere told of a *duck hawk*, noted for striking down ducks on wing, though flying with their usual rapidity. Many extravagances were mingled with these accounts, particularly that it always struck the ducks with its breast bone, which was universally said to project several inches, and to be strong and sharp. From the best verbal descriptions I could obtain of this hawk, I have strong suspicions that it is no other than the *black hawk*, as its wings were said to be long and very pointed, the colour very dark, the size nearly alike, and several other traits given, that seemed particularly to belong to this species. As I have been promised specimens of this celebrated hawk next winter, a short time will enable me to determine the matter more satisfactorily. Few gunners in that quarter are unacquainted with the *duck hawk*, as it often robs them of their wounded birds before they are able to reach them.

Since writing the above, I have ascertained that the *duck hawk* is not this species, but the celebrated peregrine falcon, a figure and description of which will be given in our third volume.

RED-SHOULDERED HAWK. (*Falco lineatus.*)

PLATE LIII.—Fig. 3.

Arct. Zool. p. 206, No. 102.—*Lath.* i. 56, No. 36.—*Turt. Syst.* p. 153.—
Peale's Museum, No. 205.

BUTEO? LINEATUS.—Jardine.*

Falco (sub-genus Circus) hyemalis, *Bonap. Synop.* p. 33.—Red-shouldered Hawk,
Aud. pl. 56, male and female; *Orn. Biog.* i. p. 296.

This species is more rarely met with than either of the former. Its haunts are in the neighbourhood of the sea. It preys on larks, sandpipers, and the small ringed plover, and frequently on ducks. It flies high and irregularly, and not in the sailing manner of the long-winged hawks. I have occasionally observed this bird near Egg Harbour, in New

* This bird is certainly distinct from the *F. hyemalis* of this volume; and, independent of the distinctions of plumage, the very different habits of both pointed out by Mr Audubon can hardly be reconciled. All the characters and habits of the bird lean much more to the goshawks; it delights in woody countries, builds on trees, and is much more active. The plumage generally is that of the buzzards and *Circi;* but the under parts present a combination of the transverse barring of *Astur.* In addition to the description of Wilson, Audubon observes, that this bird is rarely observed in the middle districts, where, on the contrary, the winter falcon usually makes its appearance from the north at the approach of autumn. " It is one of the most noisy of its genus, during spring especially, when it would be difficult to walk the skirts of woods bordering a large plantation, without hearing its discordant shrill notes, *ka-hee, ka-hee,* as it sails in rapid circles at a very great elevation. The interior of the woods seems the fittest haunts for the red-shouldered hawk, where they also breed. The nest is seated near the extremity of a large branch, and is as bulky as that of a common crow. It is formed externally of dry sticks and Spanish moss, and is lined with withered grass and fibrous roots. The female lays four eggs, sometimes five; they are of a broad oval form, granulated all over, pale blue, faintly blotched with brownish red at the smaller end."—Ed.

Jersey, and once in the meadows below this city. This hawk was first transmitted to Great Britain by Mr Blackburne, from Long Island, in the State of New York. With its manner of building, eggs, &c., we are altogether unacquainted.

The red-shouldered hawk is nineteen inches in length ; the head and back are brown, seamed and edged with rusty ; bill, blue black ; cere and legs, yellow ; greater wing-coverts and secondaries, pale olive brown, thickly spotted on both vanes with white and pale rusty ; primaries, very dark, nearly black, and barred or spotted with white ; tail, rounded, reaching about an inch and a half beyond the wings, black, crossed by five bands of white, and broadly tipt with the same ; whole breast and belly, bright rusty, speckled and spotted with transverse rows of white, the shafts black ; chin and cheeks, pale brownish, streaked also with black ; iris, reddish hazel ; vent, pale ochre, tipt with rusty ; legs, feathered a little below the knees, long ; these and the feet, a fine yellow ; claws, black ; femorals, pale rusty, faintly barred with a darker tint.

In the month of April I shot a female of this species, and the only one I have yet met with, in a swamp, seven or eight miles below Philadelphia. The eggs were, some of them, nearly as large as peas ; from which circumstance I think it probable they breed in such solitary parts even in this State. In colour, size, and markings, it differed very little from the male described above. The tail was scarcely quite so black, and the white bars not so pure ; it was also something larger.

FEMALE BALTIMORE ORIOLE. (*Oriolus Baltimorus.*)

PLATE LIII.—Fig. 4.

Amer. Orn. vol. i. p. 23.

ICTERUS BALTIMORE.—Daudin.

The history of this beautiful species has been particularly detailed in the first volume of the present work ; * to this repre-

* See Vol. I. p. 16.

sentation of the female, drawn of half the size of nature, a few particulars may be added. The males generally arrive several days before the females, saunter about their wonted places of residence, and seem lonely, and less sprightly than after the arrival of their mates. In the spring and summer of 1811, a baltimore took up its abode in Mr Bartram's garden, whose notes were so singular as particularly to attract my attention ; they were as well known to me as the voice of my most intimate friend. On the 30th of April 1812, I was again surprised and pleased at hearing this same baltimore in the garden, whistling his identical old chant ; and I observed that he particularly frequented that quarter of the garden where the tree stood, on the pendant branches of which he had formed his nest the preceding year. This nest had been taken possession of by the house wren, a few days after the baltimore's brood had abandoned it ; and, curious to know how the little intruder had furnished it within, I had taken it down early in the fall, after the wren herself had also raised a brood of six young in it, and which was her second that season. I found it stript of its original lining, floored with sticks or small twigs, above which were laid feathers ; so that the usual complete nest of the wren occupied the interior of that of the baltimore.

The chief difference between the male and female baltimore oriole is the superior brightness of the orange colour of the former to that of the latter. The black on the head, upper part of the back, and throat of the female, is intermixed with dull orange ; whereas, in the male, those parts are of a deep shining black ; the tail of the female also wants the greater part of the black, and the whole lower parts are of a much duskier orange.

I have observed that these birds are rarely seen in pine woods, or where these trees generally prevail. On the ridges of our high mountains they are seldom to be met with. In orchards and on well-cultivated farms they are most numerous, generally preferring such places to build in, rather than the woods or forest.

FEMALE TOWHE BUNTING. (*Emberiza erythropthalma.*)

PLATE LIII.—Fig. 5.

Amer. Orn. vol. ii. p. 35.—*Turt. Syst.* p. 534.—*Peale's Museum,* No. 5970.

PIPILO ERYTHROPTHALMA.—Vieillot.

This bird differs considerably from the male in colour, and has, if I mistake not, been described as a distinct species by European naturalists, under the appellation of the "*Rusty Bunting.*" The males of this species, like those of the preceding, arrive several days sooner than the females. In one afternoon's walk through the woods, on the 23d of April, I counted more than fifty of the former, and did not observe any of the latter, though I made a very close search for them. This species frequents in great numbers the barrens covered with shrub oaks; and inhabits even to the tops of our mountains. They are almost perpetually scratching among the fallen leaves, and feed chiefly on worms, beetles, and gravel. They fly low, flirting out their broad white-streaked tail, and uttering their common note *tow-heé.* They build always on the ground, and raise two broods in the season. For a particular account of the manners of this species, see our history of the male, Vol. I. p. 185.

The female towhe is eight inches long, and ten inches in extent; iris of the eye, a deep blood colour; bill, black; plumage above and on the breast, a dark reddish drab, reddest on the head and breast; sides under the wings, light chestnut; belly, white; vent, yellow ochre; exterior vanes of the tertials, white; a small spot of white marks the primaries immediately below their coverts, and another slighter streak crosses them in a slanting direction; the three exterior tail-feathers are tipt with white; the legs and feet, flesh-coloured.

This species seems to have a peculiar dislike to the sea-coast, as in the most favourable situations in other respects, within several miles of the sea, it is scarcely ever to be met with.

Scarcity of its particular kinds of favourite food in such places may probably be the reason, as it is well known that many kinds of insects, on the larvæ of which it usually feeds, carefully avoid the neighbourhood of the sea.

BROAD-WINGED HAWK. (*Falco Pennsylvanicus.*)

PLATE LIV.—Fig. 1.

Peale's Museum, No. 407.

ASTUR ? LATISSIMUS.—Jardine.*

Falco latissimus, *Ord's reprint of Wilson.*—Falco (sub-genus Astur) Pennsylvanicus, *Bonap. Synop.* p. 29.—The Broad-winged Hawk, *Aud.* pl. 91, male and female ; *Orn. Biog.* i. p. 461.

This new species, as well as the rest of the figures on the same plate, is represented of the exact size of life. The hawk was shot on the 6th of May in Mr Bartram's woods, near the Schuylkill, and was afterwards presented to Mr Peale, in whose collection it now remains. It was perched on the dead limb of a high tree, feeding on something which was afterwards found to be the meadow mouse (figured in plate 50). On my approach, it uttered a whining kind of whistle, and flew off to another tree, where I followed and shot it. Its great breadth

* Mr Ord's name of *latissimus* is the most proper for this hawk. Wilson seems inadvertently to have given the name of *Pennsylvanicus* to two species, and the latter being applied to the adult plumage, and *velox* to the young, the former has been retained by Temminck and the authors of the " Northern Zoology," while Ord seems to have the merit of discriminating the large species, and giving it the title above adopted. I have taken *Astur,* on the authority of Bonaparte, for its generic appellation ; though the habits and kind of food ally it more to the buzzards, it is one of those birds with dubious and combined characters. Mr Audubon describes it as of a quiet and sluggish disposition, allowing itself to be tormented by the little sparrow hawk and tyrant flycatcher. It feeds on animals and birds, and also on frogs and snakes ; breeds on trees ; the nest is placed near the stem or trunk, and is composed of dry thistles, and lined with numerous small roots and large feathers ; the eggs are four or five, of a dull grayish white, blotched with dark brown.—Ed.

1. Broad-winged Hawk. 2. Chuck-wills-widow. 3. Cape-May Warbler. 4. Female Black-cap W.

54.

of wing, or width of the secondaries, and also of its head and body, when compared with its length, struck me as peculiarities. It seemed a remarkably strong-built bird, handsomely marked, and was altogether unknown to me. Mr Bartram, who examined it very attentively, declared he had never before seen such a hawk. On the afternoon of the next day, I observed another, probably its mate or companion, and certainly one of the same species, sailing about over the same woods. Its motions were in wide circles, with unmoving wings, the exterior outline of which seemed a complete semicircle. I was extremely anxious to procure this also if possible; but it was attacked and driven away by a king-bird before I could effect my purpose, and I have never since been fortunate enough to meet with another. On dissection, the one which I had shot proved to be a male.

In size this hawk agrees nearly with the *Buzzardet* (*Falco albidus*) of Turton, described also by Pennant;* but either the descriptions of these authors are very inaccurate, the change of colour which that bird undergoes very great, or the present is altogether a different species. Until, however, some other specimens of this hawk come under my observation, I can only add to the figure here given, and which is a good likeness of the original, the following particulars of its size and plumage :—

Length, fourteen inches; extent, thirty-three inches; bill, black, blue near the base, slightly toothed; cere and corners of the mouth, yellow; irides, bright amber; frontlet and lores, white; from the mouth backwards runs a streak of blackish brown; upper parts, dark brown, the plumage tipt and the head streaked with whitish; almost all the feathers above are spotted or barred with white, but this is not seen unless they be separated by the hand; head, large, broad, and flat; cere very broad; the nostril also large; tail short, the exterior and interior feathers somewhat the shortest, the others rather longer, of a full black, and crossed with two bars of white,

* Arctic Zoology, No. 109.

tipt also slightly with whitish ; tail coverts, spotted with white;
wings, dusky brown, indistinctly barred with black ; greater
part of the inner vanes, snowy ; lesser coverts and upper part
of the back, tipt and streaked with bright ferruginous ; the
bars of black are very distinct on the lower side of the wing ;
lining of the wing, brownish white, beautifully marked with
small arrow-heads of brown ; chin, white, surrounded by
streaks of black ; breast and sides, elegantly spotted with
large arrow-heads of brown, centred with pale brown ; belly
and vent, like the breast, white, but more thinly marked with
pointed spots of brown ; femorals, brownish white, thickly
marked with small touches of brown and white ; vent, white ;
legs, very stout ; feet, coarsely scaled, both of a dirty orange
yellow ; claws, semicircular, strong and very sharp, hind one
considerably the largest.

While examining the plumage of this bird, a short time
after it was shot, one of those winged ticks with which many
of our birds are infested appeared on the surface of the
feathers, moving about, as they usually do, backwards or side-
ways like a crab, among the plumage with great facility.
The fish hawk, in particular, is greatly pestered with these
vermin, which occasionally leave him, as suits their convenience.
A gentleman who made the experiment assured me, that on
plunging a live fish hawk under water, several of these winged
ticks remained hovering over the spot, and, the instant the
hawk rose above the surface, darted again among his plumage.
The experiment was several times made, with the like result.
As soon, however, as these parasites perceive the dead body of
their patron beginning to become cold, they abandon it ; and,
if the person who holds it have his head uncovered, dive instantly among his hair, as I have myself frequently experienced;
and, though driven from thence, repeatedly return, till they
are caught and destroyed. There are various kinds of these
ticks : the one found on the present hawk is figured beside
him. The head and thorax were light brown ; the legs, six
in number, of a bright green, their joints moving almost hori-

zontally, and thus enabling the creature to pass with the greatest ease between the laminæ of feathers ; the wings were single, of a dark amber colour, and twice as long as the body, which widened towards the extremity, where it was slightly indented ; feet, two clawed.

This insect lived for several days between the crystal and dial-plate of a watch carried in the pocket ; but being placed for a few minutes in the sun, fell into convulsions and died.

CHUCK-WILL'S-WIDOW. *(Caprimulgus Carolinensis.)*

PLATE LIV.—Fig. 2.

Peale's Museum, No. 7723.

CAPRIMULGUS CAROLINENSIS.—Brisson.*

Caprimulgus Carolinensis, *Lath. Gen. Hist.*—Caprimulgus rufus, *Vieill.* (auct.
Bonap.) *Bonap. Synop.* p. 61.—Chuck-will's-widow, *Aud.* pl. 52, male
and female ; *Orn. Biog.* i. p. 273.

This solitary bird is rarely found to the north of James River, in Virginia, on the sea-board, or of Nashville, in the

* According to Mr Audubon, this species, when disturbed or annoyed about the nest, removes its eggs or young to a distance. This circumstance seems known to the negroes and American farmers, who give various accounts of the mode in which it is performed. Mr Audubon could not satisfy himself as to the truth of these accounts, and resolved to watch and judge for himself. What follows is the result of his observation :—

" When the chuck-will's-widow, either male or female (for each sits alternately), has discovered that the eggs have been touched, it ruffles its feathers, and appears extremely dejected for a minute or two, after which it emits a low murmuring cry, scarcely audible to me, as I lay concealed at a distance of not more than eighteen or twenty yards. At this time, I have seen the other parent reach the spot, flying so low over the ground, that I thought its little feet must have touched it as it skimmed along, and after a few low notes and some gesticulations, all indicative of great distress, take an egg in its large mouth, the other bird doing the same, when they would fly off together, skimming closely over the ground, until they disappeared among the branches and trees."—Ed.

State of Tennessee, in the interior ; and no instance has come to my knowledge of its having been seen either in New Jersey, Pennsylvania, or Maryland. On my journey south, I first met with it between Richmond and Petersburg, in Virginia, and also on the banks of the Cumberland in Tennesee.

Mr Pennant has described this bird under the appellation of the "short-winged goatsucker" (Arct. Zool., No. 336), from a specimen which he received from Dr Garden of Charleston, South Carolina ; but in speaking of its manners, he confounds it with the whip-poor-will, though the latter is little more than half the cubic bulk of the former, and its notes altogether different. "In South Carolina," says this writer, speaking of the present species, "it is called, from one of its notes, *chuck, Chuck-will's-widow*, and, in the northern provinces, *whip-poor-will*, from the resemblance which another of its notes bears to those words" (Arct. Zool., p. 434). He then proceeds to detail the manners of the common whip-poor-will, by extracts from Dr Garden and Mr Kalm, which clearly prove that all of them were personally unacquainted with that bird, and had never seen or examined any other than two of our species, the short-winged or chuck-will's-widow, and the long-winged or night hawk, to both of which they indiscriminately attribute the notes and habits of the whip-poor-will.

The chuck-will's-widow, so called from its notes, which seem exactly to articulate those words, arrives on the seacoast of Georgia about the middle of March, and in Virginia early in April. It commences its singular call generally in the evening, soon after sunset, and continues it, with short occasional interruptions, for several hours. Towards morning these repetitions are renewed, and continue until dawn has fairly appeared. During the day it is altogether silent. This note or call instantly attracts the attention of a stranger, and is strikingly different from that of the whip-poor-will. In sound and articulation it seems plainly to express the words which have been applied to it (*chuck-will's-widow*), pronouncing each syllable leisurely and distinctly, putting the principal emphasis on the last word. In a still evening it may be heard at the distance of nearly a mile, the tones of its voice being stronger and more full than those of the whip-poor-will, who utters his with much greater rapidity. In the Chickasaw country, and throughout the whole Mississippi territory, I found the present species very numerous in the months of April and May, keeping up a continual noise during the whole evening, and, in moonlight, throughout the whole of the night.

The flight of this bird is low, skimming about at a few feet above the surface of the ground, frequently settling on old logs, or on the fences, and from thence sweeping around in pursuit of various winged insects that fly in the night. Like the whip-poor-will, it prefers the declivities of glens and other deeply shaded places, making the surrounding mountains ring with echoes the whole evening. I several times called the attention of the Chickasaws to the notes of this bird, on which occasions they always assumed a grave and thoughtful aspect; but it appeared to me that they made no distinction between the two species; so that whatever superstitious notions they may entertain of the one, are probably applied to the other.

This singular genus of birds, formed to subsist on the superabundance of nocturnal insects, are exactly and surprisingly fitted for their peculiar mode of life. Their flight is low, to accommodate itself to their prey; silent, that they may be the better concealed, and sweep upon it unawares; their sight, most acute in the dusk, when such insects are abroad; their evolutions, something like those of the bat, quick and sudden; their mouths, capable of prodigious expansion, to seize with more certainty, and furnished with long branching hairs or bristles, serving as palisadoes to secure what comes between them. Reposing so much during the heats of day, they are much invested with vermin, particularly about the head, and are provided with a comb on the inner edge of the middle claw, with which they are often employed in ridding themselves of these pests, at least when

in a state of captivity. Having no weapons of defence
except their wings, their chief security is in the solitude of
night, and in their colour and close retreats by day; the
former so much resembling that of dead leaves of various
hues, as not to be readily distinguished from them even when
close at hand.

The chuck-will's-widow lays its eggs, two in number, on
the ground generally, and, I believe, always in the woods; it
makes no nest; the eggs are of a dull olive colour, sprinkled
with darker specks; are about as large as those of a pigeon,
and exactly oval. Early in September they retire from the
United States.

This species is twelve inches long, and twenty-six in
extent; bill, yellowish, tipt with black; the sides of the mouth
are armed with numerous long bristles, strong, tapering, and
furnished with finer hairs branching from each; cheeks and
chin, rust colour, specked with black; over the eye extends
a line of small whitish spots; head and back, very deep
brown, powdered with cream, rust, and bright ferruginous,
and marked with long ragged streaks of black; scapulars,
broadly spotted with deep black, bordered with cream, and
interspersed with whitish; the plumage of that part of the
neck which falls over the back, is long, something like that
of a cock, and streaked with yellowish brown; wing quills,
barred with black and bright rust; tail, rounded, extending
about an inch beyond the tips of the wings; it consists of ten
feathers, the four middle ones are powdered with various
tints of ferruginous, and elegantly marked with fine zigzag
lines, and large herring-bone figures of black; exterior edges
of the three outer feathers, barred like the wings; their
interior vanes, for two-thirds of their length, are pure snowy
white, marbled with black, and ferruginous at the base; this
white spreads over the greater part of the three outer feathers
near their tips; across the throat is a slight band or mark of
whitish; breast, black, powdered with rust; belly and vent,
lighter; legs, feathered before nearly to the feet, which are of

a dirty purplish flesh colour ; inner side of the middle claw, deeply pectinated.

The female differs chiefly in wanting the pure white on the three exterior tail-feathers, these being more of a brownish cast.

CAPE MAY WARBLER. (*Sylvia maritima.*)

PLATE LIV.—FIG. 3.

SYLVICOLA MARITIMA.—JARDINE.[*]

Sylvia maritima, *Bonap. Synop.* p. 79.—The Carbonated Warbler? *Aud.* pl. 60, male ; *Orn. Biog.* i. p. 308.

THIS new and beautiful little species was discovered in a maple swamp in Cape May county, not far from the coast, by Mr George Ord of Philadelphia, who accompanied me on a shooting excursion to that quarter in the month of May last (1811). Through the zeal and activity of this gentleman I succeeded in procuring many rare and elegant birds among the sea islands and extensive salt marshes that border that part of the Atlantic, and much interesting information relative to their nests, eggs, and particular habits. I have also at various times been favoured with specimens of other birds from the same friend ; for all which I return my grateful acknowledgments.

The same swamp that furnished us with this elegant little stranger, and indeed several miles around it, were ransacked by us both for another specimen of the same, but without success. Fortunately it proved to be a male,[†] and being in excellent plumage, enabled me to preserve a faithful portrait of the original.

Whether this be a summer resident in the lower parts of

[*] The Prince of Musignano first directed my attention to the identity of this bird of Wilson and Audubon's carbonated warbler. I cannot perceive any essential difference, that is, judging from the two plates and descriptions. Mr Audubon procured his species in the State of Kentucky.—ED.

[†] Female figured Vol. III.

New Jersey, or merely a transient passenger to a more northern climate, I cannot with certainty determine. The spring had been remarkably cold, with long and violent north-east storms, and many winter birds, as well as passengers from the south, still lingered in the woods as late as the 20th of May, gleaning, in small companies, among the opening buds and infant leaves, and skipping nimbly from twig to twig, which was the case with the bird now before us when it was first observed. Of its notes or particular history I am equally uninformed.

The length of this species is five inches and a half; extent, eight and a half; bill and legs, black; whole upper part of the head, deep black; line from the nostril over the eye, chin, and sides of the neck, rich yellow; ear-feathers, orange, which also tints the back part of the yellow line over the eye; at the anterior and posterior angle of the eye is a small touch of black; hind head and whole back, rump, and tail-coverts, yellow olive, thickly streaked with black; the upper exterior edges of several of the greater wing-coverts are pure white, forming a broad bar on the wing, the next superior row being also broadly tipt with white; rest of the wing, dusky, finely edged with dark olive yellow; throat and whole breast, rich yellow, spreading also along the sides under the wings, handsomely marked with spots of black running in chains; belly and vent, yellowish white; tail, forked, dusky black, edged with yellow olive, the three exterior feathers on each side marked on their inner vanes with a spot of white. The yellow on the throat and sides of the neck reaches nearly round it, and is very bright.

FEMALE BLACK-POLL WARBLER. *(Sylvia striata.)*

PLATE LIV.—Fig. 4.

Amer. Orn. vol. iv. p. 40.

SYLVICOLA STRIATA.—Swainson.

This bird was shot in the same excursion with the preceding, and is introduced here for the purpose of preventing future collectors, into whose hands specimens of it may chance to fall, from considering it as another and a distinct species. Its history, as far as was then known, has been detailed in a preceding part of this work, supra, p. 32. Of its nest and eggs I am still ignorant. It doubtless breeds both here and in New Jersey, having myself found it in both places during the summer. From its habit of keeping on the highest branches of trees, it probably builds in such situations, and its nest may long remain unknown to us.

Pennant, who describes this species, says that it inhabits, during summer, Newfoundland and New York, and is called in the last, *sailor*. This name, for which, however, no reason is given, must be very local, as the bird itself is one of those silent, shy, and solitary individuals, that seek the deep retreats of the forest, and are known to few or none but the naturalist.

Length of the female black-cap, five inches and a quarter, extent, eight and a quarter; bill, brownish black; crown, yellow olive, streaked with black; back, the same, mixed with some pale slate; wings, dusky brown, edged with olive; first and second wing-coverts, tipt with white; tertials, edged with yellowish white; tail-coverts, pale gray; tail, dusky, forked, the two exterior feathers marked on their inner vanes with a spot of white; round the eye is a whitish ring; cheeks and sides of the breast, tinged with yellow, and slightly spotted with black; chin, white, as are also the belly and vent; legs and feet, dirty orange.

The young bird of the first season and the female, as is usually the case, are very much alike in plumage. On their arrival early in April, the black feathers on the crown are frequently seen coming out, intermixed with the former ash-coloured ones.

This species has all the agility and many of the habits of the flycatcher.

[Parts VII. and VIII. of this work, commencing with the next description (ring-tailed eagle), seem to have been finished more hurriedly, and contain greater mistakes in the nomenclature, than any of the preceding ones ; the descriptions, however, are alike vivid and well drawn. In 1824 Mr Ord, the personal friend of Wilson, undertook, at the request of the publisher, to improve these two parts, and they were accordingly republished with that gentleman's additions. We have thought it better to print from the original edition, as showing the true opinions of its author, but have occasionally inserted, at the conclusion of the descriptions, the observations of Mr Ord, taken from his reprint. —Ed.]

RING-TAILED EAGLE. (*Falco fulvus.*)

PLATE LV.—Fig. 1.

Linn. Syst. 125.—Black Eagle, *Arct. Zool.* p. 195, No. 87.—*Lath.* i. 32, No. 6. —White-tailed Eagle, *Edw.* i. 1.—L'Aigle commun, *Buff.* i. 86, *Pl. enl.* 409. —*Bewick*, i. p. 49.—*Turt. Syst.* p. 145.—*Peale's Museum*, No. 84.

AQUILA CHRYSAETUS.—Willoughby.*

Ayle royal, *Temm. Man. d'Orn.* i. p. 38.—Aquila chrysaëtos, *Flem.* 138.—*Zool.* p. 52.—Golden Eagle, *Selby, Illust. Br. Orn.* pl. 1 and 2, the young and adult, part i. p. 4.—Aquila chrysaëtos? *North. Zool.* ii. p. 12.—*Bonap. Synop.* p. 24.

THE reader is now presented with a portrait of this celebrated eagle, drawn from a fine specimen shot in the county of Montgomery, Pennsylvania. The figure here given, though

* Wilson, like many other ornithologists, imagined that the ring-tailed and golden eagles constituted two species. Temminck, I believe, first asserted the fact of their being identical, and the attention of naturalists in this country was attracted to the circumstance by the different opinions entertained by Mr James Wilson and Mr Selby. The latter gentleman has long since satisfactorily proved their identity from observation, and the numerous specimens kept alive in various parts of Britain have set the question completely at rest. The ring-tail is the young of

1.Ring-tail Eagle. 2.Sea Eagle.

55

reduced to one-third the size of life, is strongly characteristic of its original. With respect to the habits of the species, such particulars only shall be selected as are well authenticated, rejecting whatever seems vague, or savours too much of the marvellous.

This noble bird, in strength, spirit, and activity, ranks among the first of its tribe. It is found, though sparingly, dispersed over the whole temperate and arctic regions, particularly the latter; breeding on high precipitous rocks, always preferring a mountainous country. In its general appearance, it has great resemblance to the golden eagle, from which, however, it differs in being rather less, as also in the colours and markings of the tail, and, as it is said, in being less noisy. When young, the colour of the body is considerably lighter, but deepens into a blackish brown as it advances in age.

The tail-feathers of this bird are highly valued by the various tribes of American Indians for ornamenting their calumets or pipes of peace. Several of these pipes, which were brought from the remote regions of Louisiana by Captain

the first year, and as such is correctly figured by our author. In a wild state, three years are required to complete the clouded barring, the principal mark of the adults, and which, even after that period, increase in darkness of colour. When kept in confinement, the change is generally longer in taking place; and I have seen it incomplete at six years. It commences by an extension of the bar at the end of the tail, and by additional cloudings on the white parts, which increase yearly until perfected. This bird does not seem very common in any part of America, and is even more rarely met with in the adult plumage. It was found on the borders of the Rocky Mountains by the Overland Arctic Expedition, and is known also on the plains of the Saskatchewan.

The noble bearing and aspect of the eagles and falcons have always associated them, among rude nations and in poetical comparisons, with the true courage of the warrior and the magnanimity of the prince or chief. The young Indian warrior glories in his eagle's plume, as the most honourable ornament with which he can adorn himself; the dress of a Highland chieftain is incomplete without this badge of high degree. The feathers of the war eagle are also used at the propitiatory sacrifices, and so highly are they prized, that a valuable horse is sometimes exchanged for the tail of a single eagle.—Ed.

Lewis, are now deposited in Mr Peale's Museum, each of which
has a number of the tail-feathers of this bird attached to it.
The northern as well as the southern Indians seem to follow
the like practice, as appears by the numerous calumets, formerly belonging to different tribes, to be seen in the same
magnificent collection.

Mr Pennant informs us that the independent Tartars train
this eagle for the chase of hares, foxes, wolves, antelopes, &c.,
and that they esteem the feathers of the tail the best for pluming
their arrows. The ring-tail eagle is characterised by all as a
generous, spirited, and docile bird ; and various extraordinary
incidents are related of it by different writers, not, however,
sufficiently authenticated to deserve repetition. The truth
is, the solitary habits of the eagle now before us, the vast
inaccessible cliffs to which it usually retires, united with the
scarcity of the species in those regions inhabited by man, all
combine to render a particular knowledge of its manners
very difficult to be obtained. The author has once or twice
observed this bird sailing along the alpine declivities of the
White Mountains of New Hampshire, early in October, and
again over the Highlands of Hudson's river, not far from
West Point. Its flight was easy, in high circuitous sweeps ;
its broad white tail, tipped with brown, expanded like a fan.
Near the settlements on Hudson's Bay, it is more common,
and is said to prey on hares, and the various species of grouse
which abound there. Buffon observes, that though other
eagles also prey upon hares, this species is a more fatal enemy
to those timid animals, which are the constant object of their
search, and the prey which they prefer. The Latins, after
Pliny, termed the eagle *Valeria quasi valens viribus,* because
of its strength, which appears greater than that of other eagles
in proportion to its size.

The ring-tail eagle measures nearly three feet in length ;
the bill is of a brownish horn colour ; the cere, sides of the
mouth, and feet, yellow ; iris of the eye, reddish hazel, the eye
turned considerably forwards ; eyebrow, remarkably prominent,

projecting over the eye, and giving a peculiar sternness to the aspect of the bird; the crown is flat; the plumage of the head, throat, and neck, long and pointed; that on the upper part of the head and neck, very pale ferruginous; fore part of the crown, black; all the pointed feathers are shafted with black; whole upper parts, dark blackish brown; wings, black; tail, rounded, long, of a white or pale cream colour, minutely sprinkled with specks of ash and dusky, and ending in a broad band of deep dark brown, of nearly one-third its length; chin, cheeks, and throat, black; whole lower parts, a deep dark brown, except the vent and inside of the thighs, which are white, stained with brown; legs, thickly covered to the feet with brownish white down or feathers; claws, black, very large, sharp, and formidable, the hind one full two inches long.

The ring-tail eagle is found in Russia, Switzerland, Germany, France, Scotland, and the northern parts of America. As Marco Polo, in his description of the customs of the Tartars, seems to allude to this species, it may be said to inhabit the whole circuit of the arctic regions of the globe. The golden eagle, on the contrary, is said to be found only in the more warm and temperate countries of the ancient continent.[*] Later discoveries, however, have ascertained it to be also an inhabitant of the United States.

SEA EAGLE. *(Falco ossifragus.)*

PLATE LV.—Fig. 2.

Arct. Zool. p. 194, No. 86.—*Linn. Syst.* 124.—*Lath.* i. 30.—L'Orfraie, *Buff.* i. 112, pl. 3, *Pl. enl.* 12, 415.—*Br. Zool.* i. No. 44.—*Bewick,* i. 53.—*Turt. Syst.* p. 144.—*Peale's Museum,* No. 80.

HALIÆETUS LEUCOCEPHALUS.—Savigny.[†]

Bald Eagle, Falco leucocephalus, young, *Ord's reprint.*

This eagle inhabits the same countries, frequents the same situations, and lives on the same kind of food, as the bald eagle, with whom it is often seen in company. It resembles

* Buffon, vol. i. p. 56, Trans.

† See note to the adult, in this Volume, p. 89, for synoymns, &c.

this last so much in figure, size, form of the bill, legs, and
claws, and is so often seen associating with it, both along the
Atlantic coast and in the vicinity of our lakes and large rivers,
that I have strong suspicions, notwithstanding ancient and
very respectable authorities to the contrary, of its being the
same species, only in a different stage of colour.

That several years elapse before the young of the bald eagle
receive the white head, neck, and tail, and that, during the
intermediate period, their plumage strongly resembles that of
the sea eagle, I am satisfied from my own observation on
three several birds kept by persons of Philadelphia. One of
these, belonging to the late Mr Enslen, collector of natural
subjects for the Emperor of Austria, was confidently believed
by him to be the black or sea eagle until the fourth year,
when the plumage on the head, tail, and tail-coverts began
gradually to become white ; the bill also exchanged its dusky
hue for that of yellow ; and, before its death, this bird, which
I frequently examined, assumed the perfect dress of the full-
plumaged bald eagle. Another circumstance, corroborating
these suspicions, is the variety that occurs in the colours of
the sea eagle. Scarcely two of these are found to be alike,
their plumage being more or less diluted with white. In
some the chin, breast, and tail-coverts are of a deep brown ;
in others nearly white ; and in all evidently unfixed, and
varying to a pure white. Their place and manner of building, on high trees, in the neighbourhood of lakes, large rivers,
or the ocean, exactly similar to the bald eagle, also strengthens
the belief. At the celebrated Cataract of Niagara, great
numbers of these birds, called there gray eagles, are continually
seen sailing high and majestically over the watery tumult, in
company with the bald eagles, eagerly watching for the
mangled carcasses of those animals that have been hurried
over the precipice, and cast up on the rocks below by the
violence of the rapids. These are some of the circumstances
on which my suspicions of the identity of those two birds are
founded. In some future part of the work, I hope to be able
to speak with more certainty on this subject.

Were we disposed, after the manner of some, to substitute
for plain matters of fact all the narratives, conjectures, and
fanciful theories of travellers, voyagers, compilers, &c., relative
to the history of the eagle, the volumes of these writers, from
Aristotle down to his admirer, the Count de Buffon, would
furnish abundant materials for this purpose. But the author
of the present work feels no ambition to excite surprise and
astonishment at the expense of truth, or to attempt to elevate
and embellish his subject beyond the plain realities of nature.
On this account he cannot assent to the assertion, however
eloquently made, in the celebrated parallel drawn by the
French naturalist between the lion and the eagle, viz., that
the eagle, like the lion, "disdains the possession of that property which is not the fruit of his own industry, and rejects
with contempt the prey which is not procured by his own
exertions ; " since the very reverse of this is the case in the
conduct of the bald and the sea eagle, who, during the summer
months, are the constant robbers and plunderers of the osprey,
or fish hawk, by whose industry alone both are usually fed.
Nor that, " *though famished for want of prey, he disdains to
feed on carrion,*" since we have ourselves seen the bald eagle,
while seated on the dead carcass of a horse, keep a whole
flock of vultures at a respectful distance until he had fully
sated his own appetite. The Count has also taken great pains
to expose the ridiculous opinion of Pliny, who conceived that
the ospreys formed no separate race, and that they proceeded
from the intermixture of different species of eagles, the young
of which were not ospreys, only sea eagles ; " *which sea eagles,*"
says he, " *breed small vultures, which engender great vultures,
that have not the power of propagation.*" * But, while labour-
ing to confute these absurdities, the Count himself, in his
belief of an occasional intercourse between the osprey and the
sea eagle, contradicts all actual observation, and one of the
most common and fixed laws of nature ; for it may be safely
asserted, that there is no habit more universal among the

* Hist. Nat. lib. x. c. 3.

feathered race, in their natural state, than that chastity of attachment which confines the amours of individuals to those of their own species only. That perversion of nature produced by domestication is nothing to the purpose. In no instance have I ever observed the slightest appearance of a contrary conduct. Even in those birds which never build a nest for themselves, nor hatch their young, nor even pair, but live in a state of general concubinage,—such as the cuckoo of the old, and the cow bunting of the new continent,—there is no instance of a deviation from this striking habit. I cannot, therefore, avoid considering the opinion above alluded to, that " the male osprey, by coupling with the female sea eagle, produces sea eagles; and that the female osprey, by pairing with the male sea eagle, gives birth to ospreys," * or fish hawks, as altogether unsupported by facts, and contradicted by the constant and universal habits of the whole feathered race in their state of nature.

The sea eagle is said by Salerne to build on the loftiest oaks a very broad nest, into which it drops two large eggs, that are quite round, exceedingly heavy, and of a dirty white colour. Of the precise time of building we have no account; but something may be deduced from the following circumstance :—In the month of May, while on a shooting excursion along the sea-coast, not far from Great Egg Harbour, accompanied by my friend Mr Ord, we were conducted about a mile into the woods to see an eagle's nest. On approaching within a short distance of the place, the bird was perceived slowly retreating from the nest, which, we found, occupied the centre of the top of a very large yellow pine. The woods were cut down and cleared off for several rods around the spot, which, from this circumstance, and the stately, erect trunk, and large crooked, wriggling branches of the tree, surmounted by a black mass of sticks and brush, had a very singular and picturesque effect. Our conductor had brought an axe with him, to cut down the tree; but my companion, anxious to

* Buffon, vol. i. p. 80, Trans.

save the eggs or young, insisted on ascending to the nest, which he fearlessly performed, while we stationed ourselves below, ready to defend him in case of an attack from the old eagles. No opposition, however, was offered ; and on reaching the nest, it was found, to our disappointment, empty. It was built of large sticks, some of them several feet in length ; within which lay sods of earth, sedge, grass, dry reeds, &c., piled to the height of five or six feet, by more than four in breadth. It was well lined with fresh pine tops, and had little or no concavity. Under this lining lay the recent exuviæ of the young of the present year, such as scales of the quill-feathers, down, &c. Our guide had passed this place late in February, at which time both male and female were making a great noise about the nest ; and, from what we afterwards learnt, it is highly probable it contained young, even at that early time of the season.

A few miles from this is another eagle's nest, built also on a pine tree, which, from the information received from the proprietor of the woods, had been long the residence of this family of eagles. The tree on which the nest was originally built, had been, for time immemorial, or at least ever since he remembered, inhabited by these eagles. Some of his sons cut down this tree to procure the young, which were two in number ; and the eagles soon after commenced building another nest on the very next adjoining tree, thus exhibiting a very particular attachment to the spot. The eagles, he says, make it a kind of *home* and *lodging place* in all seasons. This man asserts that the gray or sea eagles are the young of the bald eagle, and that they are several years old before they begin to breed. It does not drive its young from the nest, like the osprey or fish hawk, but continues to feed them long after they leave it.

The bird from which the figure in the plate was drawn, and which is reduced to one-third the size of life, measured three feet in length, and upwards of seven feet in extent. The bill was formed exactly like that of the bald eagle, but of a dusky brown colour; cere and legs, bright yellow; the latter, as in the bald eagle, feathered a little below the knee; irides, a bright straw colour; head above, neck, and back, streaked with light brown, deep brown, and white, the plumage being white, tipt and centred with brown; scapulars, brown; lesser wing-coverts, very pale, intermixed with white; primaries, black, their shafts brownish white; rump, pale brownish white; tail, rounded, somewhat longer than the wings, when shut, 'brown on the exterior vanes, the inner ones white, sprinkled with dirty brown; throat, breast, and belly, white, dashed and streaked with different tints of brown and pale yellow; vent, brown, tipt with white; femorals, dark brown, tipt with lighter; auriculars, brown, forming a bar from below the eye backwards; plumage of the neck, long, narrow, and pointed, as is usual with eagles, and of a brownish colour, tipt with white.

The sea eagle is said, by various authors, to hunt at night as well as during the day, and that, besides fish, it feeds on chickens, birds, hares, and other animals. It is also said to catch fish during the night; and that the noise of its plunging into the water is heard at a great distance. But, in the descriptions of these writers, this bird has been so frequently confounded with the osprey, as to leave little doubt that the habits and manners of the one have been often attributed to both, and others added that are common to neither.

[The following addition is made by Mr Ord, but I have in many instances found the reverse. I have had the golden eagle and peregrine perfectly tame, and even playful. Three sea eagles with me now are very savage:—" The bald eagle may be tamed, so as to become quite sociable, permitting one to handle it at pleasure, and even seeming pleased with such familiarities. The hawks, on the contrary, are apt to retain their savage nature under the kindest treatment; and, like the cat, will frequently remind one, on the slightest provocation, to beware of those powerful weapons with which nature has provided them."]

Drawn from Nature by A. Wilson.

Engraved by W.H.Lizars.

1 Esquimaux Curlew. 2 Red backed Snipe. 3 Semipalmated S. 4 Marbled Godwit.

56.

ESQUIMAUX CURLEW. (*Scolopax borealis.*)

PLATE LVI.—Fig. 1.

Arct. Zool. p. 461, No. 364.—*Lath.* iii.—*Turt. Syst.* p. 392.—*Peale's Museum,*
No. 4003.

NUMENIUS BOREALIS.—Latham.*

Numenius borealis, *Lath. Ind. Orn.* ii. p. 712.—*Bonap. Synop.* No. 244.—*North.
Zool.* ii. p. 378, pl. 55.

In prosecuting our researches among the feathered tribes
of this extensive country, we are at length led to the shores
of the ocean, where a numerous and varied multitude, sub-
sisting on the gleanings of that vast watery magazine of nature,
invite our attention, and, from their singularities and numbers,
promise both amusement and instruction. These we shall, as
usual, introduce in the order we chance to meet with them in
their native haunts. Individuals of various tribes thus pro-
miscuously grouped together, the peculiarities of each will
appear more conspicuous and striking, and the detail of their
histories less formal, as well as more interesting.

The Esquimaux curlew, or, as it is called by our gunners
on the sea-coast, the short-billed curlew, is peculiar to the
new continent. Mr Pennant, indeed, conceives it to be a
mere variety of the English whimbrel (*S. phæopus*); but,
among the great numbers of these birds which I have myself
shot and examined, I have never yet met with one corresponding to the descriptions given of the whimbrel, the
colours and markings being different, the bill much more

* This species has been by some supposed to be identical with the *N.
phæopus* of Europe, but I believe later investigations have proved that
it is entirely distinct, the whimbrel having not yet been found to
inhabit any part of America. The "Northern Zoology" mentions it as
inhabiting the barren lands within the arctic circle in summer, where it
feeds on insects and the berries of *Empetrum nigrum*. The Copper
Indians believe that this bird, and some others, betray the approach
of an enemy. Their nests and habits while breeding resemble those of
the common curlew.—Ed.

bent, and nearly an inch and a half longer; and the manners, in certain particulars, very different : these reasons have determined its claim to that of an independent species.

The short-billed curlew arrives in large flocks on the seacoast of New Jersey early in May, from the south, frequents the salt marshes, muddy shores and inlets, feeding on small worms and minute shell-fish. They are most commonly seen on mud flats at low water, in company with various other waders, and at high water roam along the marshes. They fly high, and with great rapidity. A few are seen in June, and as late as the beginning of July, when they generally move off towards the north. Their appearance on these occasions is very interesting : they collect together from the marshes as if by premeditated design, rise to a great height in the air, usually about an hour before sunset, and forming in one vast line, keep up a constant whistling on their way to the north, as if conversing with one another to render the journey more agreeable. Their flight is then more slow and regular, that the feeblest may keep up with the line of march ; while the glittering of their beautifully speckled wings, sparkling in the sun, produces altogether a very pleasing spectacle.

In the month of June, while the dew-berries are ripe, these birds sometimes frequent the fields, in company with the long-billed curlews, where brambles abound ; soon get very fat, and are at that time excellent eating. Those who wish to shoot them fix up a shelter of brushwood in the middle of the field, and by that means kill great numbers. In the early part of spring, and indeed during the whole time that they frequent the marshes feeding on shell-fish, they are much less esteemed for the table.

Pennant informs us that the Esquimaux curlews "were seen in flocks innumerable on the hills about Chatteaux Bay, on the Labrador coast, from August the 9th to September the 6th, when they all disappeared, being on their way from their northern breeding place." He adds, "They kept on the open grounds, fed on the *Empetrum nigrum*, and were very fat and delicious. They arrive at Hudson's Bay in April or early in May; pair and breed to the north of Albany Fort among the woods; return in August to the marshes, and all disappear in September." * About this time they return in accumulated numbers to the shores of New Jersey, whence they finally depart for the south early in November.

The Esquimaux curlew is eighteen inches long, and thirty-two inches in extent; the bill, which is four inches and a half long, is black towards the point, and a pale purplish flesh colour near the base; upper part of the head, dark brown, divided by a narrow stripe of brownish white; over each eye extends a broad line of pale drab; iris, dark coloured; hind part of the neck, streaked with dark brown; fore part and whole breast, very pale brown; upper part of the body, pale drab, centred and barred with dark brown, and edged with spots of white on the exterior vanes; three first primaries, black, with white shafts; rump and tail-coverts, barred with dark brown; belly, white; vent, the same, marked with zigzag lines of brown; whole lining of the wing, beautifully barred with brown on a dark cream ground; legs and naked thighs, a pale lead colour.

The figure of this bird, and of all the rest on the same plate, are reduced to exactly one half the size of life.

[Mr Ord adds, in his reprint, " I have some doubts whether or not this species is the Esquimaux curlew (*N. borealis*) of Dr Latham, as this ornithologist states his bird to be only thirteen inches in length, and in breadth twenty-one; whilst that above described is eighteen inches long, and thirty-two in breadth. Besides, Latham's species has a bill of two inches in length, and the bill of mine is four inches and a half long. I am aware, however, that the bills of some birds increase greatly with age; and if it should turn out hereafter that the two birds are identical, the specimen from which Latham took his description must have been quite immature."]

* Arct. Zool. vol. ii. p. 163; Phil. Trans. lxii. 411.

RED-BACKED SANDPIPER. (*Tringa alpina.*)

PLATE LVI.—Fig. 2.

Arct. Zool. p. 476, No. 391.—*Bewick*, ii. p. 113.—La Brunette, *Buff.* vii. 493. — *Peale's Museum*, No. 4094.

TRINGA ALPINA.—Pennant.*

Dunlin, *Mont. Orn. Dict.*—The Dunlin, *Bew. Br. Birds*, ii. p. 113.—Purre, *Id.* ii. p. 115.—Bécasseau brunette ou variable, *Temm* ii. 612.—Tringa alpina, *Flem. Br. Zool.* p. 108.—*Bonap. Synop.* p. 25.—Tringa alpina, the American Dunlin, *North. Zool.* ii. p. 383.

THIS bird inhabits both the old and new continents, being known in England by the name of the Dunlin, and in the United States, along the shores of New Jersey, by that of the

---

* This species is again figured, on the next plate, in the plumage of the winter, and the decided change undergone at the different ages and seasons has caused great multiplication and confusion among the synonyms. Wilson's two figures show very well the distinctions between the nuptial dress and that of winter ; and, in the bird of the first year, the plumage assumes a ruddy tinge on the upper parts, but wants the greater part of the black, so conspicuous during the love season.

On the coasts of Great Britain, the purre is the most common of the whole race, and may generally be met with, no matter what is the character of the shore. Before they have been much driven about and annoyed, they are also one of the most familiar. During winter, the flocks are sometimes immense, and will allow a person to approach very near, looking, and running a few steps, or stretching their wings in preparation for flight, listlessly, and indicative of little alarm ; a few shots, however, render them as timorous and wary as they were before careless. In spring, they separate into pairs, when some perform a migration to a considerable extent northward, while others retire to the nearer marshes and sea-merses, a few to the shores of inland lakes, and still fewer to the higher inland muirs. Having there performed the duties of incubation, they return again in autumn to the shore, where they may be found in small parties, the amount of the broods, and which gradually congregate as the season advances and more distant travellers arrive, until many hundreds are thus joined. Their nests are formed beneath or at the side of any small bush or tuft of grass, rather neatly scraped, and with a few straws of grass round the sides. The male is generally in attendance, perched on some near elevation,

Red-back. Its residence here is but transient, chiefly in April
and May, while passing to the arctic regions to breed, and
in September and October, when on its return southward to

and, on any danger approaching, runs round, uttering, at quick intervals, his shrill monotonous whistle. The female, when raised from the
nest, flutters off for a few yards, and then assumes the same manners
with the male. The young sit and squat among the grass or reeds,
and at that time the parents will come within two yards of the person
in search of them. The purre seems extensively distributed over both
the European and American continents. I have not, however, received
it from the Asiatic side, or any part of India, where so many of this
tribe are commonly found.

The genus *Pelinda* has been instituted and adopted by several
naturalists for the purre, the little sandpiper, and a few others, with
the exclusion of the pigmy curlew and knots. Though an advocate,
generally, for subdivisions, wherever any character can be seized upon,
I cannot reconcile that of these birds. I can fix upon no character
which is not equally applicable ; and the habits, the changes of plumage, and the form, are so similar, that, with the exception of modifications essential to every group, they compose one whole. The differences
in form will be noticed under the respective species; and, for the present,
I prefer retaining these birds under the generic name of *Tringa*.

The following species, not noticed by Wilson, have been added to the
American list by different ornithologists :—

T. Schinzii, Breh. On the authority of Bonaparte, identical with
the *Pelinda cinclus* var. of Say's expedition to the Rocky Mountains,
and met with by the Arctic Expedition on the borders of the lakes
which skirt the Saskatchewan plains. So nearly allied to *T. alpina*,
as to be confounded with it ; differs in size, and the distribution of
markings.

Tringa pectoralis, Bonap. ; *Pelinda pectoralis* of Say. This seems to
have been first noticed in the valuable notes to Major Laing's expedition
to the Rocky Mountains. The following description is there given by
Say :—

P. pectoralis, Say. Bill, black, reddish yellow at base; upper
mandible, with a few indented punctures near the tip ; head above,
black, plumage margined with ferruginous, a distinct brown line from
the eye to the upper mandible ; cheeks and neck beneath, cinereous,
very slightly tinged with rufous, and lineate with blackish ; orbits and
lineover the eye, white ; chin, white ; neck above, dusky, plumage
margined with cinereous ; scapulars, interscapulars, and wing-coverts,
black, margined with ferruginous, and near the exterior tips with
whitish ; primaries, dusky, slightly edged with whitish ; outer quill-

winter quarters. During their stay, they seldom collect in separate flocks by themselves, but mix with various other species of strand birds, among whom they are rendered conspicuous by the red colour of the upper part of their plumage. They frequent the muddy flats and shores of the salt marshes at low water, feeding on small worms and other insects, which generally abound in such places. In the month of May they are extremely fat.

This bird is said to inhabit Greenland, Iceland, Scandinavia, the Alps of Siberia, and, in its migrations, the coasts of the Caspian Sea.* It has not, till now, been recognised by naturalists as inhabiting this part of North America. Wherever its breeding place may be, it probably begins to lay at a late period of the season, as, in numbers of females which I examined on the 1st of June, the eggs were no larger than grains of mustard seed.

shaft white; back (beneath the interscapulars), rump, and tail-coverts, black, immaculate; tail-feathers, dusky, margined with white at tip, two intermediate ones longest, acute, attaining the tip of the wings, black, edged with ferruginous; breast, venter, vent, and inferior tail-coverts, white, plumage blackish at base; sides, white, the plumage towards the tail slightly lineate with dusky; feet, greenish yellow; toes, divided to the base; length, nearly nine inches; bill, 11–8.

T. Douglasii, Swainson. Described in the " Northern Zoology," from a specimen killed on the Saskatchewan, and is not uncommon in the Fur Countries up to the 60th parallel. The authors express a kind of doubt regarding this species, having been unable to compare it with a specimen of Bonaparte's *T. himantopus ;* but mention the tail as even with the central feathers alone, longest, and not barred with ferruginous; with chestnut coloured ear-feathers, and somewhat smaller in size.

To these nearly undescribed species, the Prince of Musignano mentions in his catalogue, *T. Temminckii*, Leisler; *T. minuta*, Leisler; *Numenius pygmæus*, Latham; the *Tringa platyrhyncha*, Temminck, and pigmy curlew of our shores; and the *T. maritima*, Brunnich, our purple sandpiper. The latter has been met with by most of the late arctic expeditions, and breeds abundantly on Melville Island and the shores of Hudson's Bay, and *T. subarquata, Becasseau corcoli*, Temm. ; and we may add, the *T. rufescens* of Vieillot, lately taken in this country.—ED.

* Pennant.

Length of the red-back, eight inches and a half; extent, fifteen inches; bill, black, longer than the head (which would seem to rank it with the snipes), slightly bent, grooved on the upper mandible, and wrinkled at the base; crown, back, and scapulars, bright reddish rust, spotted with black; wing-coverts, pale olive; quills, darker; the first tipt, the latter crossed with white; front, cheeks, hind head, and sides of the neck, quite round; also the breast, grayish white, marked with small specks of black; belly, white, marked with a broad crescent of black; tail, pale olive, the two middle feathers centred with black; legs and feet, ashy black; toes, divided to their origin, and bordered with a slightly scalloped membrane; irides, very dark.

The males and females are nearly alike in one respect, both differing greatly in colour, even at the same season, probably owing to difference of age; some being of a much brighter red than others, and the plumage dotted with white. In the month of September many are found destitute of the black crescent on the belly; these have been conjectured to be young birds.

SEMI-PALMATED SNIPE. *(Scolopax semipalmata.)*

PLATE LVI.—Fig. 3.

Arct. Zool. p. 469, No. 380.—*Peale's Museum,* No. 3942.

TOTANUS SEMIPALMATUS.—Temminck.*

Chevalier semipalmé, Totanus semipalmatus, *Temm. Man. d'Orn.* ii. p. 637.—Totanus crassirostris, *Vieill.* winter plumage, auct. *Bonap.*—*Bonap. Cat.* p. 26.

THIS is one of the most noisy and noted birds that inhabit our salt marshes in summer. Its common name is the Willet,

* Wilson has figured the winter dress of this curious species, and the Prince of Musignano has signified his intention of representing its other states. It is admitted as an accidental straggler among the species of Europe by Temminck.—ED.

by which appellation it is universally known along the shores
of New York, New Jersey, Delaware, and Maryland,—in all
of which places it breeds in great numbers.

The willet is peculiar to America. It arrives from the
south on the shores of the middle States about the 20th April
or beginning of May, and from that time to the last of July,
its loud and shrill reiterations of *pill-will-willet, pill-will-willet*,
resound almost incessantly along the marshes, and may be
distinctly heard at the distance of more than half a mile.
About the 20th of May, the willets generally begin to lay.*
Their nests are built on the ground, among the grass of the
salt marshes, pretty well towards the land or cultivated fields,
and are composed of wet rushes and coarse grass, forming a
slight hollow or cavity in a tussock. This nest is gradually
increased during the period of laying and sitting to the height
of five or six inches. The eggs are usually four in number,
very thick at the great end, and tapering to a narrower point
at the other than those of the common hen; they measure two
inches and one-eighth in length, by one and a half in their
greatest breadth, and are of a dark dingy olive, largely blotched
with blackish brown, particularly at the great end. In some,
the ground colour has a tinge of green; in others, of bluish.
They are excellent eating, as I have often experienced when
obliged to dine on them in my hunting excursions through the
salt marshes. The young are covered with a gray-coloured
down; run off soon after they leave the shell; and are led
and assisted in their search of food by the mother, while the
male keeps a continual watch around for their safety.

The anxiety and affection manifested by these birds for their
eggs and young are truly interesting. A person no sooner
enters the marshes, than he is beset by the willets, flying around
and skimming over his head, vociferating with great violence
their common cry of *pill-will-willet*, and uttering at times a
loud clicking note as he approaches nearer to their nest. As

---

* From some unknown cause, the height of laying of these birds is
said to be full two weeks later than it was twenty years ago.

they occasionally alight, and slowly shut their long white wings speckled with black, they have a mournful note expressive of great tenderness. During the term of incubation, the female often resorts to the sea-shore, where, standing up to the belly in water, she washes and dresses her plumage, seeming to enjoy great satisfaction from these frequent immersions. She is also at other times seen to wade more in the water than most of her tribe ; and when wounded in the wing, will take to the water without hesitation, and swims tolerably well.

The eggs of the willet, in every instance which has come under my observation, are placed, during incubation, in an almost upright position, with the large end uppermost; and this appears to be the constant practice of several other species of birds that breed in these marshes. During the laying season, the crows are seen roaming over the marshes in search of eggs, and wherever they come, spread consternation and alarm among the willets, who, in united numbers, attack and pursue them with loud clamours. It is worthy of remark, that among the various birds that breed in these marshes, a mutual respect is paid to each other's eggs ; and it is only from intruders from the land side, such as crows, jays, weasels, foxes, minx, and man himself, that these affectionate tribes have most to dread.

The willet subsists chiefly on small shell-fish, marine worms, and other aquatic insects ; in search of which it regularly resorts to the muddy shores and flats at low water, its general rendezvous being the marshes.

This bird has a summer and also a winter dress, its colours differing so much in these seasons as scarcely to appear to be the same species. Our figure in the plate exhibits it in its spring and summer plumage, which in a good specimen is as follows :—

Length, fifteen inches ; extent, thirty inches ; upper parts, dark olive brown ; the feathers, streaked down the centre, and crossed with waving lines of black ; wing-coverts, light olive ash, and the whole upper parts sprinkled with touches of dull yellowish white; primaries, black, white at the root half; secondaries, white, bordered with brown; rump, dark brown; tail, rounded, twelve feathers pale olive, waved with bars of black; tail-coverts, white, barred with olive; bill, pale lead colour, becoming black towards the tip; eye, very black; chin, white; breast, beautifully mottled with transverse spots of olive on a cream ground; belly and vent, white, the last barred with olive; legs and feet, pale lead colour; toes, half webbed.

Towards the fall, when these birds associate in large flocks, they become of a pale dun colour above, the plumage being shafted with dark brown, and the tail white, or nearly so. At this season they are extremely fat, and esteemed excellent eating. Experienced gunners always select the lightest coloured ones from a flock, as being uniformly the fattest.

The female of this species is generally larger than the male. In the months of October and November, they gradually disappear.

GREAT MARBLED GODWIT. (*Scolopax fedoa.*)

PLATE LVI.—Fig. 4, Female.

Arct. Zool. p. 465, No. 371.—La Barge Rousse de Baie d'Hudson, *Buff.* vii. 507.—
Peale's Museum, No. 4019.

LIMOSA FEDOA.—Vieillot.

Limosa fedoa, *Ord's edit. of Wils.*—*Bonap. Synop.* p. 328.

This is another transient visitant of our sea-coasts in spring and autumn, to and from its breeding place in the north. Our gunners call it the *straight-billed curlew*, and sometimes the *red curlew*. It is a shy, cautious, and watchful bird; yet so strongly are they attached to each other, that on wounding one in a flock, the rest are immediately arrested in their flight, making so many circuits over the spot where it lies fluttering and screaming, that the sportsman often makes great destruction among them. Like the curlew, they may also be enticed within shot by imitating their call or whistle, but can seldom be approached without some such manœuvre. They are much less numerous than the short-billed curlews, with whom, however, they not unfrequently associate. They are found among the salt marshes in May, and for some time in June, and also on their return in October and November; at which last season they are usually fat, and in high esteem for the table.

The female of this bird having been described by several writers as a distinct species from the male, it has been thought proper to figure the former; the chief difference consists in the undulating bars of black with which the breast of the male is marked, and which are wanting in the female.

The male of the great marbled godwit is nineteen inches long, and thirty-four inches in extent; the bill is nearly six inches in length, a little turned up towards the extremity, where it is black, the base is of a pale purplish flesh colour; chin and upper part of the throat, whitish; head and neck, mottled with dusky brown and black on a ferruginous ground; breast, barred with wavy lines of black; back and scapulars, black, marbled with pale brown; rump and tail-coverts, of a very light brown, barred with dark brown; tail, even, except the two middle feathers, which are a little the longest; wings, pale ferruginous, elegantly marbled with dark brown, the four first primaries black on the outer edge; whole lining and lower parts of the wings, bright ferruginous; belly and vent, light rust colour, with a tinge of lake.

The female differs in wanting the bars of black on the breast. The bill does not acquire its full length before the third year.

About fifty different species of the scolopax genus are enumerated by naturalists. These are again by some separated into three classes or sub-genera: viz., the straight-billed, or snipes; those with bills bent downwards, or the curlews; and those whose bills are slightly turned upwards, or godwits. The whole are a shy, timid, and solitary tribe, frequenting

those vast marshes, swamps, and morasses that frequently
prevail in the vicinity of the ocean and on the borders of
large rivers. They are also generally migratory, on account
of the periodical freezing of those places in the northern
regions where they procure their food. The godwits are
particularly fond of salt marshes, and are rarely found in
countries remote from the sea.

TURNSTONE.   (*Tringa interpres.*)

PLATE LVII.—Fig. 1.

Hebridal Sandpiper, *Arct. Zool.* p. 472, No. 382.—Le Tourne-pierre, *Buff.* vii.
    130, *Pl. enl.* 130.—*Bewick*, ii. p. 119, 121.—*Catesby*, i. 72.—*Peale's Museum*,
    No. 4044.
            STREPSILAS INTERPRES.—Illiger.*
Tourne-pierre à collier (Strepsilas collaris), *Temm. Man. d'Orn.* ii. p. 553.—Strep-
    silas interpres, *Flem. Br. Zool.* p. 110.—*North. Zool.* ii. p. 371.—Strepsilas
    collaris, *Bonap. Synop.*

This beautifully variegated species is common to both Europe
and America, consequently extends its migrations far to the
north. It arrives from the south on the shores of New Jersey
in April; leaves them early in June; is seen on its return to
the south in October; and continues to be occasionally seen
until the commencement of cold weather, when it disappears
for the season. It is rather a scarce species in this part of the

---

* This is the only species of turnstone known, and it is apparently
distributed over the whole world. Its breeding places, according to
the "Northern Zoology," are the shores of Hudson's Bay and the Arctic
Sea, probably in the most northern districts. On the Scotch and
English coasts they arrive in small flocks about the beginning of August,
and, as the season advances, congregate into larger assemblies. The
greater proportion of these are still in their young dress, and it is not
until the ensuing spring that this is completely changed; in this state
they have been frequently described as a second species. Early in
spring, a few straggling birds, in perfect breeding plumage, may be
observed on most of our shores, which have either been left at the
general migration, or remain during the year in a state of barrenness.
It is then that the finest specimens for stuffing are obtained.—Ed.

Drawn from Nature by A. Wilson.

1. Turnstone. 2. Ash Coloured Sandpiper. 3. The Purre. 4. Black bellied Plover. 5. Red breasted Sandpiper.

Engraved by W.H. Lizars.

57.

world, and of a solitary disposition, seldom mingling among
the large flocks of other sandpipers; but either coursing the
sands alone, or in company with two or three of its own species.
On the coast of Cape May and Egg Harbour this bird is well
known by the name of the horse-foot snipe, from its living, during the months of May and June, almost wholly on the eggs,
or spawn, of the great king crab, called here by the common
people the horse-foot. This animal is the *Monoculus poly-
phemus* of entomologists. Its usual size is from twelve to
fifteen inches in breadth, by two feet in length, though sometimes it is found much larger. The head, or forepart, is semicircular, and convex above, covered with a thin, elastic, shelly
case. The lower side is concave, where it is furnished with
feet and claws resembling those of a crab. The posterior
extremity consists of a long, hard, pointed, dagger-like tail,
by means of which, when overset by the waves, the animal
turns itself on its belly again. The male may be distinguished from the female by his two large claws having only a
single hook each, instead of the forceps of the female. In the
Bay of Delaware, below Egg Island, and in what is usually
called Maurice River Cove, these creatures seem to have
formed one of their principal settlements. The bottom of
this cove is generally a soft mud, extremely well suited to
their accommodation. Here they are resident, burying themselves in the mud during the winter; but, early in the month
of May, they approach the shore in multitudes, to obey the
great law of nature, in depositing their eggs within the influence of the sun, and are then very troublesome to the fishermen, who can scarcely draw a seine for them, they are so
numerous. Being of slow motion, and easily overset by the
surf, their dead bodies cover the shore in heaps, and in such
numbers, that for ten miles one might walk on them without
touching the ground.

The hogs from the neighbouring country are regularly
driven down, every spring, to feed on them, which they do
with great avidity; though by this kind of food their flesh

acquires a strong disagreeable fishy taste. Even the small turtles, or terrapins, so eagerly sought after by our epicures, contract so rank a taste by feeding on the spawn of the king crab, as to be at such times altogether unpalatable. This spawn may sometimes be seen lying in hollows and eddies in bushels, while the snipes and sandpipers, particularly the turnstone, are hovering about feasting on the delicious fare. The dead bodies of the animals themselves are hauled up in wagons for manure, and when placed at the hills of corn in planting time, are said to enrich the soil, and add greatly to the increase of the crop.

The turnstone derives its name from another singularity it possesses, of turning over with its bill small stones and pebbles in search of various marine worms and insects. At this sort of work it is exceedingly dexterous; and even when taken and domesticated, is said to retain the same habit.* Its bill seems particularly well constructed for this purpose, differing from all the rest of its tribe, and very much resembling in shape that of the common nuthatch. We learn from Mr Pennant that these birds inhabit Hudson's Bay, Greenland, and the arctic flats of Siberia, where they breed, wandering southerly in autumn. It is said to build on the ground, and to lay four eggs, of an olive colour, spotted with black, and to inhabit the isles of the Baltic during summer

The turnstone flies with a loud twittering note, and runs with its wings lowered; but not with the rapidity of others of its tribe. It examines more completely the same spot of ground, and, like some of the woodpeckers, will remain searching in the same place, tossing the stones and pebbles from side to side for a considerable time.

These birds vary greatly in colour; scarcely two individuals are to be found alike in markings. These varieties are most numerous in autumn when the young birds are about, and are less frequently met with in spring. The most perfect specimens I have examined are as follows :—

* Catesby.

Length eight inches and a half ; extent, seventeen inches ; bill, blackish horn ; frontlet, space passing through the eyes, and thence dropping down and joining the under mandible, black, enclosing a spot of white ; crown, white, streaked with black ; breast, black, from whence it turns up half across the neck ; behind the eye, a spot of black ; upper part of the neck, white, running down and skirting the black breast as far as the shoulder ; upper part of the back, black, divided by a strip of bright ferruginous ; scapulars, black, glossed with greenish, and interspersed with rusty red ; whole back below this, pure white, but hid by the scapulars ; rump, black ; tail-coverts, white ; tail, rounded, white at the base half, thence black to the extremity ; belly and vent, white ; wings, dark dusky, crossed by two bands of white ; lower half of the lesser coverts, ferruginous ; legs and feet, a bright vermilion, or red lead ; hind toe, standing inwards, and all of them edged with a thick warty membrane. The male and female are alike variable, and when in perfect plumage nearly resemble each other.

Bewick, in his " History of British Birds," has figured and described what he considers to be two species of turnstone ; one of which, he says, is chiefly confined to the southern, and the other to the northern parts of Great Britain. The difference, however, between these two appears to be no greater than commonly occurs among individuals of the same flock, and evidently of the same species, in this country. As several years probably elapse before these birds arrive at their complete state of plumage, many varieties must necessarily appear, according to the different ages of the individuals.

ASH-COLOURED SANDPIPER. (*Tringa cinerea.*)

PLATE LVII.—Fig. 2.

Arct. Zool. p. 474, No. 386.—*Bewick,* ii. p. 102.—*Peale's Museum,* No. 4060.

TRINGA CANUTUS.—Linnæus.—Plumage of the young.*

Synonyms of young: Tringa calidris, *Linn.* i. 252.—Tringa nævia, *Lath. Ind. Orn.* ii. 732.—Maubeche tachete, *Buff.*—Freckled Sandpiper, *Arct. Zool.* ii. p. 480.

The regularly-disposed concentric semicircles of white and dark brown that mark the upper parts of the plumage of this species, distinguish it from all others, and give it a very neat appearance. In activity it is superior to the preceding; and traces the flowing and recession of the waves along the sandy

---

* This beautiful sandpiper has also from its changes been described under various names, and our author has well represented the states of the young and summer plumage in his ash-coloured and red-breasted sandpipers of the present plate. In the winter plumage of the adult, the upper parts are of a uniform gray, and want the black and light edges represented in fig. 2.

America and Europe seem the only countries of the Knot. I have never seen it from India, but have a single specimen of a knot from New Holland, very similar, and which I considered identical, until a closer examination has led me to have doubts on the subject. Like the other migratory species, they only appear on our coasts in autumn, on their return with their broods, or more sparingly in spring, when on their way north. The young possess a good deal of the rufous colour on the under parts, which leaves them as the winter approaches. I once met a large flock on the east side of Holy Island, in the month of September, which were so tame as to allow me to kill as many as I wanted with stones from the beach: it may have been on their first arrival, when they were fatigued. I have a specimen, in full plumage, killed by a boy on Portobello sands by the same means. In general they are rather shy, and it is only in their wheeling round that a good shot can be obtained. Before the severity of the winter sets in, they are fat, and are sought after by persons who *know them,* for the table.

There is a peculiarity in the gregarious *Tringæ,* and most of the *Charadriadæ,* which is very nearly confined to these tribes,—the simultaneous flight, and the acting as it were by concert in their wheels and evolutions. Among none is it more conspicuous than in this species; and every one who has been on the shore during winter, on a day

beach with great nimbleness, wading and searching among the loosened particles for its favourite food, which is a small thin oval bivalve shell-fish, of a white or pearl colour, and not larger than the seed of an apple. These usually lie at a short depth below the surface; but in some places are seen at low water in heaps, like masses of wet grain, in quantities of more than a bushel together. During the latter part of summer and autumn, these minute shell-fish constitute the food of almost all those busy flocks that run with such activity along the sands, among the flowing and retreating waves. They are universally swallowed whole; but the action of the bird's stomach, assisted by the shells themselves, soon reduces them to a pulp. If we may judge from their effects, they must be extremely nutritious, for almost all those tribes that feed on them are at this season mere lumps of fat. Digging for these in the hard sand would be a work of considerable labour, whereas, when the particles are loosened by the flowing of the sea, the birds collect them with great ease and dexterity. It is amusing to observe with what adroitness they follow and elude the tumbling surf, while at the same time they seem wholly intent on collecting their food.

The ash-coloured sandpiper, the subject of our present account, inhabits both Europe and America. It has been seen in great numbers on the Seal Islands near Chatteaux Bay; is said to continue the whole summer in Hudson's Bay, and breeds there. Mr Pennant suspects that it also breeds in Denmark; and says, that they appear in vast flocks on the Flintshire shore during the winter season.* With us they are also migratory, being only seen in spring and autumn. They

gleaming and cloudy, may have seen the masses of these birds at a distance, when the whole were only visible, appear like a dark and swiftly moving cloud, suddenly vanish, but in a second appear at some distance, glowing with a silvery light almost too intense to gaze upon, the consequences of the simultaneous motions of the flock, at once changing their position, and showing the dark gray of their backs, or the pure white of their under parts.—ED.

* Arctic Zoology, p. 474.

are plump birds ; and, by those accustomed to the sedgy taste of this tribe, are esteemed excellent eating.

The length of this species is ten inches, extent twenty ; bill black, straight, fluted to nearly its tip, and about an inch and a half long; upper parts, brownish ash, each feather marked near the tip with a narrow semicircle of dark brown, bounded by another of white ; tail-coverts, white, marbled with olive ; wing-quills, dusky, shafts, white ; greater coverts, black, tipt with white ; some of the primaries edged also with white ; tail, plain pale ash, finely edged and tipt with white ; crown and hind head, streaked with black, ash, and white ; stripe over the eye, cheeks, and chin, white, the former marked with pale streaks of dusky, the latter pure ; breast, white, thinly specked with blackish ; belly and vent, pure white ; legs, a dirty yellowish clay colour ; toes, bordered with a narrow, thick, warty membrane ; hind toe, directed inwards, as in the turn-stone ; claws and eye, black.

These birds vary a little in colour, some being considerably darker above, others entirely white below ; but, in all, the concentric semicircles on the back, scapulars, and wing-coverts, are conspicuous.

I think it probable that these birds become much lighter coloured during the summer, from the circumstance of having shot one late in the month of June at Cape May, which was of a pale drab or dun colour. It was very thin and emaciated ; and on examination appeared to have been formerly wounded, which no doubt occasioned its remaining behind its companions.

Early in December I examined the same coast every day for nearly two weeks, without meeting with more than one solitary individual of this species, although in October they were abundant. How far to the southward they extend their migrations, we have no facts that will enable us to ascertain, though it is probable that the shores of the West India islands afford them shelter and resources during our winter.

THE PURRE. (*Tringa cinclus.*)

PLATE LVII.—FIG. 3.

Linn. Syst. 251.—*Arct. Zool.* p. 475, No. 390.—*Bewick*, ii. p. 115.—L'Alouette de Mer, *Buff.* vii. 548.—*Peale's Museum*, No. 4126.

TRINGA ALPINA.—PENNANT.

THIS is one of the most numerous of our strand birds, as they are usually called, that frequent the sandy beach on the frontiers of the ocean. In its habits it differs so little from the preceding, that, except in being still more active and expert in running and searching among the sand on the reflux of the waves, as it nimbly darts about for food, what has been said of the former will apply equally to both, they being pretty constant associates on these occasions.

The purre continues longer with us, both in spring and autumn, than either of the two preceding; many of them remain during the very severest of the winter, though the greater part retire to the more genial regions of the south, where I have seen them at such seasons, particularly on the sea-coasts of both Carolinas, during the month of February, in great numbers.

These birds, in conjunction with several others, sometimes collect together in such flocks, as to seem, at a distance, a large cloud of thick smoke, varying in form and appearance every instant, while it performs its evolutions in air. As this cloud descends and courses along the shores of the ocean, with great rapidity, in a kind of waving serpentine flight, alternately throwing its dark and white plumage to the eye, it forms a very grand and interesting appearance. At such times the gunners make prodigious slaughter among them; while, as the showers of their companions fall, the whole body often alight, or descend to the surface with them, till the sportsman is completely satiated with destruction. On some of those occasions, while crowds of these victims are fluttering along the sand, the small pigeon-hawk, constrained by necessity, ventures to make a sweep among the dead in presence of the proprietor, but as suddenly pays for his temerity with his life. Such a tyrant is man, when vested with power, and unrestrained by the dread of responsibility !

The purre is eight inches in length, and fifteen inches in extent; the bill is black, straight, or slightly bent downwards, about an inch and a half long, very thick at the base, and tapering to a slender blunt point at the extremity; eye, very small ; iris, dark hazel ; cheeks, gray ; line over the eye, belly, and vent, white; back and scapulars, of an ashy brown, marked here and there with spots of black, bordered with bright ferruginous; sides of the rump, white ; tail-coverts, olive, centred with black; chin, white; neck below, gray ; breast and sides, thinly marked with pale spots of dusky, in some pure white ; wings, black, edged and tipt with white ; two middle tail-feathers, dusky, the rest, brown ash, edged with white; legs and feet, black; toes, bordered with a very narrow scalloped membrane. The usual broad band of white crossing the wing forms a distinguishing characteristic of almost the whole genus.

On examining more than a hundred of these birds, they varied considerably in the black and ferruginous spots on the back and scapulars ; some were altogether plain, while others were thickly marked, particularly on the scapulars, with a red rust colour, centred with black. The females were uniformly more plain than the males ; but many of the latter, probably young birds, were destitute of the ferruginous spots. On the 24th of May, the eggs in the females were about the size of partridge-shot. In what particular regions of the north these birds breed is altogether unknown.

BLACK-BELLIED PLOVER. (*Charadrius apricarius.*)

PLATE LVII.—Fig. 4.

Alwagrim Plover, *Arct. Zool.* p. 483, No. 398.—Le Pluvier Doré à gorge noire, *Buff.* viii. 85.—*Peale's Museum*, No. 4196.

SQUATAROLA CINEREA.—Fleming.*

Pluvialis cinerea, *Wil. Orn.* 229.—Gray Squatarola, Squatarola grisea, *Steph. Cont. Sh. Zool.* vol. xi. p. 505.—Le Vanneau Gris, *Cuv. Reg. Anim.* vol. i. p. 467.— Squatarola cinerea, *Flem. Br. Zool.* p. 3.—Vanellus melanogaster, *North. Zool.* ii. p. 370.

THIS bird is known in some parts of the country by the name of the large whistling field plover. It generally makes its first appearance in Pennsylvania late in April; frequents the countries towards the mountains; seems parti-

---

* This species, with some others, forms the division *Vanneau pluviers*, the genus *Squatarola* of Cuvier, and, according to modern ornithologists, has been separated from the *Charadrii* on account of the presence of a hinder toe.

In the arrangement of this group, as in many others, I fear the characteristic marks have been taken in a manner too arbitrary. Those birds known by the name of *Plovers* form a small but apparently distinct group ; they contain the *C. pluvialis, Virginianus*, &c., and, but for the rudimentary toe, the gray plover would also enter it : they agree in their manners, their incubation, and changes of plumage. We, again, have another well-defined group, which is called the *Dotterels*, agreeing in similar common habitudes ; but, in one species, bearing according to arrangement the name of *Squatarola*, we have all the marks and form of plumage, but the hinder toe much developed. It therefore becomes a question whether the presence or want of this appendage should be brought into the generic character (as it always has been), or should be looked upon as one of the connections of forms. In the latter way the plovers should form the genus *Squatarola*, the dotterels *Charadrius*, and the two birds in question be placed opposite in their respective circles.

Vanellus, or the *Lapwings*, again, form another group, as well marked in their different habits, and intimately connected with *Pluvianus ;* neither of these, however, have any representative in North America.

Many gray plovers breed in the English fens, and, like the migratory sandpipers, flocks appear on the shores at the commencement of winter, where they mingle with the other species. The plate is that of the summer or breeding plumage.—ED.

cularly attached to newly-ploughed fields, where it forms its
nest of a few slight materials, as slightly put together. The
female lays four eggs, large for the size of the bird, of a light
olive colour dashed with black, and has frequently two broods
in the same season. It is an extremely shy and watchful
bird, though clamorous during breeding time. The young are
without the black colour on the breast and belly until the
second year, and the colours of the plumage above are likewise imperfect till then. They feed on worms, grubs, winged
insects, and various kinds of berries, particularly those usually
called dew-berries, and are at such times considered exquisite eating. About the beginning of September they descend
with their young to the sea-coast, and associate with the
numerous multitudes then returning from their breeding
places in the north. At this season they abound on the
plains of Long Island. They have a loud whistling note ;
often fly at a great height ; and are called by many gunners
along the coast the black-bellied killdeer. The young of the
first year have considerable resemblance to those of the golden
plover ; but may be easily distinguished from this last by the
largeness of their head and bill, and in being at least two
inches more in length. The greater number of those which
I have examined have the rudiments of a hind toe ; but the
character and manners of the plover are so conspicuous in the
bird, as to determine, at the first glance, the tribe it belongs
to. They continue about the sea-coast until early in November, when they move off to the south.

This same bird, Mr Pennant informs us, inhabits all the
north of Europe, Iceland, Greenland, and Hudson's Bay, and
all the arctic part of Siberia. It is said that at Hudson's Bay
it is called the Hawk's-eye, on account of its brilliancy. It
appears, says the same author, in Greenland, in the spring,
about the southern lakes, and feeds on worms and berries of
the heath.

This species is twelve inches long, and twenty-four inches
in extent ; the bill is thick, deeply grooved on the upper

mandible, an inch and a quarter in length, and of a black colour ; the head and globe of the eye are both remarkably large, the latter deep bluish black ; forehead, white ; crown and hind head, black, spotted with golden yellow ; back and scapulars, dusky, sprinkled with the same golden or orange coloured spots, mixed with others of white ; breast, belly, and vent, black ; sides of the breast, whitish ; wing-quills, black ; middle of the shafts, white ; greater coverts, black, tipt with white ; lining of the wing, black ; tail, regularly barred with blackish and pure white ; tail-coverts, pure white ; legs and feet, a dusky lead colour ; the exterior toe joined to the middle by a broad membrane ; hind toe, very small.

From the length of time which these birds take to acquire their full colours, they are found in very various stages of plumage. The breast and belly are at first white, gradually appear mottled with black, and finally become totally black. The spots of orange or golden on the crown, hind head, and back are at first white, and sometimes even the breast itself is marked with these spots, mingled among the black. In every stage, the seemingly disproportionate size of the head and thickness of the bill will distinguish this species.

RED-BREASTED SANDPIPER. (*Tringa rufa.*)

PLATE LVII.—Fig. 5.

Peale's Museum, No. 4050.

TRINGA CANUTUS.—Linnæus.

Tringa Islandica, *Linn.* and *Lath.*—Red Sandpiper, *Mont. Orn. Dict. Supp.*— Aberdeen Sandpiper, *Penn. Brit. Zool.* ii. No. 203.

OF this prettily-marked species I can find no description. The *Tringa Icelandica*, or Aberdeen sandpiper of Pennant and others, is the only species that has any resemblance to it ; the descriptions of that bird, however, will not apply to the present.

The common name of this species on our sea-coast is the gray-back, and among the gunners it is a particular favourite, being generally a plump, tender, and excellent bird for the table ; and, consequently, brings a good price in market.

The gray-backs do not breed on the shores of the middle States. Their first appearance is early in May. They remain a few weeks, and again disappear until October. They usually keep in small flocks, alight in a close body together on the sand flats, where they search for the small bivalve shells already described. On the approach of the sportsman, they frequently stand fixed and silent for some time ; do not appear to be easily alarmed, neither do they run about in the water as much as some others, or with the same rapidity, but appear more tranquil and deliberate. In the month of November they retire to the south.

This species is ten inches long, and twenty in extent ; the bill is black, and about an inch and a half long ; the chin, eyebrows, and whole breast are a pale brownish orange colour ; crown, hind head from the upper mandible backwards, and neck, dull white, streaked with black ; back, a pale slaty olive, the feathers tipt with white, barred and spotted with black and pale ferruginous ; tail-coverts, white, elegantly barred with black ; wings, plain, dusky black towards the extremity ; the greater coverts, tipt with white ; shafts of the primaries, white ; tail, pale ashy olive, finely edged with white, the two middle feathers somewhat the longest ; belly and vent, white, the latter marked with small arrow-heads of black ; legs and feet, black ; toes, bordered with a narrow membrane ; eye, small and black.

In some specimens, both of males and females, the red on the breast was much paler, in others it descended as far as the thighs. Both sexes seemed nearly alike.

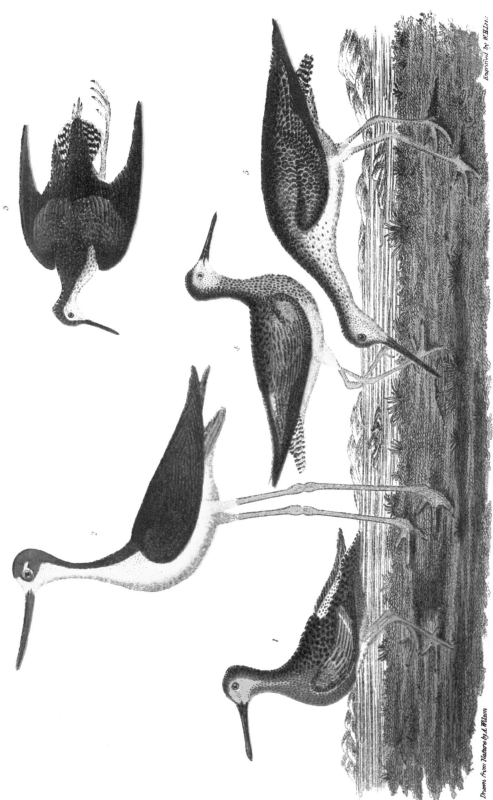

Drawn from Nature by A. Wilson.

Engraved by W.H.Lize.

1. Red-breasted Snipe. 2. Long-legged Avoset. 3. Solitary Sandpiper. 4. Yellow Shanks Snipe. 5. Tell tale Snipe.

58.

RED-BREASTED SNIPE. (*Scolopax noveboracensis.*)

PLATE LVIII.—Fig. 1.

Arct. Zool. p. 464, No. 368.—*Peale's Museum,* No. 3932.

MACRORHAMPUS GRISEUS.—Leach.*

Macrorhampus griseus, *Steph. Cont. Sh. Zool.* vol. xii. p. 61.—Scolopax grisea,
Flem. Br. Zool. p. 106.—*Bonap. Cat.* p. 27.—Le Becassine Grise, Scolopax
leucophœa, *Vieill. Gal. des Ois.* pl. 241.—Limosa scolopacea, *Say's Exped. to
Rocky Mount.* i. p. 170, 171, note.—Brown Snipe, *Mont. Orn. Dict.*—Becassine
Ponctuée, *Temm. Man.* ii. p. 679.—Brown Snipe, *Selby's Illust. Br. Orn.* pl.
24, fig. 2.

THIS bird has a considerable resemblance to the common
snipe, not only in its general form, size, and colours, but
likewise in the excellence of its flesh, which is in high estimation. It differs, however, greatly from the common snipe

* This bird will stand in the rank of a sub-genus. It was first indicated by Leach, in the Catalogue to the British Museum, under the
above title. It is one of those beautifully connecting forms which it is
impossible to place without giving a situation to themselves, and intimately connects the snipes with *Totanus* and *Limosa.* The bill is
truly that of *Scolopax,* while the plumage and changes ally it to the
other genera; from these blending characters it had been termed *Limosa
scolopacea* by Say, who gave the characters of the form without applying the name. He has the following observations in the work above
quoted :—
"Several specimens were shot in a pond near the Bowyer Creek.
Corresponds with the genus *Scolopax,* Cuvier, in having the dorsal
grooves at the tip of the upper mandible, and in having this part dilated
and rugose ; but the eye is not large, nor is it placed far back upon the
head ; which two latter characters, combined with its more elevated
and slender figure, and the circumstance of the thighs being denudated
of feathers high above the knee, and the exterior toe being united to
the middle toe by a membrane which extends as far as the first joint,
and the toes being also margined, combine to distinguish this species
from those of the genus to which the form and characters of its bill
would refer it, and approach it more closely to *Limosa.* In one specimen, the two exterior primaries on each wing were light brown, but
the quills were white. It may, perhaps, with propriety be considered
as the type of a new genus, and, under the following characters, be
placed between the genera *Scolopax* and *Limosa.* Bill, longer than the

in its manners, and in many other peculiarities, a few of which, as far as I have myself observed, may be sketched as follows :—

The red-breasted snipe arrives on the sea-coast of New Jersey early in April; is seldom or never seen inland : early in May it proceeds to the north to breed, and returns by the latter part of July or beginning of August. During its stay here, it flies in flocks, sometimes very high, and has then a loud and shrill whistle, making many evolutions over the marshes, forming, dividing, and reuniting. They sometimes settle in such numbers, and so close together, that eighty-five have been shot at one discharge of a musket. They spring from the marshes with a loud twirling whistle, generally rising high, and making several circuitous manœuvres in air before they descend. They frequent the sandbars and mud flats at low water in search of food ; and being less suspicious of a boat than of a person on shore, are easily approached by this medium, and shot down in great numbers. They usually keep by themselves, being very numerous ; are in excellent order for the table in September ; and on the approach of winter retire to the south.

I have frequently amused myself with the various action of these birds. They fly very rapidly, sometimes wheeling, coursing, and doubling along the surface of the marshes ; then shooting high in air, there separating and forming in various bodies, uttering a kind of quivering whistle. Among many which I opened in May, were several females that had very little rufous below, and the backs were also much lighter, and less marbled with ferruginous. The eggs contained in their ovaries were some of them as large as garden peas. Their

head, dilated, and rugose at tip, slightly curved downwards, and with a dorsal groove ; nasal groove, elongated ; feet, long, an extensive naked space above the knee ; toes, slightly margined, a membrane connecting the joints of the exterior toes ; first of the primaries, rather longest."

It is of rare occurrence in Europe, a few specimens only being mentioned, and a solitary instance of its appearance on the coast of Britain is recorded by Montagu.—Ed.

stomachs contained masses of those small snail-shells that lie in millions on the salt marshes; the wrinkles at the base of the bill, and the red breast, are strong characters of this species, as also the membrane which unites the outer and middle toes together.

The red-breasted snipe is ten inches and a half long, and eighteen inches in extent; the bill is about two inches and a quarter in length, straight, grooved, black towards the point, and of a dirty eel-skin colour at the base, where it is tumid and wrinkled; lores, dusky; cheeks and eyebrows, pale yellowish white, mottled with specks of black; throat and breast, a reddish buff colour; sides, white, barred with black; belly and vent, white, the latter barred with dusky; crown, neck above, back, scapulars, and tertials, black, edged, mottled, and marbled with yellowish white, pale and bright ferruginous, much in the same manner as the common snipe; wings, plain olive, the secondaries, centred and bordered with white; shaft, of the first quill, very white; rump, tail-coverts, and tail (which consists of twelve feathers), white, thickly spotted with black; legs and feet, dull yellowish green; outer toe united to the middle one by a small membrane; eye, very dark. The female, which is paler on the back, and less ruddy on the breast, has been described by Mr Pennant as a separate species.*

These birds, doubtless, breed not far to the northward of the United States, if we may judge from the lateness of the season when they leave us in spring, the largeness of the eggs in the ovaries of the females before they depart, and the short period of time they are absent. Of all our seaside snipes, it is the most numerous, and the most delicious for the table. From these circumstances, and the crowded manner in which it flies and settles, it is the most eagerly sought after by our gunners, who send them to market in great numbers.

* See his brown snipe, Arct. Zool., No. 369.

LONG-LEGGED AVOSET. (*Recurvirostra himantopus.*)

PLATE LVIII.—Fɪɢ. 2.

Long-legged Plover, *Arct. Zool.* p. 487, No. 405.—*Turton*, p. 416.—*Bewick*, ii. 21.—L'Echasse, *Buff.* viii. 114, *Pl. enl.* 878.—*Peale's Museum,* No. 4210.

HIMANTOPUS NIGRICOLLIS.—Vɪᴇɪʟʟᴏᴛ.*

Himantopus Mexicanus, *Ord's edit. of Wils.*—Himantopus nigricollis, *Bonap. Synop.* p. 322.

Nᴀᴛᴜʀᴀʟɪsᴛs have most unaccountably classed this bird with the genus *Charadrius,* or plover, and yet affect to make the particular confirmation of the bill, legs, and feet, the rule of their arrangement. In the present subject, however, excepting the trivial circumstance of the want of a hind toe, there is no resemblance whatever of those parts to the bill, legs, or feet, of the plover; on the contrary, they are so entirely different, as to create no small surprise at the adoption and general acceptation of a classification evidently so absurd and unnatural. This appears the more reprehensible, when we consider the striking affinity there is between this bird and the common avoset, not only in the particular form of the bill, nostrils, tongue, legs, feet, wings, and tail, but extending to the voice, manners, food, place of breeding, form of the nest, and even the very colour of the eggs of both, all of which are strikingly alike, and point out at once, to the actual observer of Nature, the true relationship of these remarkable birds.

Strongly impressed with these facts, from an intimate

Wilson confounded this species with the long-legged plover of Europe, and ranged it with the Avosets. Mr Ord, in his reprint, placed it in the genus *Himantopus,* properly established for these birds, but under the name *Mexicanus.* The Prince of Musignano is of opinion that it cannot range under this, being much smaller, and refers it to the *H. nigricollis* of Vieillot. The genus contains only a few species, all so closely allied, that near examination is necessary to distinguish them. They are all remarkable for the great disproportion of their legs. —Eᴅ.

acquaintance with the living subjects in their native wilds, I have presumed to remove the present species to the true and proper place assigned it by Nature, and shall now proceed to detail some particulars of its history.

This species arrives on the sea-coast of New Jersey about the 25th of April, in small detached flocks of twenty or thirty together. These sometimes again subdivide into lesser parties; but it rarely happens that a pair is found solitary, as, during the breeding season, they usually associate in small companies. On their first arrival, and, indeed, during the whole of their residence, they inhabit those particular parts of the salt marshes, pretty high up towards the land, that are broken into numerous shallow pools, but are not usually overflowed by the tides during the summer. These pools or ponds are generally so shallow, that, with their long legs, the avosets can easily wade them in every direction; and as they abound with minute shell-fish, and multitudes of aquatic insects and their larvæ, besides the eggs and spawn of others deposited in the soft mud below, these birds find here an abundant supply of food, and are almost continually seen wading about in such places, often up to the breast in water.

In the vicinity of these *bald places,* as they are called by the country people, and at the distance of forty or fifty yards off, among the thick tufts of grass, one of these small associations, consisting perhaps of six or eight pair, takes up its residence during the breeding season. About the first week in May they begin to construct their nests, which are at first slightly formed, of a small quantity of old grass, scarcely sufficient to keep the eggs from the wet marsh. As they lay and sit, however, either dreading the rise of the tides, or for some other purpose, the nest is increased in height with dry twigs of a shrub very common in the marshes, roots of the salt grass, seaweed, and various other substances, the whole weighing between two and three pounds. This habit of adding materials to the nest after the female begins sitting is common to almost all other birds that breed in the marshes. The eggs are four in number, of a dark yellowish clay colour, thickly marked with large blotches of black. These nests are often placed within fifteen or twenty yards of each other; but the greatest harmony seems to prevail among the proprietors.

While the females are sitting, the males are either wading through the ponds, or roaming over the adjoining marshes; but should a person make his appearance, the whole collect together in the air, flying with their long legs extended behind them, keeping up a continual yelping note of *click, click, click.* Their flight is steady, and not in short, sudden jerks, like that of the plover. As they frequently alight on the bare marsh, they drop their wings, stand with their legs half bent, and trembling, as if unable to sustain the burden of their bodies. In this ridiculous posture they will sometimes stand for several minutes, uttering a curring sound, while, from the corresponding quiverings of their wings and long legs, they seem to balance themselves with great difficulty. This singular manœuvre is, no doubt, intended to induce a belief that they may be easily caught, and so turn the attention of the person from the pursuit of their nests and young to themselves. The red-necked avoset, whom we have introduced in the present volume, practises the very same deception, in the same ludicrous manner, and both alight indiscriminately on the ground or in the water. Both will also occasionally swim for a few feet, when they chance, in wading, to lose their depth, as I have had several times an opportunity of observing.

The name by which this bird is known on the sea-coast is the stilt, or tilt, or long-shanks. They are but sparingly dispersed over the marshes, having, as has been already observed, their particular favourite spots, while in large intermediate tracts there are few or none to be found. They occasionally visit the shore, wading about in the water and in the mud in search of food, which they scoop up very dexterously with their delicately-formed bills. On being wounded while in the water, they attempt to escape by diving, at which they are by no means expert. In autumn, their flesh is tender and well tasted. They seldom raise more than one brood in the season, and depart for the south early in September. As they are well known in Jamaica, it is probable some of them may winter in that and other of the West India islands.

Mr Pennant observes that this bird is not a native of northern Europe, and there have been but few instances where it has been seen in Great Britain. It is common, says Latham, in Egypt, being found there in the marshes in October. It is likewise plentiful about the salt lakes, and is often seen on the shores of the Caspian Sea, as well as by the rivers which empty themselves into it, and in the southern deserts of Independent Tartary. The same author adds, on the authority of Ray, that it is known at Madras in the East Indies.

All the figures and descriptions which I have seen of this curious bird represent the bill as straight, and of almost an equal thickness throughout, but I have never found it so in any of the numerous specimens I have myself shot and examined. Many of these accounts, as well as figures, have been taken from dried and stuffed skins, which give but an imperfect, and often erroneous, idea of the true outlines of nature. The dimensions, colours, and markings of a very beautiful specimen, newly shot, were as follows :—

Length, from the point of the bill to the end of the tail, fourteen inches, to the tips of the wings, sixteen ; extent, twenty-eight inches ; bill, three inches long, slightly curved upwards, tapering to a fine point, the upper mandible rounded above, the whole of a deep black colour ; nostrils, an oblong slit, pervious ; tongue, short, pointed ; forehead, spot behind the eye, lower eyelid, sides of the neck, and whole lower parts, pure white ; back, rump, and tail-coverts, also white, but so concealed by the scapulars as to appear black ; tail, even, or very slightly forked, and of a dingy white ; the vent-feathers reach to the tip of the tail below ; line before the eye, auriculars, back part of the neck, scapulars, and whole wings, deep black, richly glossed with green ; legs and naked thighs, a fine pale carmine ; the latter measures three, the former four inches and a half in length, exceedingly thin, and so flexible that they may be bent considerably without danger of breaking. This thinness of the leg enables the bird to wade with expedition, and without fatigue. Feet, three-toed, the outer toe connected to the middle one by a broad membrane ; wings, long, extending two inches beyond the tail, and sharp pointed ; irides, a bright rich scarlet ; pupil, black. In some the white from the breast extends quite round the neck, separating the black of the hind neck from that of the body ; claws, blackish horn.

The female is about half an inch shorter, and differs in having the plumage of the upper back and scapulars, and also the tertials, of a deep brown colour. The stomach or gizzard was extremely muscular, and contained fragments of small snail-shells, winged bugs, and a slimy matter, supposed to be the remains of some aquatic worms. In one of these females I counted upwards of one hundred and fifty eggs, some of them as large as buckshot. The singular form of the legs and feet, with the exception of the hind toe and one membrane of the foot, is exactly like those of the avoset. The upper curvature of the bill, though not quite so great, is also the same as in the other, being rounded above, and tapering to a delicate point in the same manner. In short, a slight comparison of the two is sufficient to satisfy the most scrupulous observer that Nature has classed these two birds together ; and so believing, we shall not separate them.

SOLITARY SANDPIPER. (*Tringa solitaria.*)

PLATE LVIII.—Fig. 3.

Peale's Museum, No. 7763.

TOTANUS CHLOROPIGIUS.—Vieillot.*

Totanus glareolus, *Ord's reprint*, p. 57.—Totanus chloropigius, *Vieill.*—*Bonap. Cat.* p. 26.—*Synop.* p. 325.

This new species inhabits the watery solitudes of our highest mountains during the summer, from Kentucky to New York; but is nowhere numerous, seldom more than one or two being seen together. It takes short low flights; runs nimbly about among the mossy margins of the mountain springs, brooks, and pools, occasionally stopping, looking at you, and perpetually nodding the head. It is so unsuspicious, or so little acquainted with man, as to permit one to approach within a few yards of it, without appearing to take any notice or to be the least alarmed. At the approach of cold weather, it descends to the muddy shores of our large rivers, where it is occasionally met with singly, on its way to the south. I have made many long and close searches for the nest of this bird without success. They regularly breed on Pocano Mountain

* In the second edition of the seventh part, under the inspection of Mr Ord, this bird is described as new, by the name of *T. glareolus*. Ord thought it identical with the *T. glareolus* of Europe, and named it as such; his synonyms are, therefore, all wrong. The Prince of Musignano thus points out the differences: " *T. chloropigius* differs from *T. glareola*, not only as regards the characters of the tail-feathers, but also in being more minutely speckled, the white spots being smaller; by its longer tarsus; by the lineation of all the tail-feathers, but especially the lateral ones, the bands being broader, purer, and much more regular, whilst the latter tail-feathers of the European species are almost pure white on the inner webs; by having the shaft of the exterior primary black, whilst that of the *glareolus* is white."

The two specimens which Mr Ord shot, in which all the tail-feathers were barred, and which corresponded with *T. glareola*, may have been in fact that species. The Prince of Musignano is of opinion that it is also a native of North America.—Ed.

between Easton and Wilkesbarre, in Pennsylvania, arriving there early in May, and departing in September. It is usually silent, unless when suddenly flushed, when it utters a sharp whistle.

This species has considerable resemblance, both in manners and markings, to the green sandpiper of Europe (*Tringa ochropus*) ; but differs from that bird in being nearly one-third less, and in wanting the white rump and tail-coverts of that species ; it is also destitute of its silky olive green plumage. How far north its migrations extend I am unable to say

The solitary sandpiper is eight inches and a half long, and fifteen inches in extent ; the bill is one inch and a quarter in length, and dusky ; nostrils, pervious ; bill, fluted above and below ; line over the eye, chin, belly, and vent, pure white ; breast, white, spotted with pale olive brown ; crown and neck above, dark olive, streaked with white ; back, scapulars, and rump, dark brown olive, each feather marked along the edges with small round spots of white ; wings, plain, and of a darker tint ; under tail-covert, spotted with black ; tail, slightly rounded, the five exterior feathers on each side, white, broadly barred with black ; the two middle ones, as well as their coverts, plain olive ; legs, long, slender, and of a dusky green. Male and female alike in colour.

YELLOW-SHANKS SNIPE. (*Scolopax flavipes.*)

PLATE LVIII.—Fɪɢ. 4.

Arct. Zool. p. 463, No. 878.—*Turt. Syst.* 395.—*Peale's Museum,* No. 3938.

TOTANUS FLAVIPES.—Vᴇɪʟʟᴏᴛ.*

Totanus flavipes, *Ord's edit.* p. 59.—*Bonap. Cat.* p. 26.

Oғ this species I have but little to say. It inhabits our sea-coasts and salt marshes during summer ; frequents the flats at low water, and seems particularly fond of walking among the

* *T. flavipes* seems exclusively American.—Eᴅ.

mud, where it doubtless finds its favourite food in abundance. Having never met with its nest, nor with any person acquainted with its particular place or manner of breeding, I must reserve these matters for further observation. It is a plentiful species, and great numbers are brought to market in Boston, New York, and Philadelphia, particularly in autumn. Though these birds do not often penetrate far inland, yet, on the 5th of September, I shot several dozens of them in the meadows of Schuylkill, below Philadelphia. There had been a violent north-east storm a day or two previous, and a large flock of these, accompanied by several species of *Tringa*, and vast numbers of the short-tailed tern, appeared at once among the meadows. As a bird for the table, the yellow-shanks, when fat, is in considerable repute. Its chief residence is in the vicinity of the sea, where there are extensive mud flats. It has a sharp whistle, of three or four notes, when about to take wing and when flying. These birds may be shot down with great facility, if the sportsman, after the first discharge, will only lie close, and permit the wounded birds to flutter about without picking them up; the flock will generally make a circuit, and alight repeatedly, until the greater part of them may be shot down.

Length of the yellow-shanks, ten inches; extent, twenty; bill, slender, straight, an inch and a half in length, and black; line over the eye, chin, belly, and vent, white; breast and throat, gray; general colour of the plumage above, dusky brown olive, inclining to ash, thickly marked with small triangular spots of dull white; tail-coverts, white; tail, also white, handsomely barred with dark olive; wings, plain dusky, the secondaries edged, and all the coverts edged and tipt with white; shafts, black; eye, also black; legs and naked thighs, long and yellow; outer toe, united to the middle one by a slight membrane; claws, a horn colour. The female can scarcely be distinguished from the male.

TELL-TALE GODWIT, OR SNIPE. (*Scolopax vociferus.*)

PLATE LVIII.—Fig. 5.

Stone Snipe, *Arct. Zool.* p. 468, No. 376.—*Turt. Syst.* p. 396.—*Peale's Museum,* No. 3940.

TOTANUS MELANOLEUCUS.—Vieillot.*

T. melanoleucus, *Ord's reprint of Wils.* p. 61.—*Bonap. Synop.* p. 324.

This species and the preceding are both well known to our duck-gunners along the sea-coast and marshes, by whom they are detested, and stigmatised with the names of the greater and lesser tell-tale, for their faithful vigilance in alarming the ducks with their loud and shrill whistle on the first glimpse of the gunner's approach. Of the two, the present species is by far the most watchful; and its whistle, which consists of four notes rapidly repeated, is so loud, shrill, and alarming, as instantly to arouse every duck within its hearing, and thus disappoints the eager expectations of the marksman. Yet the cunning and experience of the latter are frequently more than a match for all of them; and before the poor tell-tale is aware, his warning voice is hushed for ever, and his dead body mingled with those of his associates.

* Bonaparte in his "Nomenclature" remarks, "This bird is undoubtedly the *S. melanoleuca* of Gmelin and Latham, first made known by Pennant. Why Wilson, who was aware of this, should have changed the name, we are at a loss to conceive. Mr Ord was, therefore, right in restoring it."

The species has not been discovered out of North America, and will take the place in that country of the European greenshank.

Totanus is a genus of Bechstein, now generally acknowledged as the proper place for the sandpipers of this form. Many of them do not undergo so decided a change during the breeding season, breed more inland, and, during winter, are as frequently found on the banks of rivers and lakes, or in inland marshes, as upon the shores. They are extremely noisy when first disturbed; a single individual readily gives the note of alarm; and when their nests are approached, they display more of the habit of the *Plovers.*—Ed.

This bird arrives on our coast early in April, breeds in the marshes, and continues until November, about the middle of which month it generally moves off to the south. The nest, I have been informed, is built in a tuft of thick grass, generally on the borders of a bog or morass. The female, it is said, lays four eggs of a dingy white, irregularly marked with black.

These birds appear to be unknown in Europe. They are simply mentioned by Mr Pennant as having been observed in autumn, feeding on the sands on the lower part of Chatteaux Bay, continually nodding their heads; and were called there stone curlews.*

The tell-tale seldom flies in large flocks, at least during summer. It delights in watery bogs and the muddy margins of creeks and inlets; is either seen searching about for food, or standing in a watchful posture, alternately raising and lowering the head, and, on the least appearance of danger, utters its shrill whistle, and mounts on wing, generally accompanied by all the feathered tribes that are near. It occasionally penetrates inland along the muddy shores of our large rivers, seldom higher than tide-water, and then singly and solitary. They sometimes rise to a great height in the air, and can be distinctly heard when beyond the reach of the eye. In the fall, when they are fat, their flesh is highly esteemed, and many of them are brought to our markets. The colours and markings of this bird are so like those of the preceding, that, unless in point of size and the particular curvature of the bill, the description of one might serve for both.

The tell-tale is fourteen inches and a half long, and twenty-five inches in extent; the bill is two inches and a quarter long, of a dark horn colour, and slightly bent upwards; the space round the eye, chin, and throat, pure white; lower part of the neck, pale ashy white, speckled with black; general colour of the upper parts, an ashy brown, thickly spotted with black and dull white, each feather being bordered and spotted on the edge

* Arctic Zoology, p. 468.

with black; wing-quills, black; some of the primaries, and all the secondaries, with their coverts, spotted round the margins with black and white; head and neck above, streaked with black and white; belly and vent, pure white; rump white, dotted with black; tail, also white, barred with brown; the wings, when closed, reach beyond the tail; thighs, naked nearly two inches above the knees; legs, two inches and three-quarters long; feet, four-toed, the outer joined by a membrane to the middle, the whole of a rich orange yellow. The female differs little in plumage from the male; sometimes the vent is slightly dotted with black, and the upper parts more brown.

Nature seems to have intended this bird as a kind of spy or sentinel for the safety of the rest; and so well acquainted are they with the watchful vigilance of this species, that, while it continues silent among them, the ducks feed in the bogs and marshes without the least suspicion. The great object of the gunner is to escape the penetrating glance of this guardian, which it is sometimes extremely difficult to effect. On the first whistle of the tell-tale, if beyond gunshot, the gunner abandons his design, but not without first bestowing a few left-handed blessings on the author of his disappointment.

[Mr Ord adds, "Pennant's spotted snipe is undoubtedly this species. He states that it arrives at Hudson's Bay in the spring; feeds on small shellfish and worms, and frequents the banks of rivers; called there by the natives, from its noise, *Sa-sa-shew.** This Indian word, pronounced with rapidity, gives a tolerable idea of the whistle of the tell-tale; and is a proof of the advantage of recording the vulgar names of animals, when these names are expressive of any peculiarity of voice or habit."]

* Arctic Zoology, vol. ii. p. 170.

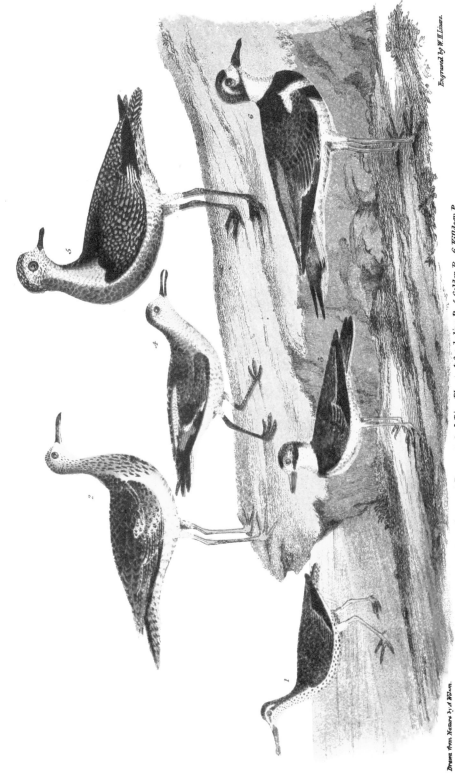

Drawn from Nature by A Wilson.

1. Spotted Sandpiper. 2. Bartram's S. 3. Ring Plover. 4. Sanderling P. 5. Golden P. 6. Killdeer P.

59.

Engraved by W. H. Lizars.

SPOTTED SANDPIPER. (*Tringa macularia.*)

PLATE LIX.—Fig. 1.

Arct. Zool. p. 473, No. 385.—La Grive d'Eau, *Buff.* viii. 140.—*Edw.* 277.—*Peale's Museum*, No. 4056.

TOTANUS MACULARIUS.—Temminck.*

Ord's reprint of Wils. part vii. p. 64.—*Temm. Man. d'Orn.* ii. p. 656.—*Bonap. Synop.* p. 325.—*Flem. Br. Zool.* p. 102.—Spotted Sandpiper, *Mont. Orn. Dict.* ii. and *Supp. Selby's Illust. of Br. Orn.* w. b. pl. 17.

This very common species arrives in Pennsylvania about the 20th of April, making its first appearance along the shores of our large rivers, and, as the season advances, tracing the courses of our creeks and streams towards the interior. Along the rivers Schuylkill and Delaware, and their tributary waters, they are in great abundance during the summer. This species is as remarkable for perpetually wagging the tail, as some others are for nodding the head ; for, whether running on the ground or on the fences, along the rails or in the water, this motion seems continual ; even the young, as soon as they are freed from the shell, run about constantly wagging the tail. About the middle of May they resort to the adjoining cornfields to breed, where I have frequently found and examined their nests. One of these now before me, and which was built at the root of a hill of Indian-corn, on high ground, is composed wholly of short pieces of dry straw. The eggs are four, of a pale clay or cream colour, marked with large irregular spots of black, and more thinly with others of a paler tint. They are large in proportion to the size of the bird, measuring an inch and a

* This is one of the most beautiful and most delicately marked among the smaller *Totani.* Closely allied to our common sand lark, *T. hypoleucos,* it is at once distinguished by the spotted marking on the under parts, which contrasts finely with their pure white. They frequent the banks of rivers more than the larger species, and have all a peculiar motion of the body and tail while running. The spotted sandpiper is common to both continents, and has been once or twice killed in Great Britain.—Ed.

quarter in length, very thick at the great end, and tapering suddenly to the other. The young run about with wonderful speed as soon as they leave the shell, and are then covered with down of a dull drab colour, marked with a single streak of black down the middle of the back, and with another behind each ear. They have a weak, plaintive note. On the approach of any person, the parents exhibit symptoms of great distress, counterfeiting lameness, and fluttering along the ground with seeming difficulty. On the appearance of a dog, this agitation is greatly increased; and it is very interesting to observe with what dexterity she will lead him from her young, by throwing herself repeatedly before him, fluttering off, and keeping just without his reach, on a contrary direction from her helpless brood. My venerable friend Mr William Bartram informs me, that he saw one of these birds defend her young for a considerable time from the repeated attacks of a ground-squirrel. The scene of action was on the river shore. The parent had thrown herself, with her two young behind her, between them and the land; and at every attempt of the squirrel to seize them by a circuitous sweep, raised both her wings in an almost perpendicular position, assuming the most formidable appearance she was capable of, and rushed forwards on the squirrel, who, intimidated by her boldness and manner, instantly retreated; but presently returning, was met, as before, in front and on flank by the daring and affectionate bird, who, with her wings and whole plumage bristling up, seemed swelled to twice her usual size. The young crowded together behind her, apparently sensible of their perilous situation, moving backwards and forwards as she advanced or retreated. This interesting scene lasted for at least ten minutes; the strength of the poor parent began evidently to flag, and the attacks of the squirrel became more daring and frequent, when my good friend, like one of those celestial agents who, in Homer's time, so often decided the palm of victory, stepped forward from his retreat, drove the assailant back to his hole, and rescued the innocent from destruction.

The flight of this bird is usually low, skimming along the surface of the water, its long wings making a considerable angle downwards from the body, while it utters a rapid cry of *weet, weet, weet,* as it flutters along, seldom steering in a direct line up or down the river, but making a long circuitous sweep, stretching a great way out, and gradually bending in again to the shore.

These birds are found occasionally along the sea-marshes, as well as in the interior; and also breed in the cornfields there, frequenting the shore in search of food; but rarely associating with the other *Tringæ.* About the middle of October, they leave us on their way to the south, and do not, to my knowledge, winter in any of the Atlantic States.

Mr Pennant is of opinion that this same species is found in Britain; but neither his description, nor that of Mr Bewick, will apply correctly to this. The following particulars, with the figure, will enable Europeans to determine this matter to their satisfaction :—

Length of the spotted sandpiper, seven inches and a half, extent, thirteen inches; bill, an inch long, straight; the tip and upper mandible dusky; lower, orange; stripe over the eye and lower eyelid, pure white; whole upper parts, a glossy olive, with greenish reflections, each feather marked with waving spots of dark brown; wing-quills, deep dusky; bastard wing, bordered and tipt with white; a spot of white on the middle of the inner vane of each quill-feather except the first; secondaries, tipt with white; tail, rounded, the six middle feathers greenish olive, the other three on each side white, barred with black; whole lower parts, white, beautifully marked with roundish spots of black, small and thick on the throat and breast, larger and thinner as they descend to the tail; legs, a yellow clay colour; claws, black.

The female is as thickly spotted below as the male; but the young birds of both sexes are pure white below, without any spots; they also want the orange on the bill. Those circumstances I have verified on numerous individuals.

BARTRAM'S SANDPIPER. (*Tringa Bartramia.*)

PLATE LIX.—Fig. 2.

Peale's Museum, No. 4040.

TOTANUS BARTRAMIUS.—Temminck.*

Totanus Bartramius, *Ord's reprint of Wils.* vol. vii. p. 67.—Chevalier à longue
queue, *Temm. Man. d'Orn.* ii. p. 650.—Totanus Bartramius, *Bonap. Synop.*
p. 325.

THIS bird being, as far as I can discover, a new species,
undescribed by any former author, I have honoured it with
the name of my very worthy friend, near whose botanic
gardens, on the banks of the river Schuylkill, I first found it.
On the same meadows I have since shot several other individuals of the species, and have thereby had an opportunity of
taking an accurate drawing as well as description of it.

Unlike most of their tribe, these birds appeared to prefer
running about among the grass, feeding on beetles and other
winged insects. There were three or four in company ; they
seemed extremely watchful, silent, and shy, so that it was
always with extreme difficulty I could approach them.

These birds are occasionally seen there during the months
of August and September, but whether they breed near, I have
not been able to discover. Having never met with them on
the sea-shore, I am persuaded that their principal residence
is in the interior, in meadows and such like places. They
run with great rapidity, sometimes spreading their tail and

* The discovery of this species, I believe, is due to our author, who
dedicated it to his venerable friend Bartram. It is admitted by
Temminck as an occasional straggler upon the Dutch and German
coasts, and is mentioned as having been only once met with by himself.
Bonaparte asserts, on the authority of Say, that it is very common in
some districts of the extensive Missouri prairies ; thus confirming the
opinion of Wilson, that its residence is in the interior, and not on the
sea-coast, like most of its congeners. The lengthened form, more conspicuous in the wedge shape of the tail, is at variance with the greater
part of the *Totani,* and reminds us of the killdeer plover.—ED.

dropping their wings, as birds do who wish to decoy you from their nest; when they alight they remain fixed, stand very erect, and have two or three sharp whistling notes as they mount to fly. They are remarkably plump birds, weighing upwards of three-quarters of a pound; their flesh is superior, in point of delicacy, tenderness, and flavour, to any other of the tribe with which I am acquainted.

This species is twelve inches long, and twenty-one in extent; the bill is an inch and a half long, slightly bent downwards, and wrinkled at the base, the upper mandible black on its ridge, the lower, as well as the edge of the upper, of a fine yellow; front, stripe over the eye, neck, and breast, pale ferruginous, marked with small streaks of black, which, on the lower part of the breast, assume the form of arrow-heads; crown, black, the plumage slightly skirted with whitish; chin, orbit of the eye, whole belly and vent, pure white; hind head and neck above, ferruginous, minutely streaked with black; back and scapulars, black, the former slightly skirted with ferruginous, the latter with white; tertials, black, bordered with white; primaries, plain black; shaft of the exterior quill, snowy, its inner vane elegantly pectinated with white; secondaries, pale brown, spotted on their outer vanes with black, and tipt with white; greater coverts, dusky, edged with pale ferruginous, and spotted with black; lesser coverts, pale ferruginous, each feather broadly bordered with white, within which is a concentric semicircle of black; rump and tail-coverts, deep brown black, slightly bordered with white, tail, tapering, of a pale brown orange colour, beautifully spotted with black, the middle feathers centred with dusky; legs, yellow, tinged with green; the outer toe joined to the middle by a membrane; lining of the wings, elegantly barred with black and white; iris of the eye, dark, or blue black; eye, very large. The male and female are nearly alike.

RING PLOVER.　(*Tringa hiaticula.*)

PLATE LIX.—Fig. 3.

Arct. Zool. p. 485, No. 401.—Le Petit Pluvier à Collier, *Buff.* viii. 90.—*Bewick*,
i. 326.—*Peale's Museum*, No. 4150.

CHARADRIUS SEMIPALMATUS.—Bonaparte.[*]

Charadrius semipalmatus, *Bonap. Synop.* p. 296.—American Ring Plover, *North.
Zool.* ii. p. 367.—Charadrius semipalmatus? *Wagl. Syst. Av.* No. 23.

In a preceding part of this work (see Plate xxxvii. Fig. 3),
a bird by this name has been figured and described, under the
supposition that it was the ring plover, then in its summer
dress; but which, notwithstanding its great resemblance to the
present, I now suspect to be a different species. Fearful of
perpetuating error, and anxious to retract where this may
inadvertently have been the case, I shall submit to the consideration of the reader the reasons on which my present suspicions are founded.

[*] The smaller *Charadriadæ* of America have been much confused,
owing to their close alliance to each other and to those of Europe, with
some of which they were thought to be identical. The Prince of
Musignano has clearly pointed out the differences which exist between
this and the species figured at Plate XXXVII, and which bears a more
close resemblance to the little African *C. pecuarius* than either the
present species or the *hiaticula* of Europe (see also our note on that
species); and although he has not been able to point out such distinctive
characters between the latter species and that now under discussion, I
have no doubt whatever of their being eventually found quite distinct;
and it will be found, by those persons who are inclined to allow so much
for the influence of climate in rendering form, colour, and plumage
distinct, that it is comparatively of no importance, and that identical
species, running through a great variety of latitude, will in fact differ
little or nothing from each other. I have transcribed the observations
of Bonaparte from his " Nomenclature of Wilson," which will show his
opinion.

He thus observes,—" The remark made by Mr Ord, relative to the
difference between the union of the toes in American and European
specimens, is no less extraordinary than correct; I have verified it on
the specimens in my collection. This character would seem to show,

The present species, or true ring plover, and also the former or light-coloured bird, both arrive on the sea-coast of New Jersey late in April. The present kind continues to be seen in flocks until late in May, when they disappear on their way farther north; the light-coloured bird remains during the summer, forms its nest in the sand, and generally produces two broods in the season. Early in September the present species returns in flocks as before; soon after this the light-coloured kind go off to the south, but the other remain a full month later. European writers inform us that the ring plover has a sharp twittering note; and this account agrees exactly with that of the present: the light-coloured species, on the contrary, has a peculiarly soft and musical note, similar to the tone of a German flute, which it utters while running along the sand, with expanded tail and hanging wings, endeavouring to decoy you from its nest. The present species is never seen to breed here; and though I have opened great numbers of them as late as the 20th of May, the eggs which the females contained were never larger than small birdshot;

in the most positive manner, that they are distinct but allied species, differing from each other as *Tringa semipalmata* of Wilson differs from his *Tringa pusilla.*"

The synonyms of Mr Ord, who noticed one of the principal distinctions in the palmation of the feet, are consequently wrong, and they should stand as above. I have added a synonym of Wagler, *C. semipalmatus*, which he takes, without any acknowledgment, from Cont. Isis, 1825, and which seems to be this species. He also refers to the *C. hiaticula* of Wilson, Plate XXXVII., under the name of *C. Okenii.* The true *C. hiaticula* has not yet, I believe, been found in North America.

"I have been endeavouring," again writes Bonaparte, "to discover some other markings on my stuffed specimens, that might enable me to establish the species on a more solid basis; but though certain small differences are discernible, such as the somewhat smaller size, and the black narrow collar of the American, &c., yet we are aware that such trifling differences occur between individuals of the same species; we shall, therefore, not rely on them until our observations shall have been repeated on numerous recent or living specimens. In the meantime, should the species prove to be distinct, it may be distinguished by the appropriate name of *C. semipalmatus.*"—ED.

while, at the same time, the light-coloured kind had every-where begun to lay in the little cavities which they had dug in the sand on the beach. These facts being considered, it seems difficult to reconcile such difference of habit in one and the same bird. The ring plover is common in England, and agrees exactly with the one now before us; but the light-coloured species, as far as I can learn, is not found in Britain; specimens of it have indeed been taken to that country, where the most judicious of their ornithologists have concluded it to be still the ring plover, but to have changed from the effect of climate. Mr Pennant, in speaking of the true ring plover, makes the following remarks:—" Almost all which I have seen from the northern parts of North America have had the black marks extremely faint, and almost lost. The climate had almost destroyed the specific marks; yet in the bill and habit, preserved sufficient to make the kind very easily ascertained." These traits agree exactly with the light-coloured species, de-scribed in our fifth volume.* But this excellent naturalist was perhaps not aware that we have the true ring plover here in spring and autumn, agreeing in every respect with that of Britain, and at least in equal numbers; why, therefore, has not the climate equally affected the present and the former sort, if both are the same species? These inconsistencies cannot be reconciled but by supposing each to be a distinct species, which, though approaching extremely near to each other in external appearance, have each their peculiar notes, colour, and places of breeding.†

The ring plover is seven inches long, and fourteen inches in extent; bill, short, orange coloured, tipt with black; front and chin, white, encircling the neck; upper part of the breast,

* Vol. II. p. 122 of this edition.

† It is mentioned as abundant in all " Arctic America " by the authors of the " Northern Zoology," " where it breeds in similar situations to the golden plover. Mr Hutchins reports that the eggs, generally four, are dark coloured, spotted with black. The natives say, that, on the approach of stormy weather, this plover makes a chirruping noise, and claps its wings."—ED.

black ; rest of the lower parts, pure white ; fore part of the crown, black ; band from the upper mandible covering the auriculars, also black ; back, scapulars, and wing-coverts, of a brownish ash colour ; wing-quills, dusky black, marked with an oval spot of white about the middle of each ; tail, olive, deepening into black, and tipt with white ; legs, dull yellow ; eye, dark hazel ; eyelids, yellow.

This bird is said to make no nest, but to lay four eggs of a pale ash colour, spotted with black, which she deposits on the ground.* The eggs of the light-coloured species, formerly described, are of a pale cream colour, marked with small round dots of black, as if done with a pen.

The ring plover, according to Pennant, inhabits America down to Jamaica and the Brazils ; is found in summer in Greenland ; migrates from thence in autumn ; is common in every part of Russia and Siberia ; was found by the navigators as low as Owyhee, one of the Sandwich Islands, and as light coloured as those of the highest latitudes. †

[Mr Ord adds to this description in his reprint : " After writing the above I had an opportunity of examining, comparatively, two or three specimens of the European ring plover which are in Mr Peale's collection. These birds corresponded with the subject of this article, except in the feet, and here I found a difference which is worthy of note. The outer toes of both the European and the American birds were united to the middle ones by a membrane of an equal size ; but the inner toes of the latter were also united by a smaller web, while those of the former were *divided to their origin.* The naturalists of Europe state that the inner toes of their species are thus divided. Here, then, is a diversity which, if constant, would constitute a specific difference. The bottoms of the toes of the present are broad as in the sanderling.

" The plover given in our fifth volume, under the name of

* Bewick. † Arct. Zool., p. 485.

hiaticula, has its inner toes divided to their origin, and the web of the outer toes is much smaller than that of the present article. All my doubts on the subject of our two plovers being now removed, I shall take the liberty of naming that of the fifth volume, the piping plover, *Charadrius melodus.*"]

SANDERLING PLOVER. (*Charadrius calidris.*)

PLATE LIX.—Fig. 4.

Linn. Syst. 255.—*Arct. Zool.* p. 486, No. 403.—Le Sanderling, *Buff.* vii. 532.— *Bewick,* ii. 19.—*Peale's Museum,* No. 4204.

CALIDRIS ARENARIA.—Illiger.*

Charadrius calidris, *Wils.* 1st edit. vii. p. 68 ; and Ch. rubidus, *Wils.* 1st edit. vii. p. 129.—Calidris, *Illig. Prod. Mam. et Av.* p. 249.—Ruddy Plover, *Penn. Arct. Zool.* ii. p. 486, summer plumage.—Sanderling variable (Calidris arenaria), *Temm. Man. d'Orn.* ii. 524.—Tringa (Calidris) arenaria, *Bonap. Synop.*—Calidris arenaria, *Flem. Br. Zool.* p. 112.—*North. Zool.* ii. p. 366.

In this well-known bird we have another proof of the imperfection of systematic arrangement, where no attention is paid to the general habits, but where one single circumstance is sometimes considered sufficient to determine the species. The genus plover is characterised by several strong family traits, one of which is that of wanting the hind toe. The sandpipers have also their peculiar external characters of bill, general

---

* *Calidris* was established for this single species, common over the world, and of form intermediate between the plovers and sandpipers. Their make is thicker ; they are less slender than the sandpipers ; the bill stronger, but, as in that group, the feet similar to those of the *Charadrii;* and with their manner of running and walking, they possess that peculiar crouch of the head upon the back seen in the common ring plover and its allies. The ruddy plover of the plate represents it in the summer plumage, in which it more resembles the changes exhibited in the knot and pigmy curlew than those of the dunlins. On the shores of Britain, it is generally met with in winter in small flocks, or in spring and autumn when going to or returning from their breeding quarters.

By Mr Hutchins it is said to make its nest rudely of grass in the marshes, and lays four dusky coloured eggs, spotted with black.—Ed.

form, &c., by which they are easily distinguished from the former. The present species, though possessing the bill, general figure, manners, and voice of the sandpipers, feeding in the same way, and associating with these in particular, yet wanting the hind toe, has been classed with the plovers, with whom, this single circumstance excepted, it has no one characteristic in common. Though we have not, in the present instance, presumed to alter this arrangement, yet it appears both reasonable and natural that, where the specific characters in any bird seem to waver between two species, that the figure, voice, and habits of the equivocal one should always be taken into consideration, and be allowed finally to determine the class to which it belongs. Had this rule been followed in the present instance, the bird we are now about to describe would have undoubtedly been classed with the sandpipers.

The history of this species has little in it to excite our interest or attention. It makes its appearance on our sea-coasts early in September, continues during the greater part of winter, and, on the approach of spring, returns to the northern regions to breed. While here, it seems perpetually busy running along the wave-worn strand, following the flux and reflux of the surf, eagerly picking up its food from the sand amid the roar of the ocean. It flies in numerous flocks, keeping a low meandering course along the ridges of the tumbling surf. On alighting, the whole scatter about after the receding wave, busily picking up those minute bivalves already described. As the succeeding wave returns, it bears the whole of them before it in one crowded line; then is the moment seized by the experienced gunner to sweep them in flank with his destructive shot. The flying survivors, after a few aerial meanders, again alight, and pursue their usual avocation as busily and unconcernedly as before. These birds are most numerous on extensive sandy beaches in front of the ocean. Among rocks, marshes, or stones covered with seaweed, they seldom make their appearance.

The sanderling is eight inches long, and fourteen inches in

extent; the bill is black, an inch and a quarter in length, slender, straight, fluted along the upper mandible, and exactly formed like that of the sandpiper; the head, neck above, back, scapulars, and tertials, are gray white; the shafts blackish, and the webs tinged with brownish ash; shoulder of the wing, black; greater coverts, broadly tipt with white; quills, black, crossed with a transverse band of white; the tail extends a little beyond the wings, and is of a grayish ash colour, edged with white, the two middle feathers being about half an inch longer than the others; eye, dark hazel; whole lower parts of the plumage, pure white; legs and naked part of the thighs, black; feet, three-toed, each divided to its origin, and bordered with a narrow membrane.

Such are the most common markings of this bird, both of males and females, particularly during the winter; but many others occur among them, early in the autumn, thickly marked or spotted with black on the crown, back, scapulars, and tertials, so as to appear much mottled, having as much black as white on those parts. In many of these I have observed the plain gray plumage coming out about the middle of October; so that perhaps the gray may be their winter, and the spotted their summer, dress.

I have also met with many specimens of this bird, not only thickly speckled with white, and black above, but also on the neck, and strongly tinged on both with ferruginous; in which dress it has been mistaken by Mr Pennant and others for a new species—the description of his "ruddy plover" agreeing exactly with this.* A figure of the sanderling in this state of plumage will be introduced in some part of the present work.

* See Arct. Zool., p. 486, No. 404.

GOLDEN PLOVER. (*Charadrius pluvialis.*)

PLATE LIX.—FIG. 5.

Arct. Zool. p. 493, No. 399.—*Bewick,* i. 322.—Le Pluvier Doré, *Buff.* viii. 81, *Pl. enl.* 904.—*Peale's Museum,* No. 4198.

CHARADRIUS VIRGINIANUS.—BONAPARTE.*

Charadrius pluvialis, *Bonap. Synop.* p. 297.—*North. Zool.* ii. p. 369.—Charadrius Virginianus, *Bonap. Osser. Sulla,* 2d edit. *Del. Regn. Anim. Cuv.* p. 93.—Charadrius marmoratus, *Wagl. Syst. Av. Char.* No. 42.

THIS beautiful species visits the sea-coast of New York and New Jersey in spring and autumn; but does not, as far as I

* The Prince of Musignano, after the publication of his "Synopsis of North American Birds," and "Observations on Wilson's Nomenclature," pointed out the distinction of the North American and European birds. The plate of Wilson also shows every character of the northern birds. The lengthened bill and legs, the more distinct dorsal spotting, and clearer colour of the forehead, the dusky hue of the under parts, and the mention, by Ord, of the brown axillaries, all point out this bird, which can never be mistaken. The following are the principal distinctions which appear between skins of *C. Virginianus* from India and New Holland, and specimens of *C. pluvialis,* shot this forenoon :—

C. pluvialis.

1. Total length, 10½ inches.

2. Length of bill to extremity of gape, 1 inch.
3. Length of wing, from joining of bastard pinion to forearm, and tip of first or longest quill, 8 inches.
4. Length of unfeathered tibia, ⅜ inch.
5. Length of tarsus, 1⅝ inch.
6. Throat, lower part of the breast, belly, vent, and crissum, pure white.

C. Virginianus.

1. The skins are about 10 inches in length, but are much stretched; 9½, or 8, as mentioned by Wagler, nearly the true length.
2. Length of bill to extremity of gape, 1⅛ inch.
3. Length of wing, from joining of bastard pinion to forearm, and tip of first or longest quill, 6½ inches.
4. Length of unfeathered tibia, ¾ inch.
5. Length of tarsus, nearly 1⅝ inch.
6. Throat, and all under parts, dull yellowish gray, with darker tips to the feathers.

can discover, breed in any part of the United States. They are most frequently met with in the months of September and October; soon after which they disappear. The young birds of the great black-bellied plover are sometimes mistaken for this species. Hence the reason why Mr Pennant remarks his having seen a variety of the golden plover, with black breasts, which he supposed to be the young.*

The golden plover is common in the northern parts of Europe. It breeds on high and heathy mountains. The female lays four eggs, of a pale olive colour, variegated with blackish spots. They usually fly in small flocks, and have a shrill whistling note. They are very frequent in Siberia,

7. Pale markings on the upper parts, dull gamboge yellow; spotting more in oblong spots; and, on the wing and tail-coverts, take the form of bars.

8. Light markings on the tail dull and undecided, with a decided dark barring.

9. Outer tail-feathers with pale margins, the distinct and frequent barring through the whole length.

10. Under wing coverts and axillaries, pure white.

11. Lesser wing coverts, tipped with white, but otherwise of a uniform colour.

7. Pale markings on the upper parts larger, and inclining more to clear white; above, more in spots on the sides of the feathers.

8. Light markings on the tail decided, nearly white; no dark bar through it.

9. Outer tail-feathers, with white tip and outer margin, which shoot down the rachis.

10. Under wing-coverts and axillaries, wood brown gray.

11. Lesser wing-coverts tipped and rather broadly edged with white.

C. pluvialis is introduced into the "Northern Zoology," but I strongly suspect these excellent ornithologists have overlooked the other species. Both may be natives of North America; I have never, however, seen or received extra European specimens of the golden plover; I possess *C. Virginianus* from India, Arctic America, and New Holland, which seems, in all those countries, very and exclusively abundant, and has always been confounded with its ally.

In plate 85 of "Ornithological Illustrations," this bird has most unaccountably been described under the title of *C. xanthochielus*, Wagler. It is undoubtedly this species, and figured from New Holland specimens.—ED.

* Arct. Zool., p. 484.

where they likewise breed ; extend also to Kamtschatka, and as far south as the Sandwich Isles. In this latter place, Mr Pennant remarks, " they are very small."

Although these birds are occasionally found along our sea-coast from Georgia to Maine, yet they are nowhere numerous; and I have never met with them in the interior. Our mountains being generally covered with forest, and no species of heath having as yet been discovered within the boundaries of the United States, these birds are probably induced to seek the more remote arctic regions of the continent to breed and rear their young in, where the country is more open, and unencumbered with woods.

The golden plover is ten inches and a half long, and twenty-one inches in extent ; bill, short, of a dusky slate colour ; eye, very large, blue black ; nostrils, placed in a deep furrow, and half covered with a prominent membrane ; whole upper parts, black, thickly marked with roundish spots of various tints of golden yellow ; wing-coverts and hind part of the neck, pale brown, the latter streaked with yellowish ; front, broad line over the eye, chin, and sides of the same, yellowish white, streaked with small pointed spots of brown olive ; breast, gray, with olive and white ; sides under the wings, marked thinly with transverse bars of pale olive ; belly and vent, white ; wing-quills, black, the middle of the shafts marked with white ; greater coverts, black tipt with white ; tail, rounded, black, barred with triangular spots of golden yellow ; legs, dark dusky slate ; feet, three-toed, with generally the slight rudiments of a heel, the outer toe connected, as far as the first joint, with the middle one. The male and female differ very little in colour.

KILDEER PLOVER. (*Charadrius vociferus.*)

PLATE LIX.—Fig. 6.

Arct. Zool. No. 400.—*Catesby*, i. 71.—Le Kildir, *Buff.* viii. 96.—*Peale's Museum,*
No. 4174.

CHARADRIUS VOCIFERUS.—Linnæus.*

Charadrius vociferus, *Bonap. Synop. North. Zool.* ii. p. 368.

This restless and noisy bird is known to almost every inhabitant of the United States, being a common and pretty constant resident. During the severity of winter, when snow covers the ground, it retreats to the sea-shore, where it is found at all seasons; but no sooner have the rivers broke up, than its shrill note is again heard, either roaming about high in air, tracing the shore of the river, or running amidst the watery flats and meadows. As spring advances, it resorts to the newly-ploughed fields, or level plains bare of grass interspersed with shallow pools, or, in the vicinity of the sea, dry bare sandy fields. In some such situation it generally chooses to breed about the beginning of May. The nest is usually slight, a mere hollow, with such materials drawn in around it as happen to be near, such as bits of sticks, straw, pebbles, or earth. In one instance, I found the nest of this bird paved with fragments of clam and oyster shells, and very neatly surrounded with a mound or border of the same, placed in a very close and curious manner. In some cases there is no vestige whatever of a nest. The eggs are usually four, of a bright rich cream or yellowish clay colour, thickly marked with blotches of black. They are large for the size of the bird, measuring more than an inch and a half in length, and a full inch in width, tapering to a narrow point at the great end.

* An abundant and well-known species, and peculiar to both continents of America, with some of the West Indian islands. According to the " Northern Zoology," it arrives on the plains of the Saskatchewan about the 20th of April, and at that season frequents the gardens and cultivated fields of the trading post with the utmost familiarity.—Ed.

Nothing can exceed the alarm and anxiety of these birds during the breeding season. Their cries of *kildeer, kildeer,* as they winnow the air overhead, dive and course around you, or run along the ground counterfeiting lameness, are shrill and incessant. The moment they see a person approach, they fly or run to attack him with their harassing clamour, continuing it over so wide an extent of ground, that they puzzle the pursuer as to the particular spot where the nest or young are concealed; very much resembling, in this respect, the lapwing of Europe. During the evening, and long after dusk, particularly in moonlight, their cries are frequently heard with equal violence, both in the spring and fall. From this circumstance, and their flying about both after dusk and before dawn, it appears probable that they see better at such times than most of their tribe. They are known to feed much on worms, and many of these rise to the surface during the night. The prowling of owls may also alarm their fears for their young at those hours; but, whatever may be the cause, the facts are so.

The kildeer is more abundant in the southern States in winter than in summer. Among the rice-fields, and even around the planters' yards, in South Carolina, I observed them very numerous in the months of February and March. There the negro boys frequently practise the barbarous mode of catching them with a line, at the extremity of which is a crooked pin with a worm on it. Their flight is something like that of the tern, but more vigorous; and they sometimes rise to a great height in the air. They are fond of wading in pools of water, and frequently bathe themselves during the summer. They usually stand erect on their legs, and run or walk with the body in a stiff horizontal position; they run with great swiftness, and are also strong and vigorous in the wings. Their flesh is eaten by some, but is not in general esteem; though others say that, in the fall, when they become very fat, it is excellent.

During the extreme droughts of summer, these birds resort to the gravelly channel of brooks and shallow streams, where they can wade about in search of aquatic insects.   At the close of summer, they generally descend to the sea-shore, in small flocks, seldom more than ten or twelve being seen together. They are then more serene and silent, as well as difficult to be approached.

The kildeer is ten inches long, and twenty inches in extent; the bill is black; frontlet, chin, and ring round the neck, white; fore part of the crown and auriculars, from the bill backwards, blackish olive; eyelids, bright scarlet; eye, very large, and of a full black; from the centre of the eye backwards, a strip of white; round the lower part of the neck is a broad band of black; below that, a band of white, succeeded by another rounding band or crescent of black; rest of the lower parts, pure white; crown and hind head, light olive brown; back, scapulars, and wing-coverts, olive brown, skirted with brownish yellow; primary quills, black, streaked across the middle with white; bastard wing, tipt with white; greater coverts, broadly tipt with white; rump and tail-coverts, orange; tail, tapering, dull orange, crossed near the end with a broad bar of black, and tipt with orange, the two middle feathers near an inch longer than the adjoining ones; legs and feet, a pale light clay colour.   The tertials, as usual in this tribe, are very long, reaching nearly to the tips of the primaries; exterior toe, joined by a membrane to the middle one, as far as the first joint.

GREAT TERN.   (*Sterna hirundo.*)

PLATE LX.—Fig. 1.

Arct. Zool. p. 524, No. 448.—Le Pierre Garin, ou Grande Hirondelle de Mer, *Buff.* viii. 331, *Pl. enl.* 987.—*Bewick*, ii. 181.—*Peale's Museum*, No. 3485.

STERNA WILSONII.—Bonaparte.[*]

Sterna hirundo, *Bonap. Synop.* p. 354.—St. Wilsonii, *Bonap. Osserv. Sulla,* 2d edit. *Del Regn. Anim. Cuv.* p. 135.

This bird belongs to a tribe very generally dispersed over the shores of the ocean.   Their generic characters are these :—

[*] Mr Ord, in his reprint, and C. L. Bonaparte, when writing his "Synopsis and Observations on the Nomenclature of Wilson," considered

1. Great Tern. 2. Lesser T. 3. Short tailed T. 4. Black Skimmer. 5. Stormy Petrel.

60.

Bill, straight, sharp pointed, a little compressed, and strong; nostrils, linear; tongue, slender, pointed; legs, short; feet, webbed; hind toe and its nail, straight; wings, long; tail, generally forked. Turton enumerates twenty-five species of this genus, scattered over various quarters of the world; six of which, at least, are natives of the United States. From their long pointed wings, they are generally known to seafaring people, and others residing near the sea-shore, by the name of *sea-swallows;* though some few, from their near resemblance, are confounded with the gulls.

The present species, or great tern, is common to the shores of Europe, Asia, and America. It arrives on the coast of New Jersey about the middle or 20th of April, led, no doubt, by the multitudes of fish which at that season visit our shallow bays and inlets. By many it is called the Sheep's-head gull, from arriving about the same time with the fish of that name.

About the middle or 20th of May, this bird commences laying. The preparation of a nest, which costs most other birds so much time and ingenuity, is here altogether dispensed with. The eggs, generally three in number, are placed on the surface of the dry drift grass, on the beach or salt marsh, and covered by the female only during the night, or in wet, raw, or stormy weather. At all other times, the hatching of them is left to the heat of the sun. These eggs measure an inch and three-quarters in length, by about an inch and two-tenths in width, and are of a yellowish dun colour, sprinkled with dark brown and pale Indian-ink. Notwithstanding they seem thus negligently abandoned during the day, it is very different in reality. One or both of the parents are generally fishing within view of the place, and, on the near approach of any

this bird as identical with the *St. hirundo* of Europe. Later comparisons by the Prince have induced him to consider it distinct, and peculiar to America, and he has dedicated it to Wilson. That gentleman mentions, as North American, in addition to the list by Wilson, *St. cyanea*, Lath.; *St. arctica*, Temm.; *St. stolida*, Linn.—ED.

person, instantly make their appearance overhead, uttering a hoarse jarring kind of cry, and flying about with evident symptoms of great anxiety and consternation. The young are generally produced at intervals of a day or so from each other, and are regularly and abundantly fed for several weeks before their wings are sufficiently grown to enable them to fly. At first the parents alight with the fish which they have brought in their mouth or in their bill, and tearing it in pieces, distribute it in such portions as their young are able to swallow. Afterwards they frequently feed them without alighting, as they skim over the spot; and, as the young become nearly ready to fly, they drop the fish among them, where the strongest and most active have the best chance to gobble it up. In the meantime, the young themselves frequently search about the marshes, generally not far apart, for insects of various kinds; but so well acquainted are they with the peculiar language of their parents that warn them of the approach of an enemy, that, on hearing their cries, they instantly squat, and remain motionless until the danger be over.

The flight of the great tern, and, indeed, of the whole tribe, is not in the sweeping shooting manner of the land swallows, notwithstanding their name; the motions of their long wings are slower, and more in the manner of the gull.

They have, however, great powers of wing and strength in the muscles of the neck, which enable them to make such sudden and violent plunges, and that from a considerable height too, headlong on their prey, which they never seize but with their bills. In the evening, I have remarked, as they retired from the upper parts of the bays, rivers, and inlets to the beach for repose, about breeding time, that each generally carried a small fish in his bill.

As soon as the young are able to fly, they lead them to the sandy shoals and ripples where fish are abundant; and while they occasionally feed them, teach them by their example to provide for themselves. They sometimes penetrate a great way inland, along the courses of rivers; and are occasionally seen about all our numerous ponds, lakes, and rivers, most usually near the close of the summer.

This species inhabits Europe as high as Spitzbergen; is found on the arctic coasts of Siberia and Kamtschatka, and also on our own continent as far north as Hudson's Bay. In New England, it is called by some the Mackerel-gull. It retires from all these places at the approach of winter to more congenial seas and seasons.

The great tern is fifteen inches long, and thirty inches in extent; bill, reddish yellow, sometimes brilliant crimson, slightly angular on the lower mandible, and tipt with black; whole upper part of the head, black, extending to a point half way down the neck behind, and including the eyes; sides of the neck, and whole lower parts, pure white; wing-quills, hoary, as if bleached by the weather, long and pointed; whole back, scapulars, and wing, bluish white, or very pale lead colour; rump and tail-coverts, white; tail, long, and greatly forked, the exterior feathers being three inches longer than the adjoining ones, the rest shortening gradually for an inch and a half to the middle ones, the whole of a pale lead colour; the outer edge of the exterior ones, black; legs and webbed feet, brilliant red lead; membranes of the feet, deeply scalloped; claws, large and black, middle one the largest. The primary quill-feathers are generally dark on their inner edges. The female differs in having the two exterior feathers of the tail, considerably shorter. The voice of these birds is like the harsh jarring of an opening door rusted on its hinges. The bone of the skull is remarkably thick and strong, as also the membrane that surrounds the brain; in this respect resembling the woodpecker's. In both, this provision is doubtless intended to enable the birds to support, without injury, the violent concussions caused by the plunging of the one and the chiselling of the other.

LESSER TERN.   (*Sterna minuta.*)

PLATE LX.—Fig. 2.

Arct. Zool. No. 449.—La Petite Hirondelle de Mer, *Buff.* viii. 337, *Pl. enl.* 996.—
Bewick, ii. 183.—*Peale's Museum,* No. 3505.

STERNA MINUTA.—Linnæus.*

Sterna minuta, *Bonap. Synop.*—*Flem. Br. Zool.* p. 144.—*Temm. Man. d'Orn.*
ii. p. 75.

This beautiful little species looks like the preceding in miniature, but surpasses it far in the rich glossy satin-like white plumage with which its throat, breast, and whole lower parts are covered.   Like the former, it is also a bird of passage, but is said not to extend its migrations to so high a northern latitude, being more delicate and susceptible of cold.   It arrives on the coast somewhat later than the other, but in equal and perhaps greater numbers ; coasts along the shores, and also over the pools in the salt marshes, in search of prawns, of which it is particularly fond ; hovers, suspended in the air, for a few moments above its prey, exactly in the manner of some of our small hawks, and dashes headlong down into the water after it, generally seizing it with its bill ; mounts instantly again to the same height, and moves slowly along as before, eagerly examining the surface below.   About the 25th of May, or beginning of June, the female begins to lay.

* This species is common to Europe and the northern continent of America.   Bonaparte mentions another closely allied species, which appears to take its place in South America, and has been confounded with it.

The breeding places of this tern are somewhat different from many of those British species with which we are acquainted.   Most of the latter breed on rocky coasts and solitary islands, while the little tern prefers flat shingly beaches, where the eggs are deposited in the manner described by Wilson,—in some little hollow or footstep.   They become clamorous on approaching the nest, but seem hardly so familiar or bold as most of the others.   The young soon leave the hollow where they were hatched, and move about as far as their limited powers will allow.—Ed.

The eggs are dropped on the dry and warm sand, the heat of which, during the day, is fully sufficient for the purpose of incubation. This heat is sometimes so great, that one can scarcely bear the hand in it for a few moments without inconvenience. The wonder would, therefore, be the greater should the bird sit on her eggs during the day, when her warmth is altogether unnecessary, and perhaps injurious, than that she should cover them only during the damps of night, and in wet and stormy weather; and furnishes another proof that the actions of birds are not the effect of mere blind impulse, but of volition, regulated by reason, depending on various incidental circumstances to which their parental cares are ever awake. I lately visited those parts of the beach on Cape May where this little bird breeds. The eggs, generally four in number, were placed on the flat sands, safe beyond the reach of the highest summer tide. They were of a yellowish brown colour, blotched with rufous, and measured nearly an inch and three-quarters in length. During my whole stay, these birds flew in crowds around me, and often within a few yards of my head, squeaking like so many young pigs, which their voice strikingly resembles. A humming-bird, that had accidentally strayed to the place, appeared suddenly among this outrageous group, several of whom darted angrily at him; but he shot like an arrow from them, directing his flight straight towards the ocean. I have no doubt but the distressing cries of the terns had drawn this little creature to the scene, having frequently witnessed his anxious curiosity on similar occasions in the woods.

The lesser tern feeds on beetles, crickets, spiders, and other insects, which it picks up from the marshes, as well as on small fish, on which it plunges at sea. Like the former, it also makes extensive incursions inland along the river courses, and has frequently been shot several hundred miles from the sea. It sometimes sits for hours together on the sands, as if resting after the fatigues of flight to which it is exposed.

The lesser tern is extremely tame and unsuspicious, often

passing you on its flight, and within a few yards, as it traces the windings and indentations of the shore in search of its favourite prawns and skippers. Indeed, at such times it appears altogether heedless of man, or its eagerness for food overcomes its apprehensions for its own safety. We read in ancient authors, that the fishermen used to float a cross of wood, in the middle of which was fastened a small fish for a bait, with limed twigs stuck to the four corners, on which the bird darting was entangled by the wings. But this must have been for mere sport, or for its feathers, the value of the bird being scarcely worth the trouble, as they are generally lean, and the flesh savouring strongly of fish.

The lesser tern is met with in the south of Russia, and about the Black and Caspian Seas; also in Siberia about the Irtish.* With the former, it inhabits the shores of England during the summer, where it breeds, and migrates, as it does here, to the south as the cold of autumn approaches.

This species is nine and a half inches long, and twenty inches in extent ; bill, bright reddish yellow ; nostril, pervious ; lower mandible, angular ; front, white, reaching in two narrow points over the eye ; crown, band through the eye, and hind head, black, tapering to a point as it descends ; cheeks, sides of the neck, and whole lower parts, of the most rich and glossy white, like the brightest satin ; upper parts of the back and wings, a pale glossy ash or light lead colour ; the outer edges of the three exterior primaries, black, their inner edges white ; tail, pale ash, but darker than the back, and forked, the two outer feathers an inch longer, tapering to a point ; legs and feet, reddish yellow ; webbed feet, claws, and hind toe exactly formed like those of the preceding. The female nearly resembles the male, with the exception of having the two exterior tail-feathers shorter.

* Pennant.

SHORT-TAILED TERN. (*Sterna plumbea.*)

PLATE LX.—FIG. 3.

Peale's Museum, No. 3519.

STERNA NIGRA.—LINNÆUS.*

Sterna plumbea, *Bonap. Nomencl.* No. 244.—Sterna nigra, *Bonap. Synop.* p. 355.

A SPECIMEN of this bird was first sent me by Mr Beasley of Cape May ; but being in an imperfect state, I could form no correct notion of the species, sometimes supposing it might be a young bird of the preceding tern.   Since that time, however, I have had an opportunity of procuring a considerable number of this same kind, corresponding almost exactly with each other.   I have ventured to introduce it in this place as a new species ; and have taken pains to render the figure in the plate a correct likeness of the original.

On the 6th of September 1812, after a violent north-east storm, which inundated the meadows of Schuylkill in many places, numerous flocks of this tern all at once made their appearance, flying over those watery spaces, picking up grasshoppers, beetles, spiders, and other insects, that were floating on the surface.   Some hundreds of them might be seen at the same time, and all seemingly of one sort.   They were busy, silent, and unsuspicious, darting down after their prey without hesitation, though perpetually harassed by gunners, whom the novelty of their appearance had drawn to the place.   Several flocks of the yellow-shanks snipe, and a few purres, appeared

* C. L. Bonaparte remarks,—"*S. plumbea* is evidently, even judging only by Wilson's figure and description, no other than the young of the European *S. nigra*, of which so many nominal species had already been made.   Indeed, so evident did the matter appear to us, even before we compared the species, that we cannot conceive why this hypothesis did not strike every naturalist, particularly as the *S. nigra* is well known to inhabit these States, though not noticed by Wilson in its adult dress. It is a singular fact that we hardly observed one adult among twenty young, which were common in the latter part of summer at Long Beach, New York."—ED.

also in the meadows at the same time, driven thither doubtless by the violence of the storm.

I examined upwards of thirty individuals of this species by dissection, and found both sexes alike in colour. Their stomachs contained grasshoppers, crickets, spiders, &c., but no fish. The people on the sea-coast have since informed me that this bird comes to them only in the fall, or towards the end of summer, and is more frequently seen about the mill-ponds and fresh-water marshes than in the bays; and add, that it feeds on grasshoppers and other insects which it finds on the meadows and marshes, picking them from the grass, as well as from the surface of the water. They have never known it to associate with the lesser tern, and consider it altogether a different bird. This opinion seems confirmed by the above circumstances, and by the fact of its greater extent of wing, being full three inches wider than the lesser tern; and also making its appearance after the others have gone off.

The short-tailed tern measures eight inches and a half from the point of the bill to the tip of the tail, and twenty-three inches in extent; the bill is an inch and a quarter in length, sharp pointed, and of a deep black colour; a patch of black covers the crown, auriculars, spot before the eye and hind head; the forehead, eyelids, sides of the neck, passing quite round below the hind head, and whole lower parts, are pure white; the back is dark ash, each feather broadly tipt with brown; the wings, a dark lead colour, extending an inch and a half beyond the tail, which is also of the same tint, and slightly forked; shoulders of the wing, brownish ash; legs and webbed feet, tawny. It had a sharp shrill cry when wounded and taken.

This is probably the *brown tern* mentioned by Willoughby, of which so many imperfect accounts have already been given. The figure in the plate, like those which accompany it, is reduced to one half the size of life.

BLACK SKIMMER, OR SHEERWATER.
(*Rhynchops nigra.*)

PLATE LX.—FIG. 4.

Arct. Zool. No. 445.—*Catesby*, i. 90.—Le Bec-en-ciseaux, *Buff.* viii. 454, tab. 36.—
Peale's Museum, No. 3530.

RHYNCHOPS NIGRA.—LINNÆUS.*

Rhynchops nigra, *Steph. Cont. Sh. Zool.* vol. xiii. p. 136.—*Cuv. Reg. Anim.* i.
522.—*Bonap. Synop.*—*Less. Man. d'Orn.* ii. p. 385.

THIS truly singular fowl is the only species of its tribe hitherto discovered. Like many others, it is a bird of passage in the United States, and makes its first appearance on the shores of New Jersey early in May. It resides there, as well as along the whole Atlantic coast, during the summer ; and retires early in September. Its favourite haunts are low sandbars raised above the reach of the summer tides, and also dry flat sands on the beach in front of the ocean. On such places it usually breeds along the shores of Cape May, in New Jersey. On account of the general coldness of the spring there, the sheerwater does not begin to lay until early in June, at which time these birds form themselves into small

---

* This very curious genus is composed, according to ornithologists, of two species,—that of our author and the *R. flavirostris*, Vieillot ; though I suspect that another is involved in the birds which I have seen from the Southern Ocean. In form and plumage they bear a strong resemblance to the terns, but are at once distinguished by the bill, which will show the greatest instance of the lateral development of that member. The manners of these birds, in adaptation to the structure of the bill and mouth, are noted by our author ; and it seems generally thought that their practice of skimming and cutting the water, as it were in search of food, is their only mode of procuring subsistence. The immense flocks of this species, mingled with gulls and terns, with their peculiar mode of feeding on some bivalve shells, is thus described by Lesson, and shows that sometimes a more substantial food is required, for the procuring of which the form of their bill is no less beautifully adapted, and that the opinion of Wilson is at variance with reality :—" Il formait avec les mouettes et quelque autres oiseaux

societies, fifteen or twenty pair frequently breeding within a few yards of each other. The nest is a mere hollow formed in the sand, without any other materials. The female lays three eggs, almost exactly oval, of a clear white, marked with large round spots of brownish black, and intermixed with others of pale Indian-ink. These eggs measure one inch and three-quarters, by one inch and a quarter. Half a bushel and more of eggs have sometimes been collected from one sandbar, within the compass of half an acre. These eggs have something of a fishy taste, but are eaten by many people on the coast. The female sits on them only during the night, or in wet and stormy weather. The young remain for several weeks before they are able to fly ; are fed with great assiduity by both parents, and seem to delight in lying with loosened wings, flat on the sand, enjoying its invigorating warmth. They breed but once in the season.

The singular confirmation of the bill of this bird has excited much surprise ; and some writers, measuring the divine proportions of nature by their own contracted standards of conception, in the plenitude of their vanity have pronounced it to be " a lame and defective weapon." Such ignorant presumption, or rather impiety, ought to hide its head in the dust on a calm display of the peculiar construction of this

de mer, des bandes tellement épaisses qu'il resemblait à des longues écharpes noires et mobiles qui obscurcissaient le ciel depuis les rives de Penco jusqu' à l'île de Quiriquine, dans un espace de douze milles. Quoique le bec-en-ciseaux semble défavorisé par la forme de son bec, nous aquîmes la preuve qu'il savait s'en servir avec avantage et avec le plus grande adresse. Les plages sablonneuses de Penco sont en effet remplies de *Mactres*, coquilles bivalves, que la marée decendente laisse presque à sec dans des petites mares ; le bec-en-ciseaux tres au fait de ce phénomène, se place auprès de ces mollusques, attend que leur valve s'ent ouvre un peu, et profite aussitôt de ce mouvement en enforçant la lame inférieure, et tranchante de son bec entre les valves qui se referment. L'oiseaux enlève alors la coquille, la frappe sur la grève, coupe le ligament du molusque, et peut ensuite avaler celui-ci sans obstacle. Plusieurs fois nous avons été témoins de cet instinct très perfectionné."—ED.

singular bird, and the wisdom by which it is so admirably adapted to the purposes or mode of existence for which it was intended. The sheerwater is formed for skimming, while on wing, the surface of the sea for its food, which consists of small fish, shrimps, young fry, &c., whose usual haunts are near the shore and towards the surface. That the lower mandible, when dipt into and cleaving the water, might not retard the bird's way, it is thinned and sharpened like the blade of a knife; the upper mandible being, at such times, elevated above water, is curtailed in its length, as being less necessary, but tapering gradually to a point, that, on shutting, it may offer less opposition. To prevent inconvenience from the rushing of the water, the mouth is confined to the mere opening of the gullet, which, indeed, prevents mastication taking place there; but the stomach, or gizzard, to which this business is solely allotted, is of uncommon hardness, strength, and muscularity, far surpassing, in these respects, any other water-bird with which I am acquainted. To all these is added a vast expansion of wing, to enable the bird to sail with sufficient celerity while dipping in the water. The general proportion of the length of our swiftest hawks and swallows to their breadth is as one to two; but, in the present case, as there is not only the resistance of the air, but also that of the water, to overcome, a still greater volume of wing is given, the sheerwater measuring nineteen inches in length, and upwards of forty-four in extent. In short, whoever has attentively examined this curious apparatus, and observed the possessor, with his ample wings, long bending neck, and lower mandible, occasionally dipped into and ploughing the surface, and the facility with which he procures his food, cannot but consider it a mere playful amusement when compared with the dashing immersions of the tern, the gull, or the fish-hawk, who, to the superficial observer, appear so superiorly accommodated.

The sheerwater is most frequently seen skimming close along shore about the first of the flood, at which time the young fry, shrimp, &c., are most abundant in such places,

There are also numerous inlets among the low islands between the sea-beach and mainland of Cape May, where I have observed the sheerwaters, eight or ten in company, passing and repassing, at high water, particular estuaries of those creeks that run up into the salt marshes, dipping, with extended neck, their open bills into the water, with as much apparent ease as swallows glean up flies from the surface. On examining the stomachs of several of these, shot at the time, they contained numbers of a small fish usually called *silver-sides*, from a broad line of a glossy silver colour that runs from the gills to the tail. The mouths of these inlets abound with this fry or fish, probably feeding on the various matters washed down from the marshes.

The voice of the sheerwater is harsh and screaming, resembling that of the tern, but stronger. It flies with a slowly flapping flight, dipping occasionally, with steady expanded wings and bended neck, its lower mandible into the sea, and with open mouth receiving its food as it ploughs along the surface. It is rarely seen swimming on the water; but frequently rests in large parties on the sandbars at low water. One of these birds which I wounded in the wing, and kept in the room beside me for several days, soon became tame, and even familiar. It generally stood with its legs erect, its body horizontal, and its neck rather extended. It frequently reposed on its belly, and stretching its neck, rested its long bill on the floor. It spent most of its time in this way, or in dressing and arranging its plumage with its long scissors-like bill, which it seemed to perform with great ease and dexterity. It refused every kind of food offered it, and I am persuaded never feeds but when on the wing. As to the reports of its frequenting oyster-beds, and feeding on these fish, they are contradicted by all those persons with whom I have conversed whose long residence on the coast where these birds are common has given them the best opportunities of knowing.

The sheerwater is nineteen inches in length, from the point of the bill to the extremity of the tail; the tips of the wings, when shut, extend full four inches farther; breadth, three feet eight inches; length of the lower mandible, four inches and a half; of the upper, three inches and a half; both of a scarlet red, tinged with orange, and ending in black; the lower extremely thin; the upper grooved, so as to receive the edge of the lower; the nostril is large and pervious, placed in a hollow near the base and edge of the upper mandible, where it projects greatly over the lower; upper part of the head, neck, back, and scapulars, deep black; wings, the same, except the secondaries, which are white on the inner vanes, and also tipt with white; tail, forked, consisting of twelve feathers, the two middle ones about an inch and a half shorter than the exterior ones, all black, broadly edged on both sides with white; tail-coverts, white on the outer sides, black in the middle; front, passing down the neck below the eye, throat, breast, and whole lower parts, pure white; legs and webbed feet, bright scarlet, formed almost exactly like those of the tern. Weight, twelve ounces avoirdupois. The female weighed nine ounces, and measured only sixteen inches in length, and three feet three inches in extent; the colours and markings were the same as those of the male, with the exception of the tail, which was white, shafted, and broadly centred with black.

The birds from which these descriptions were taken were shot on the 25th of May, before they had begun to breed. The female contained a great number of eggs, the largest of which were about the size of duckshot; the stomach, in both, was an oblong pouch, ending in a remarkably hard gizzard, curiously puckered or plaited, containing the half-dissolved fragments of the small silver-sides, pieces of shrimps, small crabs, and skippers, or sandfleas.

On some particular parts of the coast of Virginia, these birds are seen on low sandbars in flocks of several hundreds together. There more than twenty nests have been found within the space of a square rod. The young are at first so exactly of a colour with the sand on which they sit, as to be with difficulty discovered unless after a close search.

The sheerwater leaves our shores soon after his young are fit for the journey. He is found on various coasts of Asia, as well as America, residing principally near the tropics, and migrating into the temperate regions of the globe only for the purpose of rearing his young. He is rarely or never seen far out at sea ; and must not be mistaken for another bird of the same name, a species of petrel,* which is met with on every part of the ocean, skimming with bended wings along the summits, declivities, and hollows of the waves.

STORMY PETREL.   *(Procellaria pelagica.)*

PLATE LX.—Fig. 6.

Arct. Zool. No. 464.—Le Petrel, ou l'Oiseaux Tempète, *Pl. enl.* 993.—*Bewick*, ii. 223.—*Peale's Museum*, No. 3034.

THALASIDROMA WILSONII.—Bonaparte.†

Thalasidroma Wilsonii, *Bonap. Synop.* p. 367.—Procellaria Wilsonii, *Steph. Cont. Sh. Zool.* xiii. p. 224.—Procellaria Wilsonii, *Ord's reprint of Wils.* p. 94.— *Journ. of the Acad. of N. S. of Philad.* iii. p. 231, pl. ix.

There are few persons who have crossed the Atlantic, or traversed much of the ocean, who have not observed these

---

* *Procellaria puffinus*, the sheerwater petrel.

† This species, confounded (and with little wonder, from its near alliance) by Wilson with the *P. pelasgica*, has been named as above by the Prince of Musignano, another tribute to the memory of our American ornithologist, and he has added the following differences and distinctive characters. Bonaparte has also added the *T. Bullockii* to the American list.

The smaller petrels of other countries are much allied to these ; they amount to a considerable number, many of which are yet undetermined, and are confused with each other, in the want of proper distinguishing characters being assigned to each. It is from this that the *P. pelasgica* has been assigned a distribution so extensive. Some species are found in most latitudes, and from their similarity most observers seem to be

solitary wanderers of the deep skimming along the surface of
the wild and wasteful ocean; flitting past the vessel like
swallows, or following in her wake, gleaning their scanty
pittance of food from the rough and whirling surges.
Habited in mourning, and making their appearance generally
in greater numbers previous to or during a storm, they have
long been fearfully regarded by the ignorant and superstitious,

unaware when they have passed the boundary of one, and entered the
opposite limits of another form.

They resemble each other in another propensity,—that of following
the course of vessels, attracted by the shelter afforded in the wake, or
retained by the small marine insects and seeds which are sucked into
it, and the subsistence they may obtain from the refuse thrown overboard. Being most commonly seen when all is gloomy above, the view
bounded by the horizon alone, or by a thick atmosphere and boisterous
waves, and when they are the only beings visible, running on the
" trough of the sea,"—

> As though they were the shadows of themselves,
> Reflected from a loftier flight through space—

it can hardly be wondered at that associations with the spirits have
arisen in the minds of men naturally prone, and sometimes wrought up
to superstition, and that they have begotten for themselves such names
as are quoted by our author. These ideas are universal. Several small
species about the Madeiras bear the name of Anhiga, conveying the
idea of their affinity to imps.

Procellaria Bullockii has been described by Bonaparte in the Journal
of the Academy of Natural Sciences of Philadelphia as an addition to
the birds of America. It is stated to be but rare throughout the Atlantic
Ocean, and to be found on the Banks of Newfoundland. It is also
European, and was first discovered by Mr Bullock breeding at St Kilda,
and ought now to stand under the name of its discoverer, *Thalasidroma
Bullockii.* They also sometimes occur on the mainland of Britain ; and
it is remarkable that all those procured there have been found in a
dead or dying state in some frequented place—often on the public road.
It is expressly mentioned by M. Frecynet, in his " Voyage Autour du
Monde," that the small petrels cannot rise from a flat surface, such as
the deck of a ship. It is possible that the specimens discovered in this
state of exhaustion may have been unable again to resume their flight,
and thus perished. Two specimens occurred in Dumfriesshire during
the last year, both found on the public road,—the one dead, the other
nearly so.—ED.

not only as the foreboding messengers of tempests and dangers to the hapless mariner, but as wicked agents, connected, somehow or other, in creating them. "Nobody," say they, "can tell anything of where they come from or how they breed, though (as sailors sometimes say) it is supposed that they hatch their eggs under their wings as they sit on the water." This mysterious uncertainty of their origin, and the circumstances above recited, have doubtless given rise to the opinion so prevalent among this class of men, that they are in some way or other connected with that personage who has been styled the Prince of the Power of the Air. In every country where they are known, their names have borne some affinity to this belief. They have been called *Witches,** *Stormy petrels*, the *Devil's birds*, *Mother Carey's chickens,†* probably from some celebrated ideal hag of that name; and their unexpected and numerous appearance has frequently thrown a momentary damp over the mind of the hardiest seaman.

It is the business of the naturalist and the glory of philosophy to examine into the reality of these things, to dissipate the clouds of error and superstition wherever they begin to darken and bewilder the human understanding, and to illustrate nature with the radiance of truth. With these objects in view, we shall now proceed, as far as the few facts we possess will permit, in our examination into the history of this celebrated species.

The *stormy petrel*, the least of the whole twenty-four species of its tribe enumerated by ornithologists, and the smallest of all palmated fowls, is found over the whole Atlantic Ocean from Europe to North America, at all distances from land, and in all weathers, but is particularly numerous near vessels immediately preceding and during a gale, when flocks of

---

* Arctic Zoology, p. 464.

† This name seems to have been originally given them by Captain Carteret's sailors, who met with these birds on the coast of Chili. See Hawkesworth's Voyages, vol. i. p. 203.

them crowd in her wake, seeming then more than usually active in picking up various matters from the surface of the water. This presentiment of a change of weather is not peculiar to the petrel alone, but is noted in many others, and common to all, even to those long domesticated. The woodpeckers, the snow-birds, the swallows, are all observed to be uncommonly busy before a storm, searching for food with great eagerness, as if anxious to provide for the privations of the coming tempest. The common ducks and geese are infallibly noisy and tumultuous before falling weather; and though, with these, the attention of man renders any extra exertions for food at such times unnecessary, yet they wash, oil, dress, and arrange their plumage with uncommon diligence and activity. The intelligent and observing farmer remarks this bustle, and wisely prepares for the issue; but he is not so ridiculously absurd as to suppose that the storm which follows is produced by the agency of these feeble creatures, who are themselves equal sufferers by its effects with man. He looks on them rather as useful monitors, who, from the delicacy of their organs, and a perception superior to his own, point out the change in the atmosphere before it has become sensible to his grosser feelings, and thus, in a certain degree, contribute to his security. And why should not those who navigate the ocean contemplate the appearance of this unoffending little bird in like manner, instead of eyeing it with hatred and execration? As well might they curse the midnight lighthouse, that, star-like, guides them on their watery way, or the buoy that warns them of the sunken rocks below, as this harmless wanderer, whose manner informs them of the approach of the storm, and thereby enables them to prepare for it.

The stormy petrels, or Mother Carey's chickens, breed in great numbers on the rocky shores of the Bahama and the Bermuda Islands, and in some places on the coast of East Florida and Cuba. They breed in communities like the bank-swallows, making their nests in the holes and cavities of the rocks above the sea, returning to feed their young only during the night, with the superabundant oily food from their stomachs. At these times they may be heard making a continued cluttering sound like frogs during the whole night. In the day they are silent, and wander widely over the ocean. This easily accounts for the vast distance they are sometimes seen from land, even in the breeding season. The rapidity of their flight is at least equal to the fleetness of our swallows. Calculating this at the rate of one mile per minute, twelve hours would be sufficient to waft them a distance of seven hundred and twenty miles; but it is probable that the far greater part confine themselves much nearer land during that interesting period.

In the month of July, while on a voyage from New Orleans to New York, I saw few or none of these birds in the Gulf of Mexico, although our ship was detained there by calms for twenty days, and carried by currents as far south as Cape Antonio, the westernmost extremity of Cuba. On entering the Gulf Stream, and passing along the coasts of Florida and the Carolinas, these birds made their appearance in great numbers, and in all weathers, contributing much by their sprightly evolutions of wing to enliven the scene, and affording me every day several hours of amusement. It is indeed an interesting sight to observe these little birds in a gale, coursing over the waves, down the declivities, up the ascents of the foaming surf that threatens to burst over their heads, sweeping along the hollow troughs of the sea as in a sheltered valley, and again mounting with the rising billow, and just above its surface, occasionally dropping its feet, which, striking the water, throws it up again with additional force ; sometimes leaping, with both legs parallel, on the surface of the roughest waves for several yards at a time. Meanwhile it continues coursing from side to side of the ship's wake, making excursions far and wide to the right hand and to the left, now a great way ahead, and now shooting astern for several hundred yards, returning again to the ship as if she were all the while stationary, though perhaps running at the rate of ten knots an hour. But the most singular peculiarity of this bird is its faculty of standing, and even running, on the surface of the water, which it performs with apparent facility. When any greasy matter is thrown overboard, these birds instantly collect around it, and facing to windward, with their long wings expanded, and their webbed feet patting the water, the lightness of their bodies and the action of the wind on their wings enable them to do this with ease. In calm weather they perform the same manœuvre, by keeping their wings just so much in action as to prevent their feet from sinking below the surface. According to Buffon,* it is from this singular habit that the whole genus have obtained the name petrel, from the apostle Peter, who, as scripture informs us, also walked on the water.

As these birds often come up immediately under the stern, one can examine their form and plumage with nearly as much accuracy as if they were in the hand. They fly with the wings forming an almost straight horizontal line with the body, the legs extended behind, and the feet partly seen stretching beyond the tail. Their common note of "*weet, weet,*" is scarcely louder than that of a young duck of a week old, and much resembling it. During the whole of a dark, wet, and boisterous night which I spent on deck, they flew about the after-rigging, making a singular hoarse chattering, which in sound resembled the syllables *patrèt tu cuk cuk tu tu,* laying the accent strongly on the second syllable *tret.* Now and then I conjectured that they alighted on the rigging, making then a lower curring noise.

Notwithstanding the superstitious fears of the seamen, who dreaded the vengeance of the survivors, I shot fourteen of these birds one calm day in lat. 33,° eighty or ninety miles off the coast of Carolina, and had the boat lowered to pick them up. These I examined with considerable attention, and found the most perfect specimens as follow :—

---

* Buffon, tome xxiii. p. 299.

Length, six inches and three-quarters; extent, thirteen inches and a half; bill, black; nostrils, united in a tubular projection, the upper mandible grooved from thence, and overhanging the lower like that of a bird of prey; head, back, and lower parts, brown sooty black; greater wing-coverts, pale brown, minutely tipt with white; sides of the vent, and whole tail-coverts, pure white; wings and tail, deep black, the latter nearly even at the tip, or very slightly forked; in some specimens, two or three of the exterior tail-feathers were white for an inch or so at the root; legs and naked part of the thighs, black; feet, webbed, with the slight rudiments of a hind toe; the membrane of the foot is marked with a spot of straw yellow, and finely serrated along the edges; eyes, black. Male and female differing nothing in colour.

On opening these, I found the first stomach large, containing numerous round semi-transparent substances of an amber colour, which I at first suspected to be the spawn of some fish, but on a more close and careful inspection, they proved to be a vegetable substance, evidently the seeds of some marine plant, and about as large as mustard seed. The stomach of one contained a fish, half digested, so large that I should have supposed it too bulky for the bird to swallow; another was filled with the tallow which I had thrown overboard; and all had quantities of the seeds already mentioned both in their stomachs and gizzards; in the latter were also numerous minute pieces of barnacle shells. On a comparison of the seeds above mentioned with those of the *Gulf-weed,* so common and abundant in this part of the ocean, they were found to be the same. Thus it appears that these seeds, floating perhaps a little below the surface, and the barnacles with which ships' bottoms usually abound, being both occasionally thrown up to the surface by the action of the vessel through the water in blowing weather, entice these birds to follow in the ship's wake at such times, and not, as some have imagined, merely to seek shelter from the storm, the greatest violence of which they seem to disregard. There is also the greasy dish-washings and other oily substances thrown over by the cook, on which they feed with avidity, but with great good nature, their manners being so gentle, that I never observed the slightest appearance of quarrelling or dispute among them.

One circumstance is worthy of being noticed, and shows the vast range they take over the ocean. In firing at these birds, a quill-feather was broken in each wing of an individual, and hung fluttering in the wind, which rendered it so conspicuous among the rest as to be known to all on board. This bird, notwithstanding its inconvenience, continued with us for nearly a week, during which we sailed a distance of more than four hundred miles to the north. Flocks continued to follow us until near Sandy Hook.

The length of time these birds remain on wing is no less surprising. As soon as it was light enough in the morning to perceive them, they were found roaming about as usual; and I have often sat in the evening, in the boat which was suspended by the ship's stern, watching their movements, until it was so dark that the eye could no longer follow them, though I could still hear their low note of *weet, weet*, as they approached near to the vessel below me.

These birds are sometimes driven by violent storms to a considerable distance inland. One was shot some years ago on the river Schuylkill near Philadelphia; and Bewick mentions their being found in various quarters of the interior of England. From the nature of their food, their flesh is rank and disagreeable, though they sometimes become so fat, that, as Mr Pennant, on the authority of Brunnich, asserts, " the inhabitants of the Feroe Isles make them serve the purposes of a candle, by drawing a wick through the mouth and rump, which, being lighted, the flame is fed by the fat and oil of the body." *

[Mr Ord adds, in his reprint, " When this work was published, its author was not aware that those birds observed by

* British Zoology, vol. ii. p. 434.

navigators in almost every quarter of the globe, and known under the name of stormy petrels, formed several distinct species ; consequently, relying on the labours of his predecessors, he did not hesitate to name the subject of this chapter the *pelagica,* believing it to be identical with that of Europe. But the investigations of later ornithologists having resulted in the conviction that Europe possessed at least two species of these birds, it became a question whether or not those which are common on the coasts of the United States would form a third species ; and an inquiry has established the fact that the American stormy petrel, hitherto supposed to be the true *pelagica,* is an entirely distinct species. For this discovery we are indebted to the labours of Mr Charles Bonaparte, from whose interesting paper on the subject, published in the Journal of the Academy of Natural Sciences of Philadelphia, we shall take the liberty of making an extract. The author of the paper in question first describes and figures the true *pelagica* of the systems ; secondly, the *Leachii,* a species described by Temminck, and restricted to the vicinity of the Island of St Kilda, but which the former found diffused over a great part of the Atlantic, east of the Banks of Newfoundland ; and thirdly, the species of our coasts. He also indicates a fourth, which inhabits the Pacific Ocean ; but whether or not this last be in reality a species different from those named, has not yet been determined.

" ' When I first procured this species,' says Mr Bonaparte, ' I considered it a nondescript, and noted it as such ; the citation of Wilson's *pelagica* among the synonyms of the true *pelagica* by the most eminent ornithologist of the age, M. Temminck, not permitting a doubt of their identity. But having an opportunity of inspecting the very individual from which Wilson took his figure and drew up his description, I was undeceived, by proving the unity of my specimens with that of Wilson, and the discrepancy of these with that of Temminck. The latter had certainly never seen an individual from America, otherwise the difference between the two species

Drawn from Nature by A. Wilson.

Engraved by W.H. Lizars.

1. Green Heron. 2. Night H. 3. Young. 4 Great White H.

61.

would not have eluded the accurate eye of this naturalist. I propose for this species the name of *Wilsonii*, as a small testimony of respect to the memory of the author of the " American Ornithology," whose loss science and America will long deplore. The yellow spot upon the membrane of the feet distinguishes this species, at first sight, from the others ; and this character remains permanent in the dried specimens.' "]

GREEN HERON. (*Ardea virescens.*)

PLATE LXI.—Fig. 1.

Arct. Zool. No. 349, 350.—*Catesby*, i. p. 80.—Le Crabier Vert, *Buff.* vii. p. 404.—*Lath. Syn.* iii. p. 68.—*Peale's Museum*, No. 3797.

ARDEA VIRESCENS.—Linnæus.*

Ardea virescens, *Bonap. Synop.* p. 307.—*Wagl. Syst. Av.* No. 36.

This common and familiar species owes little to the liberality of public opinion, whose prejudices have stigmatised it with a very vulgar and indelicate nickname, and treat it on all

* There are two or three beautiful little herons confounded under this species, in the same manner, from their near alliance, as the little bittern of Europe has been with *A. exilis* and *pusilla*. They are all, however, to be distinguished when compared together, or when attention is given to the markings. The nearest ally to *A. virescens* is the East Indian *A. scapularis;* the upper parts of both are nearly similar, but the neck and under parts differ in being of a deep vinous chestnut in the one, and rich ash grey in the other. In Wilson's plate, the chestnut colour is not represented of a deep enough tint, and too much white is shown on the fore part.

In a specimen which I have lately received from South Carolina, the colour of the neck is very deep and rich, almost approaching to that of port wine ; the lengthened feathers of the back are remarkably long, and show well the white shafts which ought to be so conspicuous in both species. The confusion in the greater part of the synonyms must have arisen by the specimens from both countries being indiscriminately compared and described.—Ed.

occasions as worthless and contemptible. Yet few birds are more independent of man than this; for it fares best, and is always most numerous, where cultivation is least known or attended to, its favourite residence being the watery solitudes of swamps, pools, and morasses, where millions of frogs and lizards "tune their nocturnal notes" in full chorus, undisturbed by the lords of creation.

The green bittern makes its first appearance in Pennsylvania early in April, soon after the marshes are completely thawed. There, among the stagnant ditches with which they are intersected, and amidst the bogs and quagmires, he hunts with great cunning and dexterity. Frogs and small fish are his principal game, whose caution and facility of escape require nice address and rapidity of attack. When on the lookout for small fish, he stands in the water, by the side of the ditch, silent and motionless as a statue, his neck drawn in over his breast, ready for action. The instant a fry or minnow comes within the range of his bill, by a stroke, quick and sure as that of the rattlesnake, he seizes his prey, and swallows it in an instant. He searches for small crabs, and for the various worms and larvæ, particularly those of the dragon-fly, which lurk in the mud, with equal adroitness. But the capturing of frogs requires much nicer management. These wary reptiles shrink into the mire on the least alarm, and do not raise up their heads again to the surface without the most cautious circumspection. The bittern, fixing his penetrating eye on the spot where they disappeared, approaches with slow stealing step, laying his feet so gently and silently on the ground as not to be heard or felt; and when arrived within reach, stands fixed, and bending forwards, until the first glimpse of the frog's head makes its appearance, when, with a stroke instantaneous as lightning, he seizes it in his bill, beats it to death, and feasts on it at his leisure.

This mode of life, requiring little fatigue where game is so plenty as is generally the case in all our marshes, must be particularly pleasing to the bird, and also very interesting, from the continual exercise of cunning and ingenuity necessary to circumvent its prey. Some of the naturalists of Europe, however, in their superior wisdom, think very differently ; and one can scarcely refrain from smiling at the absurdity of those writers who declare that the lives of this whole class of birds are rendered miserable by toil and hunger ; their very appearance, according to Buffon, presenting the image of suffering, anxiety, and indigence.[*]

When alarmed, the green bittern rises with a hollow guttural scream ; does not fly far, but usually alights on some old stump, tree, or fence adjoining, and looks about with extended neck ; though, sometimes, this is drawn in so that his head seems to rest on his breast. As he walks along the fence, or stands gazing at you with outstretched neck, he has the frequent habit of jetting the tail. He sometimes flies high, with doubled neck, and legs extended behind, flapping the wings smartly, and travelling with great expedition. He is the least shy of all our herons, and perhaps the most numerous and generally dispersed, being found far in the interior, as well as along our salt marshes, and everywhere about the muddy shores of our millponds, creeks, and large rivers.

The green bittern begins to build about the 20th of April, sometimes in single pairs in swampy woods, often in companies, and not unfrequently in a kind of association with the qua-birds or night herons. The nest is fixed among the branches of the trees ; is constructed wholly of small sticks, lined with finer twigs, and is of considerable size, though loosely put together. The female lays four eggs, of the common oblong form, and of a pale light blue colour. The young do not leave the nest until able to fly ; and, for the first season at least, are destitute of the long pointed plumage on the back ; the lower parts are also lighter, and the white on the throat broader. During the whole summer, and until late in autumn, these birds are seen in our meadows and marshes,

[*] Histoire Naturelle des Oiseaux, tome xxii. p. 343.

but never remain during winter in any part of the United States.

The green bittern is eighteen inches long, and twenty-five inches in extent ; bill black, lighter below, and yellow at the base ; chin, and narrow streak down the throat, yellowish white ; neck, dark vinaceous red ; back, covered with very long, tapering, pointed feathers, of a hoary green, shafted with white, on a dark green ground ; the hind part of the neck is destitute of plumage, that it may be the more conveniently drawn in over the breast, but is covered with the long feathers of the throat and sides of the neck, that enclose it behind ; wings and tail, dark glossy green, tipt and bordered with yellowish white ; legs and feet, yellow, tinged before with green, the skin of these thick and movable ; belly, ashy brown ; irides, bright orange ; crested head, very dark glossy green. The female, as I have particularly observed in numerous instances, differs in nothing as to colour from the male ; neither of them receive the long feathers on the back during the first season.

There is one circumstance attending this bird which, I recollect, at first surprised me. On shooting and wounding one, I carried it some distance by the legs, which were at first yellow ; but on reaching home, I perceived, to my surprise, that they were red. On letting the bird remain some time undisturbed, they again became yellow, and I then discovered that the action of the hand had brought a flow of blood into them, and produced the change of colour. I have remarked the same in those of the night heron.

NIGHT HERON, OR QUA-BIRD. (*Ardea nycticorax.*)

PLATE LXI.—Fig. 2 ; Fig. 3, Young.

Arct. Zool. No. 356.—Le Bichoreau, *Buff.* vii. 435, 439, rol. 22 ; *Pl. enl.* 758, 759, 999.—*Lath. Syn.* iii. p. 52, No. 13 ; p. 53, young, called there the female.—*Peale's Museum*, No. 3728 ; young, No. 3729.

*NYCTICORAX GARDENII.**

Ardea nycticorax, *Temm. Man.* ii. p. 577.—Gardenian Heron, *Mont. Orn. Dict.* i.— *Bonap. Synop.* p. 306.—*Wagl. Syst. Av.* Ardea, No. 31.

THIS species, though common to both continents, and known in Europe for many centuries, has been so erroneously described by all the European naturalists whose works I have examined, as to require more than common notice in this place. For this purpose, an accurate figure of the male is given, and also another of what has till now been universally considered the female, with a detail of so much of their history as I am personally acquainted with.

* *Nycticorax*, or night raven, has been adopted to designate this from among the *Ardeadœ*, from the circumstance of their feeding by night, and remaining in a state of comparative rest and inactivity during the day. New Holland and Africa each possess a species. Europe and North America have one in common to both countries ; in the former, abundantly distributed, while, in the latter, it is of rare occurrence even towards the south, and in the northern parts of Great Britain, only a few instances have occurred of its capture.

In form, they are intermediate between the bitterns and true herons ; the bill is short, and stronger in proportion than in either ; the feathers on the sides of the neck are lengthened, and cover the hinder part, which is bare to a certain extent ; and in all the species, the hind head is adorned with (generally three) narrow feathers, in the form of a crest. They feed by twilight, or in clear nights ; and take their prey by watching, in the manner of the herons. They are gregarious, build on trees, and during the season of incubation are noisy and restless.

The colours in the adults of the true species are ash grey or pale fawn ; the crown and hind head and the back, or that part called by the French *manteau*, in the ash grey species, dark glossy green ; in the fawn coloured, deep chestnut. The young are always of a duskier tinge, and have the centre and tips of each feather white, giving the plumage a spotted appearance.—ED.

The night heron arrives in Pennsylvania early in April, and immediately takes possession of his former breeding place, which is usually the most solitary and deeply shaded part of a cedar swamp. Groves of swamp oak, in retired and inundated places, are also sometimes chosen, and the males not unfrequently select tall woods, on the banks of the river, to roost in during the day. These last regularly direct their course, about the beginning of evening twilight, towards the marshes, uttering, in a hoarse and hollow tone, the sound *qua,* which by some has been compared to that produced by the retchings of a person attempting to vomit. At this hour, also, all the nurseries in the swamps are emptied of their inhabitants, who disperse about the marshes, and along the ditches and river shore, in quest of food. Some of these breeding places have been occupied every spring and summer for time immemorial, by from eighty to one hundred pairs of qua-birds. In places where the cedars have been cut down for sale, the birds have merely removed to another quarter of the swamp; but when personally attacked, long teased, and plundered, they have been known to remove from an ancient breeding place in a body, no one knew where. Such was the case with one on the Delaware, near Thompson's Point, ten or twelve miles below Philadelphia, which having been repeatedly attacked and plundered by a body of crows, after many severe rencounters the herons finally abandoned the place. Several of these breeding places occur among the red cedars on the sea-beach of Cape May, intermixed with those of the little egret, green bittern, and blue heron. The nests are built entirely of sticks, in considerable quantities, with frequently three and four nests on the same tree. The eggs are generally four in number, measuring two inches and a quarter in length, by one and three-quarters in thickness, and of a very pale light blue colour. The ground or marsh below is bespattered with their excrements, lying all around like whitewash, with feathers, broken eggshells, old nests, and frequently small fish, which they have dropt by accident, and neglected to pick up.

On entering the swamp in the neighbourhood of one of these breeding places, the noise of the old and the young would almost induce one to suppose that two or three hundred Indians were choking or throttling each other. The instant an intruder is discovered, the whole rise in the air in silence, and remove to the tops of the trees in another part of the woods, while parties of from eight to ten make occasional circuits over the spot to see what is going on. When the young are able, they climb to the highest part of the trees ; but knowing their inability, do not attempt to fly. Though it is probable that these nocturnal birds do not see well during the day, yet their faculty of hearing must be exquisite, as it is almost impossible, with all the precautions one can use, to penetrate near their residence without being discovered. Several species of hawks hover around, making an occasional sweep among the young; and the bald eagle himself has been seen reconnoitring near the spot, probably with the same design.

Contrary to the generally received opinion, the males and females of these birds are so alike in colour as scarcely to be distinguished from each other ; both have also the long slender plumes that flow from the head. These facts I have exhibited by dissection on several subjects to different literary gentlemen of my acquaintance, particularly to my venerable friend Mr William Bartram, to whom I have also often shown the young, represented at fig. 3. One of these last, which was kept for some time in the botanic garden of that gentleman, by its voice instantly betrayed its origin, to the satisfaction of all who examined it. These young certainly receive their full coloured plumage before the succeeding spring, as, on their first arrival, no birds are to be seen in the dress of fig. 3 ; but, soon after they have bred, these become more numerous than the others. Early in October they migrate to the south. According to Buffon, these birds also inhabit Cayenne, and are found widely dispersed over Europe, Asia, and America. The European species, however, is certainly much smaller than the American, though in other respects corresponding exactly to it. Among a great number which I examined with attention, the following description was carefully taken from a common-sized full-grown male :—

Length of the night heron, two feet four inches; extent, four feet; bill, black, four inches and a quarter long from the corners of the mouth to the tip; lores, or space between the eye and bill, a bare bluish white skin; eyelids also large and bare, of a deep purple blue; eye, three-quarters of an inch in diameter; the iris of a brilliant blood red; pupil, black; crested crown, and hind head deep dark blue, glossed with green; front and line over the eye, white; from the hind head proceed three very narrow, white, tapering feathers, between eight and nine inches in length; the vanes of these are concave below, the upper one enclosing the next, and that again the lower; though separated by the hand, if the plumage be again shook several times, these long flowing plumes gradually enclose each other, appearing as one; these the bird has the habit of erecting when angry or alarmed; the cheeks, neck, and whole lower parts, are white, tinctured with yellowish cream, and under the wings with very pale ash; back and scapulars, of the same deep dark blue, glossed with green, as that of the crown; rump and tail-coverts, as well as the whole wings and tail, very pale ash; legs and feet, a pale yellow cream colour; inside of the middle claw, serrated.

The female differed in nothing as to plumage from the male, but in the wings being of rather a deeper ash, having not only the dark deep green blue crown and back, but also the long pendant white plumes from the hind head. Each of the females contained a large cluster of eggs of various sizes.

The young (fig. 3) was shot soon after it had left the nest, and differed very little from those which had been taken from the trees, except in being somewhat larger. This measured twenty-one inches in length, and three feet in extent; the general colour above, a very deep brown, streaked with reddish white, the spots of white on the back and wings being triangular, from the centre of the feather to the tip; quills, deep dusky, marked on the tips with a spot of white; eye, vivid orange; belly, white, streaked with dusky, the feathers being pale dusky, streaked down their centres with white; legs and feet, light green; inside of the middle claw slightly pectinated; body and wings exceedingly thin and limber; the down still stuck in slight tufts to the tips of some of the feathers.

The birds also breed in great numbers in the neighbourhood of New Orleans; for being in that city in the month of June, I frequently observed the Indians sitting in market with the dead and living young birds for sale; also numbers of gray owls (*Strix nebulosa*), and the white ibis (*Tantalus albus*), for which nice dainties I observed they generally found purchasers.

The food of the night heron or qua-bird is chiefly composed of small fish, which it takes by night. Those that I opened had a large expansion of the gullet immediately under the bill, that narrowed from thence to the stomach, which is a large oblong pouch, and was filled with fish. The teeth of the pectinated claw were thirty-five or forty in number, and as they contained particles of the down of the bird, showed evidently, from this circumstance, that they act the part of a comb, to rid the bird of vermin in those part which it cannot reach with its bill.

GREAT WHITE HERON. (*Ardea egretta.*)

PLATE LXI.—Fig. 4.

EGRETTA LEUCE.—Jardine.*

Ardea leuce, *Illig.*—Ardea alba, *Bonap. Synop.* p. 304.—Ardea egretta, *Wagl. Syst. Av.* No. 7.—*Bonap. Monog. del Gruppo Egretta, Osserv. Sulla,* 2d edit. *Del Reg. Anim. Cuv.*

This tall and elegant bird, though often seen during the summer in our low marshes and inundated meadows, yet, on

---

* Among no birds has there occurred so much confusion as among the white herons, or those more particularly forming the division *Egretts*. They are distributed over every country of the world, are not very different in size, the young are chiefly distinguished by the want of the crest, and are in many instances of a plumage similar to the full

account of its extreme vigilance and watchful timidity, is very difficult to be procured. Its principal residence is in the regions of the south, being found from Guiana, and probably beyond the line, to New York. It enters the territories of the United States late in February; this I conjecture from having first met with it in the southern parts of Georgia about that time. The high inland parts of the country it rarely or never visits;—its favourite haunts are vast inundated swamps, rice-fields, the low marshy shores of rivers, and such like places, where, from its size and colour, it is very conspicuous, even at a great distance.

The appearance of this bird during the first season, when it is entirely destitute of the long flowing plumes of the back, is so different from the same bird in its perfect plumage, which it obtains in the third year, that naturalists and others very generally consider them as two distinct species. The opportunities which I have fortunately had of observing them with the train in various stages of its progress, from its first appearance to its full growth, satisfies me that the great white heron with, and that without, the long plumes, are one and the same species, in different periods of age. In the museum of my friend, Mr Peale, there is a specimen of this bird in

winter dress : most of the species when mature are clothed in a garb of the purest white.

The bird with which our present species is more immediately connected is the *Ardea alba*, Gmel., a European bird, confounded with the young of *A. egretta*, and not yet, I believe, found in North America. The chief differences are presence of the crest, and much longer proportion of the legs. *A. egretta* seems to range extensively over the continents of America and some of the islands ; I am not aware of its being found elsewhere ; and the African, Asiatic, and New Holland allied species will, I suspect, turn out distinct, and most probably belong to their respective countries.

To the North American *egretta* must be added the *Ardea Pealii*, discovered by Bonaparte. It is distinguished from its allies by the flesh colour of the bill, is much smaller than *A. alba*, differs from *A. garzetta* by its large compound crest, and from *A. candissima* by the quality and texture of the ornamental feathers.—Ed.

which the train is wanting ; but on a closer examination, its rudiments are plainly to be perceived, extending several inches beyond the common plumage.

The great white heron breeds in several of the extensive cedar swamps in the lower parts of New Jersey. Their nests are built on the trees, in societies ; the structure and materials exactly similar to those of the snowy heron, but larger. The eggs are usually four, of a pale blue colour. In the months of July and August, the young make their first appearance in the meadows and marshes, in parties of twenty or thirty together. The large ditches with which the extensive meadows below Philadelphia are intersected are regularly, about that season, visited by flocks of those birds ; these are frequently shot, but the old ones are too sagacious to be easily approached. Their food consists of frogs, lizards, small fish, insects, seeds of the splatterdock (a species of *nymphœ*), and small water-snakes. They will also devour mice and moles, the remains of such having been at different times found in their stomachs.

The long plumes of these birds have at various periods been in great request on the continent of Europe, particularly in France and Italy, for the purpose of ornamenting the female head-dress. When dyed of various colours, and tastefully fashioned, they form a light and elegant duster and mosquito brush. The Indians prize them for ornamenting their hair or top-knot ; and I have occasionally observed these people wandering through the market-place of New Orleans, with bunches of those feathers for sale.

The great white heron measures five feet from the extremities of the wings, and three feet six inches from the tip of the bill to the end of the tail ; the train extends seven or eight inches farther. This train is composed of a great number of long, thick, tapering shafts, arising from the lower part of the shoulders, and thinly furnished on each side with fine flowing hairlike threads, of several inches in length, covering the lower part of the back, and falling gracefully over the tail, which it entirely conceals.   The whole plumage is of a snowy whiteness, except the train, which is slightly tinged with yellow.   The bill is nearly six inches in length, of a rich orange yellow, tipt with black; irides, a paler orange; pupil, small, giving the bird a sharp and piercing aspect; the legs are long, stout, and of a black colour, as is the bare space of four inches above the knee; the span of the foot measures upwards of six inches; the inner edge of the middle claw is pectinated; the exterior and middle toes are united at the base, for about half an inch, by a membrane.

The articulations of the vertebræ are remarkably long; the intestines measure upwards of eight feet, and are very narrow. The male and female are alike in plumage; both, when of full age, having the train equally long.

VIRGINIAN RAIL.   (*Rallus Virginianus.*)

PLATE LXII.—Fig. 1.

Arct. Zool. No. 408.—*Edw.* 279.—*Lath. Syn.* iii. p. 208, No. 1, var. A.— *Peale's Museum*, No. 4426.

RALLUS VIRGINIANUS.—LINNÆUS.*

Rallus Virginianus, *Bonap. Synop.* p. 334.

THIS species very much resembles the European water-rail (*Rallus aquaticus*), but is smaller, and has none of the slate or

* In my note upon the genus *Crex*, I mentioned the distinctions existing between that genus, *Gallinula*, and *Rallus*. The Virginian rail and that following show good examples of the latter form.   In their habits they closely agree with the aquatic species of *Crex*, are distributed over all countries of the world, and in general perform partial migrations.

When pursued or *roaded* by a dog, they may be raised once, but the second time will be a task of more difficulty; if the ground is an extensive meadow, they may be followed for an hour without success, but if there are holes or ditches they will generally seek for one of these, where they conceal themselves beneath some sod, or brow, or thicket of bushes, and may then be easily taken by the hand.   I have frequently taken our common water-rail in this manner, and sometimes with the

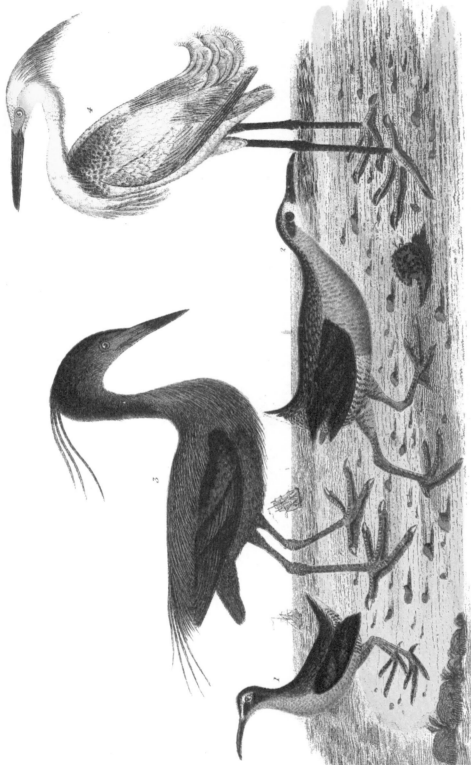

1. Virginian Rail. 2. Clapper. R. 3. Blue Crane. 4. Little Egret.

62.

lead colour on the breast which marks that of the old continent; its toes are also more than proportionably shorter, which, with a few other peculiarities, distinguish the species. It is far less numerous in this part of the United States than our common rail, and, as I apprehend, inhabits more remote northern regions. It is frequently seen along the borders of our salt marshes, which the other rarely visits ; and also breeds there, as well as among the meadows that border our large rivers. It spreads over the interior as far west as the Ohio, having myself shot it in the barrens of Kentucky early in May. The people there observe them in wet places in the groves only in spring. It feeds less on vegetable, and more on animal food, than the common rail. During the months of September and October, when the reeds and wild oats swarm with the latter species, feeding on their nutritious seeds, a few of the present kind are occasionally found ; but not one for five hundred of the others. The food of the present species consists of small snail-shells, worms, and the larvæ of insects, which it extracts from the mud ; hence the cause of its greater length of bill, to enable it the more readily to reach its food. On this account, also, its flesh is much inferior to that of the other. In most of its habits, its thin compressed form of body, its aversion to take wing, and the dexterity

head only concealed. They are easily tamed. The structure of the feathers on the forehead and crown of the rails is peculiar, and may be intended as a defence to that part from the friction of the strong grass and reeds among which they are so constantly running. The rachis of each feather is lengthened, and broadened into a flat and sharp point, having the appearance of lengthened scales ; in one or two species, the feathers consist of the rachis alone, presenting a horny appearance over the whole forehead. The bastard pinion is furnished with a spur, concealed, however, by the plumage.

The form of the Crakes and Gallinules is well adapted for their peculiar manner of life, but in this group is most conspicuous. The legs are placed far behind, the body is long, much flattened, and remarkably pliable ; and the ease and agility with which they run and thread through the long vegetation of the marshes is almost inconceivable to a person who has not witnessed it.—ED.

with which it runs or conceals itself among the grass and sedge, are exactly similar to those of the common rail, from which genus, notwithstanding the difference of its bill, it ought not to be separated.

This bird is known to some of the inhabitants along the sea-coast of New Jersey by the name of the fresh-water mud-hen, this last being the common appellation of the clapper rail, which the present species resembles in everything but size. The epithet fresh-water is given it, because of its frequenting those parts of the marsh only where fresh-water springs rise through the bogs into the salt marshes. In these places it usually constructs its nest, one of which, through the active exertions of my friend Mr Ord, while traversing with me the salt marshes of Cape May, we had the good fortune to discover. It was built in the bottom of a tuft of grass, in the midst of an almost impenetrable quagmire, and was composed altogether of old wet grass and rushes. The eggs had been floated out of the nest by the extraordinary rise of the tide in a violent north-east storm, and lay scattered about among the drift weed. The female, however, still lingered near the spot, to which she was so attached as to suffer herself to be taken by hand. She doubtless intended to repair her nest, and commence laying anew, as, during the few hours that she was in our possession, she laid one egg, corresponding in all respects with the others. On examining those floated out of the nest, they contained young, perfectly formed, but dead. The usual number of eggs is from six to ten. They are shaped like those of the domestic hen, measuring one inch and two-tenths long, by very nearly half an inch in width, and are of a dirty white or pale cream colour, sprinkled with specks of reddish and pale purple, most numerous near the great end. They commence laying early in May, and probably raise two broods in the season. I suspect this from the circumstance of Mr Ord having, late in the month of July, brought me several young ones of only a few days old, which were caught among the grass near the border of the Delaware. The parent rail showed great solicitude for their safety. They were wholly black, except a white spot on the bill; were covered with a fine down, and had a soft piping note. In the month of June of the same year, another pair of these birds began to breed amidst a boggy spring in one of Mr Bartram's meadows, but were unfortunately destroyed.

The Virginian rail is migratory, never wintering in the northern or middle States. It makes its first appearance in Pennsylvania early in May, and leaves the country on the first smart frosts, generally in November. I have no doubt but many of them linger in the low woods and marshes of the southern States during winter.

This species is ten inches long, and fourteen inches in extent; bill, dusky red; cheeks and stripe over the eye, ash; over the lores, and at the lower eyelid, white; iris of the eye, red; crown and whole upper parts, black, streaked with brown, the centre of each feather being black; wing-coverts, hazel brown, inclining to chestnut; quills, plain deep dusky; chin, white; throat, breast, and belly, orange brown; sides and vent, black, tipt with white; legs and feet, dull red brown; edge of the bend of the wing, white.

The female is about half an inch shorter, and differs from the male in having the breast much paler, not of so bright a reddish brown; there is also more white on the chin and throat.

When seen, which is very rarely, these birds stand or run with the tail erect, which they frequently jerk upwards. They fly with the legs hanging, generally but a short distance; and the moment they alight, run off with great speed.

CLAPPER RAIL. (*Rallus crepitans.*)

PLATE LXII.—FIG. 2.

Arct. Zool. No. 407.—*Lath. Syn.* iii. p. 229, No. 2.—*Ind. Orn.* p. 756, No. 2.—*Peale's Museum*, No. 4400.

RALLUS CREPITANS.—GMELIN.

Rallus crepitans, *Bonap. Synop.* p. 333.

THIS is a very numerous and well-known species, inhabiting our whole Atlantic coast from New England to Florida. It is designated by different names, such as the mud-hen, clapper rail, meadow clapper, big rail, &c., &c. Though occasionally found along the swampy shores and tide waters of our large rivers, its principal residence is in the salt marshes. It is a bird of pasage, arriving on the coast of New Jersey about the 20th of April, and retiring again late in September. I suspect that many of them winter in the marshes of Georgia and Florida, having heard them very numerous at the mouth of Savannah river in the month of February. Coasters and fishermen often hear them while on their migrations in spring, generally a little before daybreak. The shores of New Jersey, within the beach, consisting of an immense extent of flat marsh, covered with a coarse reedy grass, and occasionally overflowed by the sea, by which it is also cut up into innumerable islands by narrow inlets, seem to be the favourite breeding place for these birds, as they are there acknowledged to be more than double in number to all other marsh fowl.

The clapper rail, or, as it is generally called, the mud-hen, soon announces its arrival in the salt marshes by its loud, harsh, and incessant cackling, which very much resembles that of a guinea-fowl. This noise is most general during the night, and is said to be always greatest before a storm. About the 20th of May, they generally commence laying and building at the same time; the first egg being usually dropt in a slight cavity, lined with a little dry grass pulled for the purpose, which, as the number of the eggs increase to their usual complement, ten, is gradually added to, until it rises to the height of twelve inches or more,—doubtless to secure it from the rising of the tides. Over this the long salt grass is artfully arched, and knit at top, to conceal it from the view above ; but this very circumstance enables the experienced egg-hunter to distinguish the spot at the distance of thirty or forty yards, though imperceptible to a common eye. The eggs are of a pale clay colour, sprinkled with small spots of dark red, and measure somewhat more than an inch and a half in length, by one inch in breadth, being rather obtuse at the small end. These eggs are exquisite eating, far surpassing those of the domestic hen. The height of laying is about the 1st of June, when the people of the neighbourhood go off to the marshes *an egging*, as it is called. So abundant are the nests of this species, and so dexterous some persons at finding them, that one hundred dozen of eggs have been collected by one man in a day. At this time, the crows, the minx, and the foxes, come in for their share ; but, not content with the eggs, these last often seize and devour the parents also. The bones, feathers, wings, &c., of the poor mud-hen lie in heaps near the hole of the minx, by which circumstance, however, he himself is often detected and destroyed.

These birds are also subject to another calamity of a more extensive kind : After the greater part of the eggs are laid, there sometimes happen violent north-east tempests, that drive a great sea into the bay, covering the whole marshes ; so that at such times the rail may be seen in hundreds, floating over the marsh in great distress ; many escape to the mainland, and vast numbers perish. On an occasion of this kind, I have seen, at one view, thousands in a single meadow, walking about exposed and bewildered, while the dead bodies of the females, who had perished on or near their nests, were strewed along the shore. This last circumstance proves how strong the ties of maternal affection is in these birds ; for, of the great numbers which I picked up and opened, not one male was to be found among them—all were females ! Such as had not yet begun to sit probably escaped. These disasters do not prevent the survivors from recommencing the work of laying and building anew ; and instances have occurred where their eggs have been twice destroyed by the sea, and yet in two weeks the eggs and nests seemed as numerous as ever.

The young of the clapper rail very much resemble those of the Virginian rail, except in being larger. On the 10th of August, I examined one of these young clapper rails, caught among the reeds in the Delaware, and apparently about three weeks old ; it was covered with black down, with the exception of a spot of white on the auriculars, and a streak of the same along the side of the breast, belly, and fore part of the thigh ; the legs were of a blackish slate colour, and the bill was marked with a spot of white near the point, and round the nostril. These run with great facility among the grass and reeds, and are taken with extreme difficulty.

The whole defence of this species seems to be in the nervous vigour of its limbs and thin compressed form of its body, by which it is enabled to pass between the stalks of grass and reeds with great rapidity. There are also everywhere among the salt marshes covered ways, under the flat and matted grass, through which the rail makes its way like a rat, without a possibility of being seen. There is generally one or more of these from its nest to the water-edge, by which it may escape unseen ; and sometimes, if closely pressed, it will dive to the other side of the pond, gut, or inlet, rising and disappearing again with the silence and celerity of thought. In smooth water it swims tolerably well, but not fast ; sitting high in the water, with its neck erect, and striking with great rapidity. When on shore, it runs with the neck extended, the tail erect, and frequently flirted up. On fair ground, they run nearly as fast as a man ; having myself, with great difficulty, caught some that were wing-broken. They have also the faculty of remaining under water for several minutes, clinging close, head downwards, by the roots of the grass. In a long stretch, they fly with great velocity, very much in the manner of a duck, with extended neck, and generally low ;

but such is their aversion to take wing, that you may traverse the marshes where there are hundreds of these birds without seeing one of them; nor will they flush until they have led the dog through numerous labyrinths, and he is on the very point of seizing them.

The food of the clapper rail consists of small shellfish, particularly those of the snail form, so abundant in the marshes; they also eat small crabs. Their flesh is dry, tastes sedgy, and will bear no comparison with that of the common rail. Early in October they move off to the south; and though, even in winter, a solitary instance of one may sometimes be seen, yet these are generally such as have been weak or wounded, and unable to perform the journey.

The clapper rail measures fourteen inches in length, and eighteen in extent; the bill is two inches and a quarter long, slightly bent, pointed, grooved, and of a reddish brown colour; iris of the eye, dark red; nostril, oblong, pervious; crown, neck, and back, black, streaked with dingy brown; chin and line over the eye, brownish white; auriculars, dusky; neck before, and whole breast, of the same red brown as that of the preceding species; wing-coverts, dark chestnut; quill-feathers, plain dusky; legs, reddish brown; flanks and vent, black, tipt or barred with white. The males and females are nearly alike.

The young birds of the first year have the upper parts of an olive brown, streaked with pale slate; wings, pale brown olive; chin and part of the throat, white; breast, ash colour, tinged with brown; legs and feet, a pale horn colour. Mr Pennant, and several other naturalists, appear to have taken their descriptions from these imperfect specimens, the clapper rail being altogether unknown in Europe.

I have never met with any of these birds in the interior at a distance from lakes or rivers. I have also made diligent inquiry for them along the shores of Lakes Champlain and Ontario, but without success.*

* Mr Ord "had an opportunity of verifying the conjecture of the author as to the winter retreat of these birds, he having found them to be extremely numerous in the marshes of the coast of Georgia in the

BLUE CRANE. (*Ardea cœrulea.*)

PLATE LXII.—Fig. 3.

Arct. Zool. No. 351.—*Catesby*, i. 76.—Le Crabier Bleu, *Buff.* vii. 398.– *Sloan Jam.* ii. 315.—*Lath. Syn.* iii. p. 78, No. 45 ; p. 79, var. A.—A. cærulescens, *Turt. Syst.* p. 379.—*Planch. Enl.* 349.—*Peale's Museum*, No. 6782.

EGRETTA CÆRULEA.—Jardine.

Ardea cærulea, *Linn. Syst.*—*Bonap. Synop.* p. 300.—Ardea cærulescens, *Wagl. Syst. Av.* No. 15.

In mentioning this species in his translation of the " Systema Naturæ," Turton has introduced what he calls two varieties, one from New Zealand, the other from Brazil ; both of which, if we may judge by their size and colour, appear to be entirely different and distinct species ; the first being green with yellow legs, the last nearly one half less than the present.* By this loose mode of discrimination, the precision of science being altogether dispensed with, the whole tribe of cranes, herons, and bitterns may be styled mere varieties of the genus

month of January. In such multitudes were they along the borders of the streams or passages which separate the sea-islands from the main, that their loud and incessant noise became quite as disgusting as the monotonous cackle of that intolerable nuisance the guinea-fowl."—*Ord's Edit.*

* I have never traced this species in any Australian collection, and have little doubt that the authors of the assertion " *that it is found there,*" will turn out incorrect. This bird has all the characters of *Egretta* except the colour, and will certainly belong to that division, though it has been generally restricted to those of pure plumage. Bonaparte, in his " Nomenclature of Wilson," says, " the young birds of the year, before their first moult, are altogether pure white, and are therefore apt to be confounded with the young of *A. candidissima.*" Wagler in his excellent " Systema " confirms this, and mentions that, in their further change, the upper parts are pale cinereous tinged with purple, beneath white, the quills partly black partly white, the tail cinereous. It is curious that in a species clothed with such rich and dark plumage the young should be pure white, the colour of the true *Egretta*, while in some of those of snowy covering, the young are a dusky greyish brown. If it can be mistaken in any state for *Egretta candidissima*, it will at once show where it ought to be placed.—Ed.

Ardea. The same writer has still further increased this confusion by designating as a different species his bluish heron (*A. cœrulescens*), which agrees almost exactly with the present. Some of these mistakes may probably have originated from the figure of this bird given by Catesby, which appears to have been drawn and coloured, not from nature, but from the glimmering recollections of memory, and is extremely erroneous. These remarks are due to truth, and necessary to the elucidation of the history of this species, which seems to be but imperfectly known in Europe.

The blue heron is properly a native of the warmer climates of the United States, migrating from thence at the approach of winter to the tropical regions, being found in Cayenne, Jamaica, and Mexico. On the muddy shores of the Mississippi, from Baton Rouge downwards to New Orleans, these birds are frequently met with. In spring they extend their migrations as far north as New England, chiefly in the vicinity of the sea, becoming more rare as they advance to the north. On the sea-beach of Cape May, I found a few of them breeding among the cedars, in company with the snowy heron, night heron, and green bittern. The figure and description of the present was taken from two of these, shot in the month of May, while in complete plumage. Their nests were composed of small sticks, built in the tops of the red cedars, and contained five eggs, of a light blue colour, and of somewhat a deeper tint than those of the night heron. Little or no difference could be perceived between the colours and markings of the male and female. This remark is applicable to almost the whole genus ; though, from the circumstance of many of the yearling birds differing in plumage, they have been mistaken for females.

The blue heron, though in the northern States it be found chiefly in the neighbourhood of the ocean, probably on account of the greater temperature of the climate, is yet particularly fond of fresh-water bogs, on the edges of the salt marsh. These it often frequents, wading about in search of tadpoles, lizards, various larvæ of winged insects, and mud worms. It moves actively about in search of these, sometimes making a run at its prey; and is often seen in company with the snowy heron, figured in the same plate. Like this last, it is also very silent, intent, and watchful.

The genus *Ardea* is the most numerous of all the wading tribes, there being no less than ninety-six different species enumerated by late writers.   These are again subdivided into particular families, each distinguished by a certain peculiarity. The cranes, by having the head bald; the storks, with the orbits naked; and the herons, with the middle claw pectinated. To this last belong the bitterns.   Several of these are nocturnal birds, feeding only as the evening twilight commences, and reposing either among the long grass and reeds, or on tall trees, in sequestered places, during the day. What is very remarkable, these night wanderers often associate, during the breeding season, with the others, building their nests on the branches of the same tree; and, though differing so little in external form, feeding on nearly the same food, living and lodging in the same place, yet preserve their race, language, and manners, as perfectly distinct from those of their neighbours as if each inhabited a separate quarter of the globe.

The blue heron is twenty-three inches in length, and three feet in extent; the bill is black, but from the nostril to the eye, in both mandibles, is of a rich light purplish blue; iris of the eye, gray; pupil, black, surrounded by a narrow silvery ring; eyelid, light blue; the whole head and greater part of the neck is of a deep purplish brown; from the crested hind head shoot three narrow pointed feathers that reach nearly six inches beyond the eye; lower part of the neck, breast, belly, and whole body, a deep slate colour, with lighter reflections; the back is covered with long, flat, and narrow feathers, some of which are ten inches long, and extend four inches beyond the tail; the breast is also ornamented with a number of these long slender feathers; legs, blackish green; inner side of the middle claw pectinated. The breast and sides of the rump, under the plumage, are clothed with a mass of yellowish white unelastic cottony down, similar to that in most of the tribe, the uses of which are not altogether understood. Male and female alike in colour.

The young birds of the first year are destitute of the purple plumage on the head and neck.

SNOWY HERON. (*Ardea candidissima.*)

PLATE LXII.—Fig. 4.

Lath. Sup. i. p. 230.—No. 3748.

EGRETTA CANDIDISSIMA.—Bonaparte.*

Ardea candidissima, *Bonap. Synop.* p. 305.—*Monog. del Gruppo Egretta. Osserv. Sulla,* 2d edit. *Del Reg. Anim. Cuv.* p. 101.—*Wagl. Syst. Av.* i. No. 11.

This elegant species inhabits the sea-coast of North America from the Isthmus of Darien to the Gulf of St Lawrence, and is, in the United States, a bird of passage, arriving from the south early in April, and leaving the middle States again in October. Its general appearance, resembling so much that of the little egret of Europe, has, I doubt not, imposed on some of the naturalists of that country, as I confess it did on

---

* This species has, like the others, been also confounded with a near ally. Wagler has unravelled the confusion in his "Systema," and the Prince of Musignano in his Monograph on this group, as quoted above. To make the matter still clearer, I transcribe the Prince's observations on the "Nomenclature of Wilson." "Two closely allied species of small white-crested herons have much puzzled naturalists, who seem to have rivalled each other in confounding them, some by considering them as identical, others by making several nominal species, thus rendering their synonymy almost inextricable. The species are the *A. garzetta* of Europe and the subject of the present remarks. The latter does not inhabit Europe, but is said to be found in Asia (which we are inclined to doubt) as frequently as on this continent, where it is widely extended. Wilson is free from all the above-mentioned errors, having, as usual, admirably established the species. He was, moreover, judicious in his selection of the English and Latin names ; and it was, doubtless, after a careful investigation, that he selected the name of *candidissima*, which Mr Ord has changed to *A. Carolinensis.*"—Ed.

me.*   From a more careful comparison, however, of both
birds, I am satisfied that they are two entirely different and
distinct species.   These differences consist in the large flowing
crest, yellow feet, and singularly curled plumes of the back of
the present; it is also nearly double the size of the European
species.

The snowy heron seems particularly fond of the salt marshes
during summer, seldom penetrating far inland.   Its white
plumage renders it a very conspicuous object, either while on
wing or while wading the meadows or marshes.   Its food consists of those small crabs usually called *fiddlers*, mud worms,
snails, frogs, and lizards.   It also feeds on the seeds of some
species of *nymphœ*, and of several other aquatic plants.

On the 19th of May I visited an extensive breeding place
of the snowy heron among the red cedars of Summers's Beach,
on the coast of Cape May.   The situation was very sequestered, bounded on the land side by a fresh-water marsh or
pond, and sheltered from the Atlantic by ranges of sandhills.
The cedars, though not high, were so closely crowded together
as to render it difficult to penetrate through among them.
Some trees contained three, others four nests, built wholly of
sticks.   Each had in it three eggs, of a pale greenish blue
colour, and measuring an inch and three-quarters in length,
by an inch and a quarter in thickness.   Forty or fifty of these
eggs were cooked, and found to be well tasted; the white was
of a bluish tint, and almost transparent, though boiled for a
considerable time; the yolk very small in quantity.   The birds
rose in vast numbers, but without clamour, alighting on the
tops of the trees around, and watching the result in silent
anxiety.   Among them were numbers of the night heron, and
two or three purple-headed herons.   Great quantities of eggshells lay scattered under the trees, occasioned by the depredations of the crows, who were continually hovering about the
place.   In one of the nests I found the dead body of the bird

* "On the American continent the little egret is met with at New
York and Long Island."—*Latham,* vol. iii. p. 90.

itself, half devoured by the hawks, crows, or gulls. She had probably perished in defence of her eggs.

The snowy heron is seen at all times during summer among the salt marshes, watching and searching for food, or passing, sometimes in flocks, from one part of the bay to the other. They often make excursions up the rivers and inlets, but return regularly in the evening to the red cedars on the beach to roost. I found these birds on the Mississippi early in June, as far up as Fort Adams, roaming about among the creeks and inundated woods.

The length of this species is two feet one inch; extent, three feet two inches; the bill is four inches and a quarter long, and grooved; the space from the nostril to the eye, orange yellow, the rest of the bill black; irides, vivid orange; the whole plumage is of a snowy whiteness; the head is largely crested with loose unwebbed feathers, nearly four inches in length; another tuft of the same covers the breast; but the most distinguished ornament of this bird is a bunch of long silky plumes, proceeding from the shoulders, covering the whole back, and extending beyond the tail; the shafts of these are six or seven inches long, extremely elastic, tapering to the extremities, and thinly set with long, slender, bending threads or fibres, easily agitated by the slightest motion of the air; these shafts curl upwards at the ends. When the bird is irritated, and erects those airy plumes, they have a very elegant appearance: the legs and naked part of the thighs are black; the feet, bright yellow; claws, black, the middle one pectinated.

The female can scarcely be distinguished by her plumage, having not only the crest, but all the ornaments of the male, though not quite so long and flowing.

The young birds of the first season are entirely destitute of the long plumes of the breast and back; but as all those that have been examined in spring are found crested and ornamented as above, they doubtless receive their full dress on the first moulting. Those shot in October measured twenty-two inches in length by thirty-four in extent; the crest was beginning to form; the legs, yellowish green daubed with black; the feet, greenish yellow; the lower mandible, white at the base; the wings, when shut, nearly of a length with the tail, which is even at the end.

The little egret, or European species, is said by Latham and Turton to be nearly a foot in length. Bewick observes, that it rarely exceeds a foot and a half; has a much shorter crest, with two long feathers; the feet are black; and the long plumage of the back, instead of turning up at the extremity, falls over the rump.

The young of both these birds are generally very fat, and esteemed by some people as excellent eating.

ROSEATE SPOONBILL.   (*Platalea ajaja.*)

PLATE LXIII.—Fig. 1.

Arct. Zool. No. 338.—*Lath. Syn.* iii. p. 16, No. 2.—La Spatule Coleur de Rose, *Briss. Orn.* v. p. 3562, pl. 30.—*Buff.* vii. 456, pl. col. 116.—*Peale's Museum,* No. 3553.

PLATALEA AJAJA.—Linnæus.[*]

Platalea ajaja, *Bonap. Synop.* p. 346.

THIS stately and elegant bird inhabits the sea-shores of America from Brazil to Georgia. It also appears to wander up the Mississippi sometimes in summer, the specimen from which the figure in the plate was drawn having been sent me

---

[*] This group, remarkable for the curious development of the bill, joins a number of characters in common with the herons and tantali. They live during the breeding season in communities, and feed in twilight; their food is fish and aquatic animals, and they are said to search in the mud with their bills in the manner of ducks, where the soft and closely nervous substance enables them to detect the smaller insects. To look at the bill in a stuffed or preserved state, it is hard and horny, but when living it is remarkably tender, and has rather a fleshy and soft look and feel. The common British species is easily tamed, and, like most of its nearer allies, eats voraciously; fish will support them, and even porridge, with a little raw meat; the gape is very wide, and substances are swallowed in immediate succession, taken always crosswise, and then tossed over. The trachea in the male performs a single convolution in

1. Rose tte Spoonbill. 2. American Avoset. 3. Ruddy Plover. 4. Semipalmated Sandpiper.

from the neighbourhood of Natchez, in excellent order ; for which favour I am indebted to the family of my late benevolent and scientific friend William Dunbar, Esq., of that territory. It is now deposited in Mr Peale's Museum. This species, however, is rarely seen to the northward of the Alatamaha river, and even along the peninsula of Florida is a scarce bird. In Jamaica, several other of the West India islands, Mexico, and Guiana, it is more common, but confines itself chiefly to the sea-shore and the mouths of rivers. Captain Henderson says it is frequently seen at Honduras. It wades about in quest of shellfish, marine insects, small crabs, and fish. In pursuit of these it occasionally swims and dives.

There are few facts on record relative to this very singular bird. It is said that the young are of a blackish chestnut the first year, of the roseate colour of the present the second year, and of a deep scarlet the third.* Having never been so fortunate as to meet with them in their native wilds, I regret my present inability to throw any further light on their history and manners. These, it is probable, may resemble, in many respects, those of the European species, the white spoonbill, once so common in Holland.† To atone for this deficiency, I have endeavoured faithfully to delineate the figure of this American species, and may, perhaps, resume the subject in some future part of the present work.

the sternum. The genus contains three or four species : that of Europe, found also in India ; a species from Africa very near *P. ajaja*, peculiar to America ; and the *Spatule huppée* of Sonnerat, which Mons. Temminck thinks distinct. In all, the young do not attain full plumage till after the first moult.—ED.

   * Latham.

   † The European species breeds on trees by the seaside ; lays three or four white eggs, powdered with a few pale red spots, and about the size of those of a hen ; are very noisy during breeding time ; feed on fish, mussels, &c., which, like the bald eagle, they frequently take from other birds, frightening them by clattering their bill : they are also said to eat grass, weeds, and roots of reeds : they are migratory ; their flesh is reported to savour of that of a goose ; the young are reckoned good food.

The roseate spoonbill now before us measured two feet six inches in length, and near four feet in extent; the bill was six inches and a half long from the corner of the mouth, seven from its upper base, two inches over at its greatest width, and three-quarters of an inch where narrowest; of a black colour for half its length, and covered with hard scaly protuberances, like the edges of oyster-shells; these are of a whitish tint, stained with red; the nostrils are oblong, and placed in the centre of the upper mandible; from the lower end of each there runs a deep groove along each side of the mandible, and about a quarter of an inch from its edge; whole crown and chin, bare of plumage, and covered with a greenish skin; that below the under mandible, dilatable like those of the genus *Pelicanus;* space round the eye, orange; irides, blood red; cheeks and hind head, a bare black skin; neck, long, covered with short white feathers, some of which, on the upper part of the neck, are tipt with crimson; breast, white, the sides of which are tinged with a brown burnt colour; from the upper part of the breast proceeds a long tuft of fine hairlike plumage, of a pale rose colour; back, white, slightly tinged with brownish; wings, a pale wild rose colour, the shafts lake; the shoulders of the wings are covered with long hairy plumage, of a deep and splendid carmine; upper and lower tail-coverts, the same rich red; belly, rosy; rump, paler; tail, equal at the end, consisting of twelve feathers of a bright brownish orange, the shafts reddish; legs and naked part of the thighs, dark dirty red; feet, half webbed; toes, very long, particularly the hind one. The upper part of the neck had the plumage partly worn away, as if occasioned by resting it on the back in the manner of the ibis. The skin on the crown is a little wrinkled; the inside of the wing a much richer red than the outer.

AMERICAN AVOSET. (*Recurvirostra Americana.*)

PLATE LXIII.—Fig. 26.

Arct. Zool. No. 421.—*Lath. Syn.* iii. p. 295, No. 2.—*Peale's Museum,*
No. 4250.

RECURVIROSTRA AMERICANA.—Linnæus.*

Avocetté Isabelle, Recurvirostra Americana, *Temm. Man. d'Orn.* ii. p. 594.—
Recurvirostra Americana, *Bonap. Synop.* p. 345.

This species, from its perpetual clamour and flippancy of
tongue, is called by the inhabitants of Cape May the lawyer ;
the comparison, however, reaches no further, for our lawyer
is simple, timid, and perfectly inoffensive.

In describing the long-legged avoset of this volume, the similarity between that and the present was taken notice of. This
resemblance extends to everything but their colour. I found
both these birds associated together on the salt marshes of
New Jersey on the 20th of May. They were then breeding.
Individuals of the present species were few in respect to the
other. They flew around the shallow pools exactly in the
manner of the long-legs, uttering the like sharp note of *click,
click, click,* alighting on the marsh or in the water indiscriminately, fluttering their loose wings, and shaking their half-bent
legs, as if ready to tumble over, keeping up a continual yelping note. They were, however, rather more shy, and kept at
a greater distance. One which I wounded attempted repeatedly to dive ; but the water was too shallow to permit him to
do this with facility. The nest was built among the thick
tufts of grass, at a small distance from one of these pools. It
was composed of small twigs of a seaside shrub, dry grass, sea-

* This curious genus contains four known species ; perhaps ere long
another may be made out. They nearly resemble each other, and all
possess the turned-up bill. In their manners they assimilate generally
with the totani, feed like them, and are very clamorous when their nest
is approached. Like them, also, though possessed of partially webbed
feet, they do not swim or take the water freely, except when wading, or
by compulsion.—Ed.

weed, &c., raised to the height of several inches. The eggs were four, of a dull olive colour, marked with large irregular blotches of black, and with others of a fainter tint.

This species arrives on the coast of Cape May late in April; rears its young, and departs again to the south early in October. While here, it almost constantly frequents the shallow pools in the salt marshes ; wading about, often to the belly, in search of food, viz., marine worms, snails, and various insects that abound among the soft muddy bottoms of the pools.

The male of this species is eighteen inches and a half long, and two feet and a half in extent ; the bill is black, four inches in length, flat above, the general curvature upwards, except at the extremity, where it bends slightly down, ending in an extremely fine point ; irides, reddish hazel ; whole head, neck, and breast, a light sorrel colour ; round the eye, and on the chin, nearly white ; upper part of the back and wings, black ; scapulars, and almost the whole back, white, though generally concealed by the black of the upper parts ; belly, vent, and thighs, pure white ; tail, equal at the end, white, very slightly tinged with cinereous ; tertials, dusky brown ; greater coverts tipt with white ; secondaries, white on their outer edges and whole inner vanes ; rest of the wing, deep black ; naked part of the thighs, two and a half inches ; legs, four inches, both of a very pale light blue, exactly formed, thinned, and netted, like those of the long-legs ; feet, half webbed ; the outer membrane somewhat the broadest ; there is a very slight hind toe, which, claw and all, does not exceed a quarter of an inch in length. In these two latter circumstances alone it differs from the long-legs, but is in every other strikingly alike.

The female was two inches shorter, and three less in extent ; the head and neck a much paler rufous, fading almost to white on the breast, and separated from the black of the back by a broader band of white ; the bill was three inches and a half long ; the leg half an inch shorter ; in every other respect marked as the male. She contained a great number of eggs, some of them nearly ready for exclusion. The stomach was filled with small snails, periwinkle shellfish, some kind of mossy vegetable food, and a number of aquatic insects. The intestines were infested with tape-worms, and a number of smaller bot-like worms, some of which wallowed in the cavity of the abdomen.

In Mr Peale's collection there is one of this same species, said to have been brought from New Holland, differing little in the markings of its plumage from our own. The red brown on the neck does not descend so far, scarcely occupying any of the breast ; it is also somewhat less.

In every stuffed and dried specimen of these birds which I have examined, the true form and flexure of the bill is altogether deranged, being naturally of a very tender and delicate substance.*

RUDDY PLOVER. *(Charadrius rubidus.)*

PLATE LXIII.—Fig. 3.

Arct. Zool. No. 404.—*Lath. Syn.* iii. p. 195, No. 2.—*Turt. Syst.* p. 415.

CALIDRIS ARENARIA.—Illiger.

Tringa arenaria, *Bonap. Synop.* p. 320.

This bird is frequently found in company with the sanderling, which, except in colour, it very much resembles. It is generally seen on the sea-coast of New Jersey in May and October, on its way to and from its breeding place in the north. It runs with great activity along the edge of the flowing or retreating waves on the sands, picking up the small bivalve

* Mr Ord further observes, "It is remarkable that in the Atlantic States this species invariably affects the neighbourhood of the ocean, we never having known an instance of its having been seen in the interior ; and yet Captain Lewis met with this bird at the ponds in the vicinity of the Falls of the Missouri. That it was our species I had ocular evidence by a skin brought by Captain Lewis himself, and presented, among other specimens of natural history, to the Philadelphia Museum." See "History of Lewis and Clark's Expedition," vol. ii. p. 343.

shellfish which supplies so many multitudes of the plover and sandpiper tribes.

I should not be surprised if the present species turn out hereafter to be the sanderling itself in a different dress. Of many scores which I examined, scarce two were alike ; in some, the plumage of the back was almost plain ; in others, the black plumage was just shooting out. This was in the month of October. Naturalists, however, have considered it as a separate species ; but have given us no further particulars than that, "in Hudson's Bay it is known by the name of Mistchaychekiskaweshish," *—a piece of information certainly very instructive.

The ruddy plover is eight inches long, and fifteen in extent ; the bill is black, an inch long, and straight ; sides of the neck and whole upper parts, speckled largely with white, black, and ferruginous ; the feathers being centred with black, tipt with white, and edged with ferruginous, giving the bird a very motley appearance ; belly and vent, pure white ; wing-quills, black, crossed with a band of white ; lesser coverts, whitish, centred with pale olive, the first two or three rows black ; two middle tail-feathers, black ; the rest, pale cinereous, edged with white ; legs and feet, black ; toes, bordered with a very narrow membrane. On dissection, both males and females varied in their colours and markings.

SEMIPALMATED SANDPIPER. (*Tringa semipalmata.*)

PLATE LXIII.—Fig. 4.

Peale's Museum, No. 4025.

TRINGA SEMIPALMATA.—Wilson.

Tringa semipalmata, *Bonap. Synop.* p. 316.

This is one of the smallest of its tribe, and seems to have been entirely overlooked, or confounded with another which it much resembles (*Tringa pusilla*), and with whom it is often found associated.

* Latham.

Its half-webbed feet, however, are sufficient marks of distinction between the two. It arrives and departs with the preceding species; flies in flocks with the stints, purres, and a few others; and is sometimes seen at a considerable distance from the sea, on the sandy shores of our fresh-water lakes. On the 23d of September I met with a small flock of these birds in Burlington Bay, on Lake Champlain. They are numerous along the sea-shores of New Jersey, but retire to the south on the approach of cold weather.

This species is six inches long, and twelve in extent; the bill is black, an inch long, and very slightly bent; crown and body above, dusky brown, the plumage edged with ferruginous, and tipt with white; tail and wings, nearly of a length; sides of the rump, white; rump and tail coverts, black; wing-quills dusky black, shafted, and banded with white, much in the manner of the least snipe; over the eye a line of white; lesser coverts, tipt with white; legs and feet, blackish ash, the latter half webbed. Males and females alike in colour.

These birds varied greatly in their size, some being scarcely five inches and a half in length, and the bill not more than three-quarters; others measured nearly seven inches in the whole length, and the bill upwards of an inch. In their general appearance they greatly resemble the stints or least snipe; but unless we allow that the same species may sometimes have the toes half webbed, and sometimes divided to the origin,—and this not in one or two solitary instances, but in whole flocks, which would be extraordinary indeed,—we cannot avoid classing this as a new and distinct species.

LOUISIANA HERON. (*Ardea Ludoviciana.*)

PLATE LXIV.—FIG. 1.

Peale's Museum, No. 3750.

ARDEA LUDOVICIANA.—WILSON.

Ardea leucogaster, *Ord's reprint,* part viii. p. 1.—Ardea Ludoviciana,
Bonap. Synop. p. 304.

THIS is a rare and delicately-formed species, occasionally found on the swampy river shores of South Carolina, but more frequently along the borders of the Mississippi, particularly below New Orleans. In each of these places it is migratory; and in the latter, as I have been informed, builds its nest on trees, amidst the inundated woods. Its manners correspond very much with those of the blue heron. It is quick in all its motions, darting about after its prey with surprising agility. Small fish, frogs, lizards, tadpoles, and various aquatic insects, constitute its principal food.

There is a bird described by Latham in his "General Synopsis," vol. iii. p. 88, called the *Demi Egret,** which, from the account there given, seems to approach near to the present species. It is said to inhabit Cayenne.

Length of the Louisiana heron, from the point of the bill to the extremity of the tail, twenty-three inches; the long hair-like plumage of the rump and lower part of the back extends several inches farther; the bill is remarkably long, measuring full five inches, of a yellowish green at the base, black towards the point, and very sharp; irides, yellow; chin and throat, white, dotted with ferruginous and some blue; the rest of the neck is of a light vinous purple, intermixed on the lower part next the breast with dark slate-coloured plumage; the whole feathers of the neck are long, narrow, and pointed; head, crested, consisting first of a number of long narrow purple feathers, and under these seven or eight pendant ones, of a pure white, and twice the length of the former; upper part

*͏ See also Buffon, vol. vii. p. 378.

1. *Louisiana Heron.* 2. *Pied Oyster-catcher.* 3. *Hooping Crane.* 4. *Long billed Curlew.*

of the back and wings, light slate ; lower part of the back and rump, white, but concealed by a mass of long unwebbed hairlike plumage, that falls over the tail and tips of the wings, extending three inches beyond them ; these plumes are of a dirty purplish brown at the base, and lighten towards the extremities to a pale cream colour ; the tail is even at the tip, rather longer than the wings, and of a fine slate ; the legs and naked thighs, greenish yellow ; middle claw pectinated ; whole lower parts pure white. Male and female alike in plumage, both being crested.

PIED OYSTER-CATCHER. (*Hæmatopus ostralegus.*)

PLATE LXIV.—Fig. 2.

Arct. Zool. No. 406.—*Catesby*, i. 85.—*Bewick*, ii. 23.—*Peale's Museum*, No. 4258.

HÆMATOPUS PALLIATUS ?—Temminck.*

Hæmatopus ostralegus, *Bonap. Synop.* p. 300.—Hæmatopus palliatus ? *Jard. and Selby, Illust. Ornith.* vol. iii. plate 125.

This singular species, although nowhere numerous, inhabits almost every sea-shore both on the new and old continent,

* The oyster-catchers of Europe and America are said by Temminck and Bonaparte to be identical. Such also was the opinion of most ornithologists, and my own, until a closer comparison of American specimens with British showed a distinction. There is another, however, with which the American bird may be confounded, and I cannot decidedly say that it is distinct, the *H. palliatus*, Temm. I have not seen that species ; but from the description of the upper parts being grayish brown, it must either be distinct, or the young state of the North American bird. My specimens of the latter are of the purest black and white.

Bonaparte, in his " Nomenclature," says the species is common to both continents ; and mentions that he had specimens before him, from each country, decidedly alike. From this circumstance I should be inclined to give two species to North America, as the distinctions between them are so great as it would be impossible to overlook on an examination such as he was likely to give.

The following are the distinctive marks of the species in my posses-

but is never found inland. It is the only one of its genus
hitherto discovered, and, from the confirmation of some of its
parts, one might almost be led by fancy to suppose that it
had borrowed the eye of the pheasant, the legs and feet of the
bustard, and the bill of the woodpecker.

The oyster-catcher frequents the sandy sea-beach of New
Jersey and other parts of our Atlantic coast in summer, in
small parties of two or three pairs together. They are extremely shy, and, except about the season of breeding, will
seldom permit a person to approach within gunshot. They
walk along the shore in a watchful, stately manner, at times

sion :—The bill appears generally to be more slender ; the quills want
the white band running in a slanting direction across, being in the
American specimen entirely black ; the secondaries in the American,
except the first, are pure white ; in the British specimen, each, except
the three or four last, have a black mark near the tips, which decrease
in size as they proceed. The whole interior surface of the wing is pure
white ; in the other it is black, except where the white secondaries
appear. In the British bird, the tail-coverts and rump are pure white,
the latter running upon the back, until it is hid by the scapulary and
back feathers. In the American, the tail-coverts only are white, forming, as it were, a band of that colour, interrupted by the black tip of
the tail ; the whole rump and lower part of the back, black.

If that before us prove distinct, this genus will contain five species,
distributed over the whole world, and allied so closely, that every
member is alike, with a different distribution only of black and white
to distinguish them. They are, the common European bird, perhaps
also American, *H. ostralegus ;* the black oyster-catcher, *H. niger*, found
in Australia and Africa ; *H. palliatus*, Temm., South American, and
which may turn out to be the immature state of the species we have
mentioned ; and the *Ostralega leucopus* of Lesson, found on the Malowine
Isles, and remarkable in having white legs and feet. The species in my
possession may stand as the fifth, under the name of *H. arcticus.**

As they are allied in form, so they are in habit. They frequent low

* When this note was written, I had not seen the elaborate review of Cuvier's
"Regne Animale" by the Prince of Musignano. He is aware that the North
American and European species are distinct, and mentions that the more
northern regions produce an additional one. I believe the bird figured by
Wilson, and the skins in my possession, will prove to be this, and may stand as
I have named it above. That ornithologist also gives as a principal character to
H. palliatus, that the upper parts are "*di un color fosco invece di nero*," at
variance with the pure black and white of our specimens.—Ed.

probing it with their long wedge-like bills, in search of small shellfish. This appears evident on examining the hard sands where they usually resort, which are found thickly perforated with oblong holes, two or three inches in depth. The small crabs called fiddlers, that burrow in the mud, at the bottom of inlets, are frequently the prey of the oyster-catcher ; as are mussels, spout-fish, and a variety of other shellfish and sea-insects with which those shores abound.

The principal food, however, of this bird, according to European writers, and that from which it derives its name, is the oyster, which it is said to watch for, and snatch suddenly from the shells, whenever it surprises them sufficiently open.

sandy beaches, feeding on the shellfish during the recess of the tide, and resting while it flows. The oyster-catcher of Europe is to be found on all the sandy British coasts in immense abundance. All those which I have observed breeding have chosen low rocky coasts, and deposit their eggs on some shelf or ledge, merely baring the surface from any moss or other substance covering the rock. When approached, the parents fly round, uttering with great vehemence their clamorous note. I have never found them breeding on a sandy beach, though I have observed these birds for the last ten years, in a situation fitted in every way for that kind of incubation, and have known them retire regularly to a distance of about six or seven miles (a more populous quarter), where they had the advantage of a ledge of insulated rocks bounding the coast. A great many, both old and young birds—perhaps among the latter those of a late brood—are always to be found on these coasts, and enliven the monotony of an extensive sand-beach with their clean and lively appearance and their shrill notes, As the young begin to assemble the flocks increase ; by the month of August they consist of many thousands ; and at full tide they may be seen, like an extensive black line, at the distance of miles. They remain at rest until about half tide, when a general motion is made, and the line may be seen broken as the different parties advance close to the water's edge. After this they keep pace with the reflux, until the feeding banks begin to be uncovered, of which they seem to have an instinctive knowledge, when they leave their resting-place in small troops, taking day after day the same course. They are difficult to approach, but when one is shot, the flock will hover over it for some time without heeding the intruder. During flight they assume the ▷ wedge shape, like ducks. They feed at night when the tide is suitable, and are often very noisy. Mussels and smaller shellfish, crabs, &c., &c., are their most common food.—Ed.

In search of these, it is reported that it often frequents the oyster-beds, looking out for the slightest opening through which it may attack its unwary prey. For this purpose the form of its bill seems very fitly calculated. Yet the truth of these accounts are doubted by the inhabitants of Egg Harbour and other parts of our coast, who positively assert that it never haunts such places, but confines itself almost solely to the sands ; and this opinion I am inclined to believe correct, having myself uniformly found these birds on the smooth beach bordering the ocean, and on the higher, dry, and level sands just beyond the reach of the summer tides. On this last situation, where the dry flats are thickly interspersed with drifted shells, I have repeatedly found their nests between the middle and 25th of May. The nest itself is a slight hollow in the sand, containing three eggs, somewhat less than those of a hen, and nearly of the same shape, of a bluish cream colour, marked with large roundish spots of black, and others of a fainter tint. In some, the ground cream colour is destitute of the bluish tint, the blotches larger, and of a deep brown. The young are hatched about the 25th of May, and sometimes earlier, having myself caught them running along the beach about that period. They are at first covered with down of a greyish colour, very much resembling that of the sand, and marked with a streak of brownish black on the back, rump, and neck, the breast being dusky, where, in the old ones, it is black. The bill is at that age slightly bent downwards at the tip, where, like most other young birds, it has a hard protuberance that assists them in breaking the shell; but in a few days afterwards this falls off.* These run along the shore with great ease and swiftness.

* Latham observes that the young are said to be hatched in about three weeks ; and though they are wild when in flocks, yet are easily brought up tame, if taken young. " I have known them," says he, " to be thus kept for a long time, frequenting the ponds and ditches during the day, attending the ducks and other poultry to shelter of nights, and not unfrequently to come up of themselves as evening approaches."—*General Synopsis,* vol. iii. p. 220.

The female sits on her eggs only during the night, or in remarkably cold and rainy weather; at other times the heat of the sun and of the sand, which is sometimes great, renders incubation unnecessary. But although this is the case, she is not deficient in care or affection. She watches the spot with an attachment, anxiety, and perseverance that are really surprising, till the time arrives when her little offspring burst their prisons, and follow the guiding voice of their mother. When there is appearance of danger, they squat on the sand, from which they are with difficulty distinguished, while the parents make large circuits around the intruder, alighting sometimes on this hand, sometimes on that, uttering repeated cries, and practising the common affectionate stratagem of counterfeited lameness, to allure him from their young.

These birds run and fly with great vigour and velocity. Their note is a loud and shrill whistling *wheep-wheep-wheo,* smartly uttered. A flock will often rise, descend, and wheel in air with remarkable regularity, as if drilled to the business, the glittering white of their wings at such times being very conspicuous. They are more remarkable for this on their first arrival in the spring. Some time ago, I received a stuffed specimen of the oyster-catcher from a gentleman of Boston, an experienced sportsman, who, nevertheless, was unacquainted with this bird. He informed me that two very old men to whom it was shown called it a *hagdel.* He adds, " It was shot from a flock, which was first discovered on the beach near the entrance of Boston harbour. On the approach of the gunner, they rose, and instantly formed in line like a corps of troops, and advanced in perfect order, keeping well dressed. They made a number of circuits in the air previous to being shot at, but wheeled in line; and the man who fired into the flock observed that all their evolutions were like a regularly-organised military company."

The oyster-catcher will not only take to the water when wounded, but can also swim and dive well. This fact I can assert from my own observation, the exploits of one of them in this way having nearly cost me my life. On the sea-beach of Cape May, not far from a deep and rapid inlet, I broke the wing of one of these birds, and being without a dog, instantly pursued it towards the inlet, which it made for with great rapidity. We both plunged in nearly at the same instant; but the bird eluded my grasp, and I sunk beyond my depth; it was not until this moment that I recollected having carried in my gun along with me. On rising to the surface, I found the bird had dived, and a strong ebb current was carrying me fast towards the ocean, encumbered with a gun and all my shooting apparatus. I was compelled to relinquish my bird, and to make for the shore, with considerable mortification, and the total destruction of the contents of my powder-horn. The wounded bird afterwards rose, and swam with great buoyancy out among the breakers.

On the same day I shot and examined three individuals of this species, two of which measured each eighteen inches in length, and thirty-five inches in extent; the other was somewhat less. The bills varied in length, measuring three inches and three-quarters, three and a half, and three and a quarter, thinly compressed at the point, very much like that of the woodpecker tribe, but remarkably narrowed near the base where the nostrils are placed, probably that it may work with more freedom in the sand. This instrument, for two-thirds of its length towards the point, was evidently much worn by digging; its colour, a rich orange scarlet, somewhat yellowish near the tip; eye, large; orbits, of the same bright scarlet as the bill; irides, brilliant yellow; pupil, small, bluish black; under the eye is a small spot of white, and a large bed of the same on the wing-coverts; head, neck, scapulars, rump, wing-quills, and tail, black; several of the primaries are marked on the outer vanes with a slanting band of white; secondaries, white, part of them tipt with black; the whole lower parts of the body, sides of the rump, tail-coverts, and that portion of the tail which they cover, are pure white; the wings, when shut, cover the whole white plumage of the back and rump; legs, and naked part of the thighs, pale red ; feet, three-toed, the outer joined to the middle by a broad and strong membrane, and each bordered with a rough warty edge ; the soles of the feet are defended from the hard sand and shells by a remarkably thick and callous warty skin.

On opening these birds, the smallest of the three was found to be a male; the gullet widened into a kind of crop ; the stomach or gizzard contained fragments of shellfish, pieces of crabs, and of the great king crab, with some dark brown marine insects. The flesh was remarkably firm and muscular ; the skull, thick and strong, intended, no doubt, as in the woodpecker tribe, for the security of the brain from the violent concussions it might receive while the bird was engaged in digging. The female and young birds have the back and scapulars of a sooty brownish olive.

This species is found as far south as Cayenne and Surinam. Dampier met with it on the coast of New Holland ; the British circumnavigators also saw it on Van Diemen's Land, Tierra del Fuego, and New Zealand.

WHOOPING CRANE. (*Ardea Americana.*)

PLATE LXIV.—Fig. 3, Male.

Arct. Zool. No. 339.—*Catesby,* i. 75.—*Lath.* iii. p. 42.—La Grue d'Amerique, *Pl. enl.* 889.—*Peale's Museum,* No. 3704.

GRUS AMERICANA.—Temminck.*

Grus Americana, *Bonap. Synop.* p. 302.—*North. Zool.* ii. p. 372.

This is the tallest and most stately species of all the feathered tribes of the United States, the watchful inhabitant of exten-

* This crane has also suffered under the too general confusion of names, so that it becomes somewhat difficult to determine with precision that which should by priority be allotted to it. It is an extra European species, and seems to be the Asiatic bird generally known under the name of *G. gigantea,* Pall. Temminck, however, says that Gmelin changed this name from the original one of *G. leucogeranos,* Pall., and has figured and described it as such in the *Planches Colorées.*

sive salt marshes, desolate swamps, and open morasses in the neighbourhood of the sea.   Its migrations are regular, and of the most extensive kind, reaching from the shores and inundated tracts of South America to the arctic circle.   In these immense periodical journeys, they pass at such a prodigious height in the air as to be seldom observed.   They have, however, their resting stages on the route to and from their usual breeding places, the regions of the north.   A few sometimes make their appearance in the marshes of Cape May in December, particularly on and near Egg Island, where they are

It appears to extend over Asia to China, and specimens have been brought from Japan.   Are they all one species ?

America will also possess another majestic crane, *Grus Canadensis,* Temm., inhabiting the northern parts, but not commonly found in the middle States ; it is met with in summer in all parts of the Fur Countries to the shores of the Arctic Sea.

The birds of this genus were formerly arranged among the herons, to which they bear a certain alliance, but were, by Pallas, with propriety separated, and form a very natural division in a great class.   They are at once distinguished from *Ardea* by the bald head, and the broad, waving and pendulous form of the greater coverts.   Some extend over every part of the world, but the group is, notwithstanding, limited to only a few species.   They are majestic in appearance, and possess a strong and powerful flight, performing very long migrations, preparatory to which they assemble, and, as it were, exercise themselves before starting. They are social, and feed and migrate in troops.   Major Long, speaking of the migrations of the second American species, *G. Canadensis,* says, " They afford one of the most beautiful instances of animal motion we can anywhere meet with.   They fly at a great height, and wheeling in circles, appear to rest without effort on the surface of an aerial current, by whose eddies they are borne about in an endless series of revolutions ; each individual describes a large circle in the air, independently of his associates, and uttering loud, distinct, and repeated cries.   They continue thus to wing their flight upwards, gradually receding from the earth, until they become mere specks upon the sight, and finally altogether disappear, leaving only the discordant music of their concert to fall faintly on the ear, exploring

' Heavens not its own, and worlds unknown before.' "

The *Grus Canadensis,* or sandhill crane, will be figured and described by the Prince of Musignano in the remaining volumes of his " Continuation," which we hope ere long to receive.—ED.

known by the name of storks. The younger birds are easily distinguished from the rest by the brownness of their plumage. Some linger in these marshes the whole winter, setting out north about the time the ice breaks up. During their stay, they wander along the marshes and muddy flats of the sea-shore in search of marine worms, sailing occasionally from place to place with a low and heavy flight, a little above the surface; and have at such times a very formidable appearance. At times they utter a loud, clear, and piercing cry, which may be heard at the distance of two miles. They have also various modulations of this singular note, from the peculiarity of which they derive their name. When wounded, they attack the gunner or his dog with great resolution; and have been known to drive their sharp and formidable bill, at one stroke, through a man's hand.

During winter, they are frequently seen in the low grounds and rice plantations of the southern States, in search of grain and insects. On the 10th of February, I met with several near the Waccamau river, in South Carolina; I also saw a flock at the ponds near Louisville, Kentucky, on the 20th of March. They are extremely shy and vigilant, so that it is with the greatest difficulty they can be shot. They sometimes rise in the air spirally to a great height, the mingled noise of their screaming, even when they are almost beyond the reach of sight, resembling that of a pack of hounds in full cry. On these occasions, they fly around in large circles, as if reconnoitring the country to a vast extent for a fresh quarter to feed in. Their flesh is said to be well tasted, nowise savouring of fish. They swallow mice, moles, rats, &c., with great avidity. They build their nests on the ground, in tussocks of long grass, amidst solitary swamps, raise it to more than a foot in height, and lay two pale blue eggs, spotted with brown. These are much larger, and of a more lengthened form, than those of the common hen.

The cranes are distinguished from the other families of their genus by the comparative baldness of their heads, the broad flag of plumage projecting over the tail, and in general by their superior size. They also differ in their internal organisation from all the rest of the heron tribe, particularly in the conformation of the windpipe, which enters the breast-bone in a cavity fitted to receive it, and after several turns goes out again at the same place, and thence descends to the lungs. Unlike the herons, they have not the inner side of the middle claw pectinated, and, in this species at least, the hind toe is short, scarcely reaching the ground.

The vast marshy flats of Siberia are inhabited by a crane very much resembling the present, with the exception of the bill and legs being red ; like those of the present, the year-old birds are said also to be tawny.

It is highly probable that the species described by naturalists as the brown crane (*Ardea Canadensis*), is nothing more than the young of the whooping crane, their descriptions exactly corresponding with the latter. In a flock of six or eight, three or four are usually of that tawny or reddish brown tint on the back, scapulars, and wing-coverts ; but are evidently yearlings of the whooping crane, and differ in nothing but in that and size from the others. They are generally five or six inches shorter, and the primaries are of a brownish cast.

The whooping crane is four feet six inches in length, from the point of the bill to the end of the tail, and, when standing erect, measures nearly five feet ; the bill is six inches long, and an inch and a half in thickness, straight, extremely sharp, and of a yellowish brown colour ; the irides are yellow ; the forehead, whole crown, and cheeks, are covered with a warty skin, thinly interspersed with black hairs ; these become more thickly set towards the base of the bill ; the hind head is of an ash colour, the rest of the plumage pure white, the primaries excepted, which are black ; from the root of each wing rise numerous large flowing feathers, projecting over the tail and tips of the wings ; the uppermost of these are broad, drooping, and pointed at the extremities ; some of them are also loosely webbed, their silky fibres curling inwards, like those of the ostrich. They seem to occupy the place of the tertials. The legs and naked part of the thighs are black, very thick and strong; the hind toe seems rarely or never to reach the hard ground, though it may probably assist in preventing the bird from sinking too deep in the mire.

LONG-BILLED CURLEW. (*Numenius longirostris.*)

PLATE LXIV.—Fig. 4.

Peale's Museum, No. 3910.

NUMENIUS LONGIROSTRIS.—Wilson.*

Numenius longirostris, *Bonap. Synop.* p. 314.—*North. Zool.* ii. p. 376.

This American species has been considered by the naturalists of Europe to be a mere variety of their own, notwithstanding

* Wilson had the merit of distinguishing and separating this species from the common curlew of Europe, and giving it the appropriate name of *longirostris*, from the extraordinary length of the bill. It will fill in America the place of the common curlew in this country, and appears to have the same manners, frequenting the sea-shores in winter, and the rich dry prairies during the breeding season. *Numenius arquata*, the British prototype of *N. longirostris*, during the breeding season is entirely an inhabitant of the upland moors and sheep pastures, and in the soft and dewy mornings of May and June forms an object in their early solitude which adds to their wildness. At first dawn, when nothing can be seen but rounded hills of rich and green pasture, rising one beyond another, with perhaps an extensive meadow between, looking more boundless by the mists and shadows of morn, a long string of sheep marching off at a sleepy pace on their well-beaten track to some more favourite feeding ground, the shrill tremulous call of the curlew to his mate has something in it wild and melancholy, yet always pleasing to the associations. In such situations do they build, making almost no nest, and, during the commencement of their amours, run skulkingly among the long grass and rushes, the male rising and sailing round, or descending with the wings closed above his back, and uttering his peculiar quavering whistle. The approach of an intruder requires more demonstration of his powers, and he approaches near, buffeting and *whauping* with all his might. When the young are hatched they remain near the spot, and are for a long time difficult to raise; a pointer will stand and road them, and at this time they are tender and well

its difference of colour and superior length of bill. These differences not being accidental, or found in a few individuals, but common to all, and none being found in America corresponding with that of Europe, we do not hesitate to consider the present as a distinct species peculiar to this country.

Like the preceding, this bird is an inhabitant of marshes in the vicinity of the sea. It is also found in the interior, where, from its long bill and loud whistling note, it is generally known.

The curlews appear in the salt marshes of New Jersey about the middle of May on their way to the north, and in September on their return from their breeding places. Their food consists chiefly of small crabs, which they are very dexterous at probing for, and pulling out of their holes with their long bills; they also feed on those small sea-snails so abundant in the marshes, and on various worms and insects. They are likewise fond of bramble-berries, frequenting the fields and

flavoured. By autumn, they are nearly all dispersed to the sea-coasts, and have now lost their clear whistle. They remain here until next spring, feeding at low tide on the shore, and retiring for a few miles to inland fields at high water; on their return again at the ebb, they show a remarkable instance of the instinctive knowledge implanted in and most conspicuous in the migratory sea and water fowl. During my occasional residence on the Solway, for some years past, in the month of August, these birds, with many others, were the objects of observation. They retired regularly inland after their favourite feeding-places were covered. A long and narrow ledge of rocks runs into the Firth, behind which we used to lie concealed for the purpose of getting shots at various sea-fowl returning at ebb. None were so regular as the curlew. The more aquatic were near the sea, and could perceive the gradual reflux; the curlews were far inland, but as soon as we could perceive the top of a sharp rock standing above water, we were sure to perceive the first flocks leave the land, thus keeping pace regularly with the change of the tides. They fly in a direct line to their feeding grounds, and often in a wedge shape; on alarm, a simultaneous cry is uttered, and the next coming flock turns from its course, uttering in repetition the same alarm note. In a few days they became so wary as not to fly over the concealed station. They are one of the most difficult birds to approach, except during spring, but may be enticed by imitating their whistle.—Ed.

uplands in search of this fruit, on which they get very fat, and are then tender and good eating, altogether free from the sedgy taste with which their flesh is usually tainted while they feed in the salt marshes.

The curlews fly high, generally in a wedge-like form, somewhat resembling certain ducks, occasionally uttering their loud whistling note, by a dexterous imitation of which a whole flock may sometimes be enticed within gunshot, while the cries of the wounded are sure to detain them until the gunner has made repeated shots and great havoc among them.

This species is said to breed in Labrador, and in the neighbourhood of Hudson's Bay. A few instances have been known of one or two pairs remaining in the salt marshes of Cape May all summer. A person of respectability informed me that he once started a curlew from her nest, which was composed of a little dry grass, and contained four eggs, very much resembling in size and colour those of the mud-hen, or clapper rail. This was in the month of July. Cases of this kind are so rare, that the northern regions must be considered as the general breeding place of this species.

The long-billed curlew is twenty-five inches in length, and three feet three inches in extent, and, when in good order, weighs about thirty ounces, but individuals differ greatly in this respect; the bill is eight inches long, nearly straight for half its length, thence curving considerably downwards to its extremity, where it ends in an obtuse knob that overhangs the lower mandible; the colour black, except towards the base of the lower, where it is of a pale flesh colour; tongue, extremely short, differing in this from the snipe; eye, dark; the general colour of the plumage above is black, spotted and barred along the edge of each feather with pale brown; chin, line over the eye and round the same, pale brownish white; neck, reddish brown, streaked with black; spots on the breast more sparingly dispersed; belly, thighs, and vent, pale plain rufous, without any spots; primaries, black on the outer edges, pale brown on the inner, and barred with black; shaft of the outer one, snowy ; rest of the wing, pale reddish brown, elegantly barred with undulating lines of black ; tail, slightly rounded, of an ashy brown, beautifully marked with herring-bones of black ; legs and naked thighs, very pale light blue or lead colour, the middle toe connected with the two outer ones as far as the first joint by a membrane, and bordered along the sides with a thick warty edge ; lining of the wing, dark rufous, approaching a chestnut, and thinly spotted with black. Male and female alike in plumage. The bill continues to grow in length until the second season, when the bird receives its perfect plumage. The stomach of this species is lined with an extremely thick skin, feeling to the touch like the rough hardened palm of a sailor or blacksmith. The intestines are very tender, measuring usually about three feet in length, and as thick as a swan's quill. On the front, under the skin, there are two thick callosities, which border the upper sides of the eye, lying close to the skull. These are common, I believe, to most of the tringa and scolopax tribes, and are probably designed to protect the skull from injury while the bird is probing and searching in the sand and mud.

YELLOW-CROWNED HERON.　(*Ardea violacea.*)

PLATE LXV.—Fig. 1.

Le Crabier de Bahama, *Briss.* v. p. 481, 41.—Crested Bittern, *Catesby*, i. p. 79. —Le Crabier Gris de Fer, *Buff.* vii. p. 399.—*Arct. Zool.* No. 352.—*Peale's Museum*, No. 3738.

NYCTICORAX VIOLACEA.—Bonaparte.*

Ardea violacea, *Bonap. Synop.* p. 306.

This is one of the nocturnal species of the heron tribe, whose manners, place, and mode of building its nest, resemble greatly

* This curious species is an instance of one of those connecting links which intervene constantly among what have been defined *fixed groups.* The general form and appearance is decidedly a *Nycticorax*, and at the extremity of that form we should place it. Its manners and social manner of breeding are exactly those of the qua-bird, but it possesses

1. Yellow-crowned Heron. 2. Great Heron. 3. American Bittern. 4. Least B.

65.

those of the common night heron (*Ardea nycticorax*); the form of its bill is also similar. The very imperfect figure and description of this species by Catesby seem to have led the greater part of European ornithologists astray, who appear to have copied their accounts from that erroneous source, otherwise it is difficult to conceive why they should either have given it the name of yellow-crowned, or have described it as being only fifteen inches in length, since the crown of the perfect bird is pure white, and the whole length very near two feet. The name, however, erroneous as it is, has been retained in the present account, for the purpose of more particularly pointing out its absurdity, and designating the species.

This bird inhabits the lower parts of South Carolina, Georgia, and Louisiana in the summer season; reposing during the day among low, swampy woods, and feeding only in the night. It builds in societies, making its nest with sticks among the branches of low trees, and lays four pale blue eggs. This species is not numerous in Carolina, which, with its solitary mode of life, makes this bird but little known there. It abounds on the Bahama Islands, where it also breeds; and great numbers of the young, as we are told, are yearly taken for the table, being accounted in that quarter excellent eating. This bird also extends its migrations into Virginia, and even farther north; one of them having been shot a few years ago on the borders of Schuylkill, below Philadelphia.

The food of this species consists of small fish, crabs, and lizards, particularly the former; it also appears to have a strong attachment to the neighbourhood of the ocean.

The yellow-crowned heron is twenty-two inches in length, from the point of the bill to the end of the tail; the long flowing plumes of the back extend four inches farther; breadth, from tip to tip of the expanded wings, thirty-four

the crest and long dorsal plumes of the egrets. As far as we at present see, it will form the passage from the last-mentioned form to the night herons, which will again reach the bitterns by those confused under the name of *tiger bitterns.*—Ed.

inches; bill, black, stout, and about four inches in length, the upper mandible grooved exactly like that of the common night heron; lores, pale green; irides, fiery red; head and part of the neck, black, marked on each cheek with an oblong spot of white; crested crown and upper part of the head white, ending in two long narrow tapering plumes of pure white, more than seven inches long; under these are a few others of a blackish colour; rest of the neck and whole lower parts, fine ash, somewhat whitish on that part of the neck where it joins the black; upper parts, a dark ash, each feather streaked broadly down the centre with black, and bordered with white; wing-quills, deep slate, edged finely with white; tail, even at the end, and of the same ash colour; wing-coverts, deep slate, broadly edged with pale cream; from each shoulder proceed a number of long loosely-webbed tapering feathers, of an ash colour, streaked broadly down the middle with black, and extending four inches or more beyond the tips of the wings; legs and feet, yellow; middle claw, pectinated. Male and female, as in the common night heron, alike in plumage.

I strongly suspect that the species called by naturalists the Cayenne night heron (*Ardea Cayanensii*), is nothing more than the present, with which, according to their descriptions, it seems to agree almost exactly.

GREAT HERON. (*Ardea Herodias.*)

PLATE LXV.—Fig. 2.

Le Heron Hupé de Virginie, *Briss.* v. p. 416, 10.—Grand Heron, *Buff.* vii. p. 355; *Id.* p. 386.—Largest Crested Heron, *Catesby, App.* pl. 10, fig. 1.—*Lath. Syn.* iii. p. 85, No. 51.—*Arct. Zool.* No. 341, 342.—*Peale's Museum*, No. 3629; young, 3631.

ARDEA HERODIAS.—LINNÆUS.[*]

Ardea Herodias, *Bonap. Synop.* p. 304.—*North. Zool.* ii. p. 373.

THE history of this large and elegant bird having been long involved in error and obscurity, I have taken more than com-

[*] This may be called the representative of the European heron; it is considerably larger, but in the general colours bears a strong resemblance,

mon pains to present a faithful portrait of it in this place, and to add to that every fact and authentic particular relative to its manners which may be necessary to the elucidation of the subject.*

The great heron is a constant inhabitant of the Atlantic coast from New York to Florida ; in deep snows and severe weather seeking the open springs of the cedar and cypress swamps, and the muddy inlets occasionally covered by the

and is, moreover, the only North American bird that can rank with the genus *Ardea* in its restricted sense. In manners they are similar, feed in the evening, or early in the morning, when their prey is most active in search of its own victims ; but roost at night except during very clear moonlight. They are extremely shy and watchful, and the height they are able to overlook, with the advantage of their long legs and neck, renders them difficult of approach, unless under extensive cover. When watching their prey they may be said to resemble a cat, prying anxiously about the sides of the ditches, lake, or stream, but as soon as the least motion or indication of a living creature is seen, they are fixed and ready to make a dart almost always unerring. Mouse, frog, or fish, even rails, and the young of the larger waterfowl, are transfixed, and being carried to the nearest bank or dry ground, are immediately swallowed, always with the head downwards. Their prey appears to be often, if not always, transfixed,—a mode of capture not generally known, but admirably fitted to secure one as vigilant as the aggressor. One or two of the wild and beautiful islets on Loch Awe are occupied as breeding places by the herons, where I have climbed to many of their nests, all well supplied with trout and eels, invariably pierced or stuck through. None of the species breed on the ground, and it is a curious and rather anomalous circumstance, that the Ardeadæ, the ibis, and some allied birds, which are decidedly waders, and formed for walking, should build and roost on trees, where their motions are all awkward, and where they seem as if constantly placed in a situation contrary to their habits or abilities. A heronry, during the breeding season, is a curious and interesting, as well as picturesque object.—ED.

* Latham says of this species, that " all the upper parts of the body, the belly, tail, and legs, are brown ; " and this description has been repeated by every subsequent compiler. Buffon, with his usual eloquent absurdity, describes the heron as " exhibiting the picture of wretchedness, anxiety, and indigence ; condemned to struggle perpetually with misery and want ; sickened with the restless cravings of a famished appetite ; " a description so ridiculously untrue, that, were it possible for these birds to comprehend it, it would excite the risibility of the whole tribe.

tides. On the higher inland parts of the country, beyond the mountains, they are less numerous; and one which was shot in the upper parts of New Hampshire was described to me as a great curiosity. Many of their breeding places occur in both Carolinas, chiefly in the vicinity of the sea. In the lower parts of New Jersey, they have also their favourite places for building and rearing their young. These are generally in the gloomy solitudes of the tallest cedar swamps, where, if un-molested, they continue annually to breed for many years. These swamps are from half a mile to a mile in breadth, and sometimes five or six in length, and appear as if they occupied the former channel of some choked-up river, stream, lake, or arm of the sea. The appearance they present to a stranger is singular.

A front of tall and perfectly straight trunks, rising to the height of fifty or sixty feet without a limb, and crowded in every direction, their tops so closely woven together as to shut out the day, spreading the gloom of a perpetual twilight below. On a nearer approach, they are found to rise out of the water, which, from the impregnation of the fallen leaves and roots of the cedars, is of the colour of brandy. Amidst this bottom of congregated springs, the ruins of the former forest lie piled in every state of confusion. The roots, prostrate logs, and, in many places, the water, are covered with green mantling moss, while an undergrowth of laurel fifteen or twenty feet high intersects every opening so completely as to render a passage through laborious and harassing beyond description; at every step you either sink to the knees, clamber over fallen timber, squeeze yourself through between the stub-born laurels, or plunge to the middle in ponds made by the uprooting of large trees, which the green moss concealed from observation. In calm weather, the silence of death reigns in these dreary regions; a few interrupted rays of light shoot across the gloom; and unless for the occasional hollow screams of the herons, and the melancholy chirping of one or two species of small birds, all is silence, solitude, and desolation.

When a breeze rises, at first it sighs mournfully through the tops ; but as the gale increases, the tall mastlike cedars wave like fishing poles, and rubbing against each other, produce a variety of singular noises, that, with the help of a little imagination, resemble shrieks, groans, growling of bears, wolves, and such like comfortable music.

On the tops of the tallest of these cedars the herons construct their nests, ten or fifteen pair sometimes occupying a particular part of the swamp. The nests are large, formed of sticks, and lined with smaller twigs ; each occupies the top of a single tree. The eggs are generally four, of an oblong pointed form, larger than those of a hen, and of a light greenish blue, without any spots. The young are produced about the middle of May, and remain on the trees until they are full as heavy as the old ones, being extremely fat, before they are able to fly. They breed but once in the season. If disturbed in their breeding place, the old birds fly occasionally over the spot, sometimes honking like a goose, sometimes uttering a coarse, hollow grunting noise like that of a hog, but much louder.

The great heron is said to be fat at the full moon, and lean at its decrease ; this might be accounted for by the fact of their fishing regularly by moonlight through the greater part of the night as well as during the day ; but the observation is not universal, for at such times I have found some lean, as well as others fat. The young are said to be excellent for the table, and even the old birds, when in good order and properly cooked, are esteemed by many.

The principal food of the great heron is fish, for which he watches with the most unwearied patience, and seizes them with surprising dexterity. At the edge of the river, pond, or sea-shore, he stands fixed and motionless, sometimes for hours together. But his stroke is quick as thought, and sure as fate, to the first luckless fish that approaches within his reach ; these he sometimes beats to death, and always swallows head foremost, such being their uniform position in the stomach.

He is also an excellent mouser, and of great service to our meadows in destroying the short-tailed or meadow mouse, so injurious to the banks. He also feeds eagerly on grasshoppers, various winged insects, particularly dragonflies, which he is very expert at striking, and also eats the seeds of that species of *nymphœ* usually called splatterdocks, so abundant along our fresh-water ponds and rivers.

The heron has great powers of wing, flying sometimes very high, and to a great distance; his neck doubled, his head drawn in, and his long legs stretched out in a right line behind him, appearing like a tail, and probably serving the same rudder-like office. When he leaves the sea-coast, and traces on wing the courses of the creeks or rivers upwards, he is said to prognosticate rain; when downwards, dry weather. He is most jealously vigilant and watchful of man, so that those who wish to succeed in shooting the heron must approach him entirely unseen, and by stratagem. The same inducements, however, for his destruction, do not prevail here as in Europe. Our sea-shores and rivers are free to all for the amusement of fishing. Luxury has not yet constructed her thousands of fish-ponds, and surrounded them with steel traps, spring guns, and heron snares.* In our vast fens, meadows, and sea-marshes, this stately bird roams at pleasure, feasting on the never-failing magazines of frogs, fish, seeds, and insects with which they abound, and of which he probably considers himself the sole lord and proprietor. I have several times seen

* "The heron," says an English writer, "is a very great devourer of fish, and does more mischief in a pond than an otter. People who have kept herons have had the curiosity to number the fish they feed them with into a tub of water, and counting them again afterwards, it has been found that they will eat up fifty moderate dace and roaches in a day. It has been found, that in carp-ponds visited by this bird, one heron will eat up a thousand store carp in a year; and will hunt them so close, as to let very few escape. The readiest method of destroying this mischievous bird is by fishing for him in the manner of pike, with a baited hook. When the haunt of the heron is found out, three or four small roach or dace are to be procured, and each of them is to be baited on a wire, with a strong hook at the end, entering the wire just at the gills,

the bald eagle attack and tease the great heron ; but whether for sport, or to make him disgorge his fish, I am uncertain.

The common heron of Europe (*Ardea major*) very much resembles the present, which might, as usual, have probably been ranked as the original stock, of which the present was a mere degenerated species, were it not that the American is greatly superior in size and weight to the European species ; the former measuring four feet four inches, and weighing upwards of seven pounds ; the latter, three feet three inches, and rarely weighing more than four pounds. Yet, with the exception of size, and the rust-coloured thighs of the present, they are extremely alike. The common heron of Europe, however, is not an inhabitant of the United States.

The great heron does not receive his full plumage during the first season, nor until the summer of the second. In the first season, the young birds are entirely destitute of the white plumage of the crown, and the long pointed feathers of the back, shoulders, and breast. In this dress I have frequently shot them in autumn ; but in the third year, both males and females have assumed their complete dress, and, contrary to all the European accounts which I have met with, both are then so nearly alike in colour and markings as scarcely to be distinguished from each other, both having the long flowing crest, and all the ornamental white pointed plumage of the back and breast. Indeed, this sameness in the plumage of the males and females, when arrived at their perfect state, is a characteristic of the whole of the genus with which I am acquainted. Whether it be different with those of Europe, or

and letting it run just under the skin to the tail ; the fish will live in this manner for five or six days, which is a very essential thing ; for if it be dead, the heron will not touch it. A strong line is then to be prepared of silk and wire twisted together, and is to be about two yards long ; tie this to the wire that holds the hook, and to the other end of it there is to be tied a stone of about a pound weight ; let three or four of these baits be sunk in different shallow parts of the pond, and, in a night or two's time, the heron will not fail to be taken with one or other of them."

that the young and imperfect birds have been hitherto mistaken for females, I will not pretend to say, though I think the latter conjecture highly probable, as the night raven (*Ardea nycticorax*) has been known in Europe for several centuries, and yet, in all their accounts, the sameness of the colours and plumage of the male and female of that bird is nowhere mentioned ; on the contrary, the young or yearling bird has been universally described as the female.

On the 18th of May, I examined, both externally and by dissection, five specimens of the great heron, all in complete plumage, killed in a cedar swamp near the head of Tuckahoe river, in Cape May county, New Jersey. In this case, the females could not be mistaken, as some of the eggs were nearly ready for exclusion.

Length of the great heron, four feet four inches from the point of the bill to the end of the tail ; and to the bottom of the feet, five feet four inches ; extent, six feet ; bill eight inches long, and one inch and a quarter in width, of a yellow colour, in some, blackish on the ridge, extremely sharp at the point, the edges also sharp, and slightly serrated near the extremity ; space round the eye, from the nostril, a light purplish blue ; irides, orange, brightening into yellow where they join the pupil ; forehead and middle of the crown, white passing over the eye; sides of the crown and hind head, deep slate or bluish black, and elegantly crested, the two long, tapering black feathers being full eight inches in length ; chin, cheeks, and sides of the head, white for several inches ; throat white, thickly streaked with double rows of black ; rest of the neck, brownish ash, from the lower part of which shoot a great number of long, narrow, pointed white feathers, that spread over the breast, and reach nearly to the thighs; under these long plumes, the breast itself and middle of the belly are of a deep blackish slate, the latter streaked with white ; sides, blue ash ; vent, white ; thighs and ridges of the wings, a dark purplish rust colour ; whole upper parts of the wings, tail, and body, a fine light ash, the latter ornamented with a profusion of long, narrow, white, tapering feathers, originating on the shoulders or upper part of the back, and falling gracefully over the wings ; primaries, very dark slate, nearly black ; naked thighs, brownish yellow ; legs, brownish black, tinctured with yellow, and netted with seams of whitish ; in some, the legs are nearly black. Little difference could be perceived between the plumage of the males and females ; the latter were rather less, and the long pointed plumes of the back were not quite so abundant.

The young birds of the first year have the whole upper part of the head of a dark slate ; want the long plumes of the breast and back ; and have the body, neck, and lesser coverts of the wings considerably tinged with ferruginous.

On dissection, the gullet was found of great width from the mouth to the stomach, which has not the two strong muscular coats that form the gizzard of some birds ; it was more loose, of considerable and uniform thickness throughout, and capable of containing nearly a pint. It was entirely filled with fish, among which were some small eels, all placed head downwards ; the intestines measured nine feet in length, were scarcely as thick as a goose-quill, and incapable of being distended ; so that the vulgar story of the heron swallowing eels, which, passing suddenly through him, are repeatedly swallowed, is absurd and impossible. On the external coat of the stomach of one of these birds, opened soon after being shot, something like a blood-vessel lay in several meandering folds, enveloped in a membrane, and closely adhering to the surface. On carefully opening this membrane, it was found to contain a large, round, living worm, eight inches in length ; another, of like length, was found coiled, in the same manner, on another part of the external coat. It may also be worthy of notice, that the intestines of the young birds of the first season, killed in the month of October, when they were nearly as large as the others, measured only six feet four or five inches ; those of the full-grown ones, from eight to nine feet in length.

AMERICAN BITTERN. (*Ardea minor.*)

PLATE LXV.—Fig. 3.

Le Butor de la Baye de Hudson, *Briss.* v. p. 449, 25.—*Buff.* vii. p. 430.—
Edw. 136.—*Lath. Syn.* iii. p. 58.— *Peale's Museum*, No. 3727.

BOTAURUS MINOR.—Bonaparte.

Ardea minor, *Bonap. Synop.* p. 307.—Ardea Mokoho, *Wagl. Syst. Av.* No. 29.

This is another nocturnal species, common to all our sea and river marshes, though nowhere numerous. It rests all day among the reeds and rushes, and, unless disturbed, flies and feeds only during the night. In some places it is called the Indian-hen; on the sea-coast of New Jersey it is known by the name of *dunkadoo* a word probably imitative of its common note. They are also found in the interior, having myself killed one at the inlet of the Seneca Lake in October. It utters, at times, a hollow guttural note among the reeds, but has nothing of that loud booming sound for which the European bittern is so remarkable. This circumstance, with its great inferiority of size and difference of marking, sufficiently prove them to be two distinct species, although hitherto the present has been classed as a mere variety of the European bittern. These birds, we are informed, visit Severn River, at Hudson's Bay, about the beginning of June; make their nests in swamps, laying four cinereous green eggs among the long grass. The young are said to be at first black.

These birds, when disturbed, rise with a hollow *kwa,* and are then easily shot down, as they fly heavily. Like other night birds, their sight is most acute during the evening twilight; but their hearing is at all times exquisite.

The American bittern is twenty-seven inches long, and three feet four inches in extent; from the point of the bill to the extremity of the toes, it measures three feet; the bill is four inches long; the upper mandible, black; the lower, greenish yellow; lores and eyelids, yellow; irides, bright yellow; upper part of the head, flat, and remarkably depressed; the plumage there is of a deep blackish brown, long behind and on the neck, the general colour of which is a yellowish brown shaded with darker; this long plumage of the neck the bird can throw forward at will when irritated, so as to give him a more formidable appearance; throat, whitish, streaked with deep brown; from the posterior and lower part of the auriculars, a broad patch of deep black passes diagonally across the neck, a distinguished characteristic of this species; the back is deep brown, barred and mottled with innumerable specks and streaks of brownish yellow; quills, black, with a leaden gloss, and tipt with yellowish brown; legs and feet, yellow, tinged with pale green; middle claw, pectinated; belly, light yellowish brown, streaked with darker; vent, plain; thighs, sprinkled on the outside with grains of dark brown; male and female nearly alike, the latter somewhat less. According to Bewick, the tail of the European bittern contains only ten feathers; the American species has invariably twelve. The intestines measured five feet six inches in length, and were very little thicker than a common knitting-needle; the stomach is usually filled with fish or frogs.*

This bird, when fat, is considered by many to be excellent eating.

* I have taken an entire water-rail from the stomach of the European bittern.—Ed.

LEAST BITTERN.　(*Ardea exilis.*)

PLATE LXV.—Fig. 4, Male.

Lath. Syn. iii. p. 26, No. 28.—*Peale's Museum*, No. 3814; female, 3815.

ARDEOLA EXILIS.—Bonaparte.*

Ardeola exilis, *Bonap. Synop.* p. 309.—Ardea exilis, *Wagl. Syst. Av.* No. 43.—
　　Le Heron Rouge et Noir, *Azar. Voy.* 360.—Descript. Opt. Auct. *Wagl.*

This is the smallest known species of the whole tribe. It is commonly found in fresh-water meadows, and rarely visits the salt marshes. One shot near Great Egg Harbour was presented to me as a very uncommon bird. In the meadows of Schuylkill and Delaware, below Philadelphia, a few of these birds breed every year, making their nests in the thick tussocks of grass in swampy places. When alarmed, they seldom fly far, but take shelter among the reeds or long grass. They are scarcely ever seen exposed, but skulk during the day ; and, like the preceding species, feed chiefly in the night.

This little creature measures twelve inches in length, and sixteen in extent ; the bill is more than two inches and a quarter long, yellow, ridged with black, and very sharp pointed ; space round the eye, pale yellow ; irides, bright yellow ; whole upper part of the crested head, the back, scapulars, and tail, very deep slate, reflecting slight tints of green ; throat, white, here and there tinged with buff ; hind part of the neck, dark chestnut bay ; sides of the neck, cheeks, and line over the eye, brown buff ; lesser wing-coverts, the same ; greater wing-coverts, chestnut, with a spot of the same

---

* Bonaparte proposes the title of *Ardeola* as a subgenus for this species and the *A. minuta* of Britain. They differ from the other (*A. virescens*, &c.) small herons, in having the space above the knees plumed, and in the scapularies taking the broad form of those of the bitterns and night herons, instead of beautifully lengthened plumes.

Three species will constitute this group—that of America, *A. exilis ;* *A. minuta*, of Europe ; and *A. pusilla*, Wagl., of New Holland. They are all very similar ; the latter has been confounded hitherto with the others.—Ed.

Drawn from Nature by A. Wilson.

1. *Wood Ibis.* 2. *Scarlet I.* 3. *White I.* 4. *Flamingo.*

66.

Engraved by W.H.Lizars.

at the bend of the wing ; the primary coverts are also tipt with the same ; wing-quills, dark slate ; breast, white, tinged with ochre, under which lie a number of blackish feathers ; belly and vent, white ; sides, pale ochre ; legs, greenish on the shins, hind part and feet, yellow ; thighs, feathered to within a quarter of an inch of the knees ; middle claw, pectinated ; toes, tinged with pale green ; feet, large, the span of the foot measuring two inches and three-quarters. Male and female, nearly alike in colour. The young birds are brown on the crown and back. The stomach was filled with small fish ; and the intestines, which were extremely slender, measured in length about four feet.

The least bittern is also found in Jamaica, and several of the West India islands.

WOOD IBIS. (*Tantalus loculator.*)

PLATE LXVI.—Fig. 1.

Gmel. Syst. p. 647.—Le Grand Courly d'Amerique, *Briss.* v. p. 335, 8.—Couricaca, *Buff.* vii. p. 276, *Pl. enl.* 868.—*Catesby,* i. 81.—*Arct. Zool.* No. 360.—*Lath. Syn.* iii. p. 104.—*Peale's Museum,* No. 3832.

TANTALUS LOCULATOR.—Linnæus.*

Tantalus loculator, *Bonap. Synop,* p. 310.— *Wagl. Syst. Av.* No. 1.

The wood ibis inhabits the lower parts of Louisiana, Carolina, and Georgia ; is very common in Florida, and extends

---

* This species, I believe peculiar to the New World, is extensively dispersed over it, but migratory towards the north. The bird stated by Latham as identical with this, from New Holland, will most probably turn out the *T. lacteus* or *leucocephalus ;* at all events, distinct. The genera *Tantalus* and *Ibis* run into each other in one of those gradual marches where it is nearly impossible to mark the distinction ; yet, taking the extremes, the difference is very great. *Tantalus loculator* is the only American species of the former group, principally distinguished by the base of the bill being equal in breadth with the forehead, which, with the face, cheeks, and throat, are bare. In their general manner, they are more sluggish than the ibis, and possess more of the inactivity of the heron when gorged, or the sedate gait of the stork and adjutants.

as far south as Cayenne, Brazil, and various parts of South America. In the United States it is migratory ; but has never, to my knowledge, been found to the north of Virginia. Its favourite haunts are watery savannas and inland swamps, where it feeds on fish and reptiles. The French inhabitants of Louisiana esteem it good eating.

The known species have been limited to about five in number, natives of America, Africa, and India. The genus *Ibis* is more extensive ; they are spread over all the world, and among themselves present very considerable modifications of form. Those of Northern America are three— the two now figured, and the *I. falcinellus* of Europe, first noticed by Mr Ord as a native of that country in the Journal of the Academy, under the name of *Tantalus Mexicanus*, and afterwards recognised by the Prince of Musignano as the bird of Europe. By Wagler, in his " Systema Avium," they are put into three divisions, distinguished by the scutellation of the tarsi, and the proportion of the toes. The face is often bare ; in one or two the crown is developed into a shield, as in *I. calva ;* in a few the head and neck are unplumed, *I. sacra* and *melanocephalus ;* and in some, as that of Europe, the face and head are nearly wholly clothed, and bear close resemblance to the curlews. They are all partly gregarious, feed in small groups, and breed on trees in most extensive communities. They include birds well known for many curious particulars connected with the history and superstitions of nations, and gorgeous from the pureness and decided contrast or dazzling richness of their plumage. To the former will belong the sacred ibis of antiquity, whose bodies, *in the words of a versatile and pleasing writer,* " from the perfection of an unknown process, have almost defied the ravages of time ; and, through its interventions, the self-same individuals exist in a tangible form which wandered along the banks of the mysterious Nile in the earliest ages of the world, or, ' in dim seclusion veiled,' inhabited the sanctuary of temples, which, though themselves of most magnificent proportions, are now scarcely discernible amid the desert dust of an unpeopled wilderness." To the others will belong the brilliant species next described, no less remarkable for its unassuming garb in the dress of the first year, and the richly plumaged glossy ibis. The last-mentioned bird is more worthy of notice, holding a prominent part in the mythology of the Egyptians, and occasionally honoured by embalmment ; it is also of extensive geographical distribution, being found in India, Africa, America, Europe, and an occasional stray individual finding a devious course to the shores of Great Britain. A specimen has occurred on the Northumbrian coast within this month. —Ed.

With the particular manners of this species I am not personally acquainted ; but the following characteristic traits are given of it by Mr William Bartram, who had the best opportunities of noting them :—

"This solitary bird," he observes, "does not associate in flocks, but is generally seen alone, commonly near the banks of great rivers, in vast marshes or meadows, especially such as are covered by inundations, and also in the vast deserted rice plantations ; he stands alone, on the topmost limb of tall, dead cypress trees, his neck contracted or drawn in upon his shoulders, and his beak resting like a long scythe upon his breast ; in this pensive posture and solitary situation, they look extremely grave, sorrowful, and melancholy, as if in the deepest thought. They are never seen on the sea-coast, and yet are never found at a great distance from it. They feed on serpents, young alligators, frogs, and other reptiles." *

The figure of this bird given in the plate was drawn from a very fine specimen, sent to me from Georgia by Stephen Elliot, Esq. of Beaufort, South Carolina ; its size and markings were as follow :—

Length, three feet two inches ; bill, nearly nine inches long, straight for half its length, thence curving downwards to the extremity, and full two inches thick at the base, where it rises high in the head, the whole of a brownish horn colour ; the under mandible fits into the upper in its whole length, and both are very sharp edged ; face, and naked head, and part of the neck, dull greenish blue, wrinkled ; eye, large, seated high in the head ; irides, dark red ; under the lower jaw is a loose corrugated skin or pouch, capable of containing about half a pint ; whole body, neck, and lower parts, white ; quills, dark glossy green and purple ; tail, about two inches shorter than the wings, even at the end, and of a deep and rich violet ; legs and naked thighs, dusky green ; feet and toes, yellowish, sprinkled with black ; feet, almost semipalmated, and bordered to the claws with a narrow membrane ; some of the greater

* Travels, &c., p. 150.

wing-coverts are black at the root, and shafted with black; plumage on the upper ridge of the neck generally worn, as in the presented specimen, with rubbing on the back, while in its common position of resting its bill on its breast, in the manner of the white ibis.   (See fig. 3.)

The female has only the head and chin naked; both are subject to considerable changes of colour when young, the body being found sometimes blackish above, the belly cinereous, and spots of black on the wing-coverts; all of which, as the birds advance in age, gradually disappear, and leave the plumage of the body, &c., as has been described.

SCARLET IBIS.   (*Tantalus ruber.*)

PLATE LXVI.—FIG. 2.

Le Courly Rouge du Bresil, *Briss.* v. p. 344, pl. 29, fig. 2.—Red Curlew, *Catesby,*
    i. 84.—*Arct. Zool.* No. 366, 382.—*Peale's Museum,* No. 3864; female, 3868.

IBIS RUBRA.—VIEILLOT.

Ibis rubra, *Vieill. Bonap. Synop.* p. 311.—*Wagl. Syst. Av.* No. 4.—Ibis ruber,
    Wils. Ill. of Zool. i. pl. 7, and 36 in the plumage of second and first years.—
    Ibis rouge, *Less. Man. d'Ornith.* ii. p. 254.

THIS beautiful bird is found in the most southern parts of Carolina, also in Georgia and Florida, chiefly about the sea-shore and its vicinity.   In most parts of America within the tropics, and in almost all the West India islands, it is said to be common, also in the Bahamas.   Of its manners, little more has been collected than that it frequents the borders of the sea, and shores of the neighbouring rivers, feeding on small fry, shellfish, sea-worms, and small crabs.   It is said frequently to perch on trees, sometimes in large flocks; but to lay its eggs on the ground on a bed of leaves.   The eggs are described as being of a greenish colour; the young, when hatched, black; soon after, grey; and before they are able to fly, white; continuing gradually to assume their red colour until the third year, when the scarlet plumage is complete. It is also said that they usually keep in flocks, the young and old birds separately. They have frequently been domesticated.

One of them, which lived for some time in the museum of this city, was dexterous at catching flies, and most usually walked about in that pursuit in the position in which it is represented in the plate.

The scarlet ibis measures twenty-three inches in length, and thirty-seven in extent; the bill is five inches long, thick, and somewhat of a square form at the base, gradually bent downwards, and sharply ridged, of a black colour, except near the base, where it inclines to red; irides, dark hazel; the naked face is finely wrinkled, and of a pale red; chin, also bare and wrinkled for about an inch; whole plumage, a rich glowing scarlet, except about three inches of the extremities of the four outer quill-feathers, which are of a deep steel-blue; legs and naked part of the thighs, pale red, the three anterior toes united by a membrane as far as the first joint.

Whether the female differs in the colour of her plumage from the male, or what changes both undergo during the first and second years, I am unable to say from personal observation. Being a scarce species with us, and only found on our most remote southern shores, a sufficient number of specimens have not been procured to enable me to settle this matter with sufficient certainty.

WHITE IBIS. *(Tantalus albus.)*

PLATE LXVI.—Fig. 3.

Le Courly Blanc du Bresil, *Briss.* v. p. 339, 10.—*Buff.* viii. p. 41.—White Curlew, *Catesby*, i. pl. 82.—*Lath. Syn.* iii. p. 111, No. 9.—*Arct. Zool.* No. 363.

IBIS ALBA.—Vieillot.

Ibis alba, *Wagl. Syst. Av.* No. 5.—*Bonap. Synop.* p. 312.

This species bears in every respect, except that of colour, so strong a resemblance to the preceding, that I have been almost induced to believe it the same in its white or imperfect stage

of colour. The length and form of the bill, the size, conformation, as well as colour of the legs, the general length and
breadth, and even the steel-blue on the four outer quill-feathers,
are exactly alike in both. These suggestions, however, are
not made with any certainty of its being the same, but as circumstances which may lead to a more precise examination of
the subject hereafter.

I found this species pretty numerous on the borders of Lake
Pontchartrain, near New Orleans, in the month of June, and
also observed the Indians sitting in market with strings of
them for sale. I met with them again on the low keys or
islands off the peninsula of Florida. Mr Bartram observes
that " they fly in large flocks or squadrons, evening and
morning, to and from their feeding places or roosts, and are
usually called Spanish curlews. They feed chiefly on crayfish,
whose cells they probe, and, with their strong pinching bills,
drag them out." The low islands above mentioned abound
with these creatures and small crabs, the ground in some
places seeming alive with them, so that the rattling of their
shells against one another was incessant. My venerable friend,
in his observations on these birds, adds, " It is a pleasing sight,
at times of high winds and heavy thunderstorms, to observe
the numerous squadrons of these Spanish curlews driving to
and fro, turning and tacking about high up in the air, when, by
their various evolutions in the different and opposite currents
of the wind, high in the clouds, their silvery white plumage
gleams and sparkles like the brightest crystal, reflecting the
sunbeams that dart upon them between the dark clouds."

The white ibis is twenty-three inches long, and thirty-seven
inches in extent ; bill formed exactly like that of the scarlet
species, of a pale red, blackish towards the point ; face a reddish flesh colour, and finely wrinkled ; irides, whitish ; whole
plumage pure white, except about four inches of the tips of the
four outer quill-feathers, which are of a deep and glossy steel-
blue ; legs and feet pale red, webbed to the first joint.

These birds I frequently observed standing on the dead

limbs of trees and on the shore resting on one leg, their body in an almost perpendicular position, as represented in the figure, the head and bill resting on the breast. This appears to be its most common mode of resting, and perhaps sleeping, as, in all those which I examined, the plumage on the upper ridge of the neck and upper part of the back was evidently worn by this habit. The same is equally observable on the neck and back of the wood ibis.

The present species rarely extends its visits north of Carolina, and even in that State is only seen for a few weeks towards the end of summer. In Florida they are common, but seldom remove to any great distance from the sea.

RED FLAMINGO. (*Phœnicopterus ruber.*)

PLATE LXVI.—Fig. 4.

Le Flamant, *Briss.* vi. p. 532, pl. 47, fig. 1.—*Buff.* viii. p. 475, pl. 39, *Pl. enl.* 63.—*Lath. Syn.* iii. p. 299, pl. 93.—*Arct. Zool.* No. 422.—*Catesby*, i. pl. 73, 74.—*Peale's Museum*, No. 3545, bird of the first year ; No. 3546, bird of the second year. *PHŒNICOPTERUS RUBER.*—Linnæus.

Phœnicopterus ruber, *Bonap. Synop.* p. 348.

This very singular species, being occasionally seen on the southern frontiers of the United States and on the peninsula of East Florida, where it is more common, has a claim to a niche in our ornithological museum, although the author regrets that, from personal observation, he can add nothing to the particulars of its history already fully detailed in various European works. From the most respectable of these, the "Synopsis" of Dr Latham, he has collected such particulars as appear authentic and interesting.

"This remarkable bird has the neck and legs in a greater disproportion than any other bird ; the length from the end of the bill to that of the tail is four feet two or three inches ; but to the end of the claws, measures sometimes more than six feet. The bill is four inches and a quarter long, and of a con-

struction different from that of any other bird; the upper mandible very thin and flat, and somewhat movable; the under,
thick; both of them bending downwards from the middle; the
nostrils are linear, and placed in a blackish membrane; the
end of the bill, as far as the bend, is black; from thence to the
base, reddish yellow; round the base, quite to the eye, covered
with a flesh-coloured cere; the neck is slender, and of a great
length; the tongue, large, fleshy, filling the cavity of the bill,
furnished with twelve or more hooked papillæ on each side,
turning backwards; the tip, a sharp cartilaginous substance.
The bird, when in full plumage, is wholly of a most deep
scarlet (those of Africa said to be the deepest), except the
quills, which are black; from the base of the thigh to the
claws, measures thirty-two inches, of which the feathered part
takes up no more than three inches; the bare part above the
knee, thirteen inches; and from thence to the claws, sixteen;
the colour of the bare parts is red, and the toes are furnished
with a web, as in the duck genus, but is deeply indented.
The legs are not straight, but slightly bent, the shin rather
projecting.

"These birds do not gain their full plumage till the third
year. In the first, they are of a greyish white for the most
part; the second, of a clearer white, tinged with red, or rather
rose colour; but the wings and scapulars are red; in the third
year, a general glowing scarlet manifests itself throughout;
the bill and legs also keep pace with the gradation of colour
in the plumage, these parts changing to their colours by degrees, as the bird approaches to an adult state.

"Flamingoes prefer a warm climate; in the old continent
not often met with beyond forty degrees north or south; everywhere seen on the African coast and adjacent isles, quite
to the Cape of Good Hope,* and now and then on the coasts
of Spain,† Italy, and those of France lying on the Mediterra-

---

* In Zee Coow river.—*Philosophical Transactions.* Once plenty in
the Isle of France.— *Voyage to Mauritius*, p. 66.

† About Valencia, in the Lake Albufere.—*Dillon's Travels*, p. 374.

nean Sea, being at times met with at Marseilles, and for some way up the Rhone; in some seasons frequent Aleppo * and parts adjacent; seen also on the Persian side of the Caspian Sea, and from thence along the western coast as far as the Wolga, though this at uncertain times, and chiefly in considerable flocks, coming from the north coast mostly in October and November, but so soon as the wind changes, they totally disappear.† They breed in the Cape Verd Isles, particularly in that of Sal.‡ The nest is of a singular construction, made of mud, in shape of a hillock, with a cavity at top; in this the female lays generally two white eggs, § of the size of those of a goose, but more elongated. The hillock is of such a height as to admit of the bird's sitting on it conveniently, or rather standing, as the legs are placed one on each side at full length. || The young cannot fly till full grown, but run very fast.

"Flamingoes, for the most part, keep together in flocks, and now and then are seen in great numbers together, except in breeding time. Dampier mentions having, with two more in company, killed fourteen at once; but this was effected by secreting themselves, for they are very shy birds, and will by no means suffer any one to approach openly near enough to shoot them.¶ Kolben observes that they are very numerous at the Cape, keeping in the day on the borders of the lakes and rivers, and lodging themselves of nights in the long grass on the hills. They are also common to various places in the warmer parts of America, frequenting the same latitudes as in other quarters of the world; being met with in Peru, Chili,

---

* Russel's Aleppo, p. 69.

† Decouv. Russ. ii. p. 24.

‡ Dampier's Voy. i. p. 70.

§ They never lay more than three, and seldom fewer.—*Phil. Trans.*

|| Sometimes will lay the eggs on a projecting part of a low rock, if it be placed sufficiently convenient so as to admit of the legs being placed one on each side.—*Linnæus.*

¶ Davies talks of the gunner disguising himself in an ox-hide, and, by this means, getting within gunshot.—*Hist. of Barbadoes*, p. 88.

Cayenne,* and the coast of Brazil, as well as the various islands of the West Indies. Sloane found them in Jamaica, but particularly at the Bahama Islands, and that of Cuba, where they breed. When seen at a distance, they appear as a regiment of soldiers, being arranged alongside of one another, on the borders of the rivers, searching for food, which chiefly consists of small fish,† or the eggs of them, and of water insects, which they search after by plunging in the bill and part of the head; from time to time trampling with their feet to muddy the water, that their prey may be raised from the bottom. In feeding, are said to twist the neck in such a manner that the upper part of the bill is applied to the ground; ‡ during this, one of them is said to stand sentinel, and the moment he sounds the alarm, the whole flock take wing. This bird, when at rest, stands on one leg, the other being drawn up close to the body, with the head placed under the wing on that side of the body it stands on.

"The flesh of these birds is esteemed pretty good meat, and the young thought by some equal to that of a partridge; § but the greatest dainty is the tongue, which was esteemed by the ancients an exquisite morsel. ‖ Are sometimes caught young, and brought up tame; but are ever impatient of cold, and in this state will seldom live a great while, gradually losing their colour, flesh, and appetite, and dying for want of that food which, in a state of nature at large, they were abundantly supplied with."

* Called there by the name of Tococo.
† Small shellfish.—*Gesner.*   ‡ Linnæus, Brisson.
§ Commonly fat, and accounted delicate.—*Davies' Hist. of Barbadoes,* p. 88. The inhabitants of Provence always throw away the flesh, as it tastes fishy, and only make use of the feathers as ornaments to other birds at particular entertainments.—*Dillon's Travels,* p. 374.
‖ See Plin. ix. cap. 48.

Drawn from Nature by A.Wilson.   1.Black or Surf' Duck.   2.Buffel-headed D.   3.Female.   4.Canada Goose.   5.Tufted Duck.   6.Golden.-eye.   7.Shoveller.   Engraved by W.H.Lizars.

67

BLACK OR SURF DUCK. (*Anas perspicillata.*)

PLATE LXVII.—Fig. 2, Male.

La Grande Macreuse de la Baye de Hudson, *Briss.* vi. 425, 30.—La Macreuse à large bec, *Buff.* ix. p. 244, *Pl. enl.* 995.—*Edw.* pl. 155.—*Lath. Syn.* iii. p. 479.—*Phil. Trans.* lxii. p. 417.—*Peale's Museum*, No. 2788 ; female, 2789.

OIDEMIA PERSPICILLATA.—Stephens.

Oidemia perspicillata, *Steph. Cont. Sh. Gen. Zool.* xii. p. 219.—Oidemia, subgen. Fuligula perspicillata, *Bonap. Synop.* p. 389.—Oidemia perspicillata, *North. Zool.* ii. p. 449.—*Jard. and Selby, Illust. of Ornith.* pl. 138.

This duck is peculiar to America,* and altogether confined to the shores and bays of the sea, particularly where the waves roll over the sandy beach. Their food consists principally of those small bivalve shellfish already described, spout-fish, and others that lie in the sand near its surface. For these they dive almost constantly, both in the sandy bays and amidst the tumbling surf. They seldom or never visit the salt marshes. They continue on our shores during the winter, and leave us early in May for their breeding places in the north. Their skins are remarkably strong, and their flesh coarse, tasting of fish. They are shy birds, not easily approached, and are common in winter along the whole coast, from the river St Lawrence to Florida.

The length of this species is twenty inches ; extent, thirty-two inches ; the bill is yellowish red, elevated at the base, and marked on the side of the upper mandible with a large square patch of black, preceded by another space of a pearl colour ; the part of the bill thus marked swells or projects consider-

---

* One or two instances of this bird being killed on the shores of Great Britain have occurred ; and, as an occasional visitant, it will be figured in the concluding number of Mr Selby's " Illustrations of British Ornithology." It is also occasionally met with on the continent of Europe, but generally in high latitudes, and though unfrequent elsewhere, it is not entirely confined to America.—Ed.

ably from the common surface ; the nostrils are large and pervious ; the sides of the bill broadly serrated or toothed ; both mandibles are furnished with a nail at the extremity ; irides, white or very pale cream ; whole plumage, a shining black, marked on the crown and hind head with two triangular spaces of pure white ; the plumage on both these spots is shorter and thinner than the rest ; legs and feet, blood red ; membrane of the webbed feet, black ; the primary quills are of a deep dusky brown.

On dissection, the gullet was found to be gradually enlarged to the gizzard, which was altogether filled with broken shell-fish. There was a singular hard expansion at the commencement of the windpipe, and another much larger about three-quarters of an inch above where it separates into the two lobes of the lungs ; this last was larger than a Spanish hazel-nut, flat on one side, and convex on the other. The protuberance on each side of the bill communicated with the nostril, and was hollow. All these were probably intended to contain supplies of air for the bird's support while under water ; the last may also protect the head from the sharp edges of the shells.

The female is altogether of a sooty brown, lightest about the neck ; the prominences on the bill are scarcely observable, and its colour dusky.

This species was also found by Captain Cook at Nootka Sound, on the north-west coast of America.

BUFFEL-HEADED DUCK. (*Anas albeola.*)

PLATE LXVII.—FIG. 2, MALE ; FIG. 3, FEMALE.

Le Sarselle de Louisiane, *Briss.* vi. p. 461, pl. 41, fig. 1.—Le Petit Canard à grosse tête, *Buff.* ix. p. 249.—*Edwards*, ii. p. 100.—*Catesby*, i. 95. —*Lath. Syn.* iii. p. 533.—A. bucephala, *id.* p. 121, No. 21 ; A. rustica, *id.* p. 524, No. 24.— *Peale's Museum*, No. 2730 ; female, 2731.

CLANGULA ALBEOLA.—BOIE.

Fuligula albeola, *Bonap. Synop.* p. 394.—Clangula albeola, *North. Zool.* ii. p. 458.

THIS pretty little species, usually known by the name of the butter-box or butter-ball, is common to the sea-shores, rivers, and lakes of the United States in every quarter of the country during autumn and winter. About the middle of April or early in May they retire to the north to breed. They are dexterous divers, and fly with extraordinary velocity. So early as the latter part of February, the males are observed to have violent disputes for the females. At this time they are more commonly seen in flocks, but during the preceding part of winter they usually fly in pairs. Their note is a short *quak.* They feed much on shellfish, shrimps, &c. They are some-times exceedingly fat, though their flesh is inferior to many others for the table. The male exceeds the female in size, and greatly in beauty of plumage.

The buffel-headed duck, or rather, as it has originally been, the buffalo-headed duck, from the disproportionate size of its head, is fourteen inches long, and twenty-three inches in extent ; the bill is short, and of a light blue, or leaden colour ; the plumage of the head and half of the neck is thick, long, and velvety, projecting greatly over the lower part of the neck ; this plumage on the forehead and nape is rich glossy green, changing into a shining purple on the crown and sides of the neck ; from the eyes backward passes a broad band of pure white ; iris of the eye, dark ; back, wings, and part of the scapulars, black ; rest of the scapulars, lateral band along the wing, and whole breast, snowy white ; belly, vent, and tail-coverts, dusky white ; tail, pointed, and of a hoary colour.

The female is considerably less than the male, and entirely destitute of the tumid plumage of the head ; the head, neck, and upper parts of the body, and wings, are sooty black, darkest on the crown ; side of the head marked with a small oblong spot of white ; bill, dusky ; lower part of the neck, ash, tipt with white ; belly, dull white ; vent, cinereous ; outer edges of six of the secondaries and their incumbent coverts, white, except the tips of the latter, which are black ; legs and feet, a livid blue ; tail, hoary brown ; length of the intestines, three feet six inches ; stomach filled with small shellfish. This is the spirit-duck of Pennant, so called from its dexterity in diving (Arctic Zoology, No. 487), likewise the little brown duck of Catesby (Natural History of Carolina, pl. 98).

This species is said to come into Hudson's Bay, about Severn River, in June, and make their nests in trees in the woods near ponds.* The young males during the first year are almost exactly like the females in colour.

CANADA GOOSE. (*Anas Canadensis.*)

PLATE LXVII.—Fig. 4.

L'Oye Savage de Canada, *Briss.* vi. p. 272, 4, pl. 26.—L'Oie à cravatte, *Buff.* ix. p. 82. —*Edw.* pl. 151.—*Arct. Zool.* No. 471.—*Catesby*, i. pl. 92.—*Lath. Syn.* iii. p. 450.—*Peale's Museum*, No. 2704.

ANSER CANADENSIS.—Vieillot.*

Bernicla Canadensis, *Boie.*—Anser Canadensis, *Bonap. Synop.* p. 377.—*North. Zool.* ii. p. 468.—L'Outarde, *French Canadians.*—Bustard, *Huds. B. Settlers.*

THIS is the common wild goose of the United States, universally known over the whole country, whose regular periodical

---

* Latham.

† The appellation "*geese*" will mark, in a general way, the birds and form to which *Anser* should be generically applied. They are all of large size, possess in part the gait of a gallinaceous bird, are gregarious, except during the breeding season, mostly migratory, and are formed more for

migrations are the sure signals of returning spring or approaching winter. The tracts of their vast migratory journeys are not confined to the sea-coast or its vicinity. In their aerial voyages to and from the north, these winged pilgrims pass over the interior on both sides of the mountains, as far west, at least, as the Osage River; and I have never yet visited any quarter of the country where the inhabitants are not familiarly acquainted with the regular passing and repassing of the wild geese. The general opinion here is, that they are on their way to the lakes to breed; but the inhabitants on the confines of the great lakes that separate us from Canada are equally ignorant with ourselves of the particular breeding places of those birds. There, their journey north is but commencing; and how far it extends it is impossible for us at present to ascertain, from our little acquaintance with these frozen regions. They were seen by Hearne in large flocks within the arctic circle, and were then pursuing their way still farther north. Captain Phipps speaks of seeing wild geese feeding at the water's edge on the dreary coast of Spitzbergen, in lat. 80° 27'. It is highly probable that they extend their migrations under the very pole itself, amid the

extensive flight than for the life of a truly aquatic feeding and diving bird. Most of them, during winter, at times leave the sea or lakes, and feed on the pastures, or, when to be had, on the newly-sprung grains, while some feed entirely on aquatic plants and animals. The Canada goose is easily domesticated, and it is probable that most of the specimens killed in Great Britain have escaped from preserves; it is found, however, on the continent of Europe, and stragglers may occasionally occur.

On the beautiful piece of water at Gosford House, the seat of the Earl of Wemyss, Haddingtonshire, this and many other water-birds rear their young freely. I have never seen any artificial piece of water so beautifully adapted for the domestication and introduction of every kind of waterfowl which will bear the climate of Great Britain. Of very large extent, it is embossed in beautiful shrubbery, perfectly recluse, and, even in the nearly constant observance of a resident family, several exotic species seem to look on it as their own. The Canada and Egyptian geese both had young when I visited it, and the lovely *Ana (Dendronessa) sponsa* seemed as healthy as if in her native waters —ED.

silent desolation of unknown countries, shut out since creation from the prying eye of man by everlasting and insuperable barriers of ice. That such places abound with their suitable food, we cannot for a moment doubt; while the absence of their great destroyer, man, and the splendours of a perpetual day, may render such regions the most suitable for their purpose.

Having fulfilled the great law of nature, the approaching rigours of that dreary climate oblige these vast congregated flocks to steer for the more genial regions of the south. And no sooner do they arrive at those countries of the earth inhabited by man, than carnage and slaughter is commenced on their ranks. The English at Hudson's Bay, says Pennant, depend greatly on geese, and in favourable years kill three or four thousand, and barrel them up for use. They send out their servants, as well as Indians, to shoot these birds on their passage. It is in vain to pursue them ; they therefore form a row of huts, made of boughs, at musket-shot distance from each other, and place them in a line across the vast marshes of the country. Each stand, or hovel, as they are called, is occupied by only a single person. These attend the flight of the birds, and, on their approach, mimic their cackle so well, that the geese will answer, and wheel, and come nearer the stand. The sportsman keeps motionless, and on his knees, with his gun cocked the whole time, and never fires till he has seen the eyes of the geese. He fires as they are going from him, then picks up another gun that lies by him, and discharges that. The geese which he has killed he sets upon sticks, as if alive, to decoy others ; he also makes artificial birds for the same purpose. In a good day, for they fly in very uncertain and unequal numbers, a single Indian will kill two hundred. Notwithstanding every species of goose has a different call, yet the Indians are admirable in their imitations of every one. The autumnal flight lasts from the middle of August to the middle of October ; those which are taken in this season, when the frosts begin, are preserved in their feathers, and left to be frozen for the fresh provisions of the winter stock. The feathers constitute an article of commerce, and are sent to England.

The vernal flight of the geese lasts from the middle of April until the middle of May. Their first appearance coincides with the thawing of the swamps, when they are very lean. Their arrival from the south is impatiently attended; it is the harbinger of the spring, and the month named by the Indians the goose-moon. They appear usually at their settlements about St George's Day, O.S., and fly northward, to nestle in security. They prefer islands to the continent, as farther from the haunts of man.*

After such prodigious havoc as thus appears to be made among these birds, and their running the gauntlet, if I may so speak, for many hundreds of miles through such destructive fires, no wonder they should have become more scarce, as well as shy, by the time they reach the shores of the United States.

Their first arrival on the coast of New Jersey is early in October, and their first numerous appearance is the sure prognostic of severe weather. Those which continue all winter frequent the shallow bays and marsh islands; their principal food being the broad tender green leaves of a marine plant which grows on stones and shells, and is usually called sea-cabbage; and also the roots of the sedge, which they are frequently observed in the act of tearing up. Every few days they make an excursion to the inlets on the beach for gravel. They cross indiscriminately over land or water, generally taking the nearest course to their object, differing in this respect from the brant, which will often go a great way round by water rather than cross over the land. They swim well; and, if wing-broken, dive and go a long way under water, causing the sportsman a great deal of fatigue before he can kill them. Except in very calm weather, they rarely sleep on the water, but roost all night in the marshes. When the

* Arctic Zoology.

shallow bays are frozen, they seek the mouths of inlets near the sea, occasionally visiting the air-holes in the ice ; but these bays are seldom so completely frozen as to prevent them from feeding on the bars.

The flight of the wild geese is heavy and laborious, generally in a straight line, or in two lines approximating to a point, thus, $\succ$; in both cases the van is led by an old gander, who every now and then pipes his well-known *honk*, as if to ask how they come on, and the honk of " All's well " is generally returned by some of the party. Their course is in a straight line, with the exception of the undulations of their flight. When bewildered in foggy weather, they appear sometimes to be in great distress, flying about in an irregular manner and for a considerable time over the same quarter, making a great clamour. On these occasions, should they approach the earth and alight, which they sometimes do, to rest and recollect themselves, the only hospitality they meet with is death and destruction from a whole neighbourhood already in arms for their ruin.

Wounded geese have, in numerous instances, been completely domesticated, and readily pair with the tame grey geese. The offspring are said to be larger than either ; but the characteristic marks of the wild goose still predominate. The gunners on the sea-shore have long been in the practice of taming the wounded of both sexes, and have sometimes succeeded in getting them to pair and produce. The female always seeks out the most solitary place for her nest, not far from the water. On the approach of every spring, however, these birds discover symptoms of great uneasiness, frequently looking up into the air, and attempting to go off. Some whose wings have been closely cut have travelled on foot in a northern direction, and have been found at the distance of several miles from home. They hail every flock that passes overhead, and the salute is sure to be returned by the voyagers, who are only prevented from alighting among them by the presence and habitations of man. The gunners take one or two of these domesticated geese with them to those parts of the marshes over which the wild ones are accustomed to fly; and concealing themselves within gunshot, wait for a flight, which is no sooner perceived by the decoy geese, than they begin calling aloud, until the whole flock approaches so near as to give them an opportunity of discharging two and sometimes three loaded muskets among it, by which great havoc is made.

The wild goose, when in good order, weighs from ten to twelve, and sometimes fourteen pounds. They are sold in the Philadelphia markets at from seventy-five cents to one dollar each; and are estimated to yield half a pound of feathers apiece, which produces twenty-five or thirty cents more.

The Canada goose is now domesticated in numerous quarters of the country, and is remarked for being extremely watchful, and more sensible of approaching changes in the atmosphere than the common grey goose. In England, France, and Germany, they have also been long ago domesticated. Buffon, in his account of this bird, observes, "Within these few years many hundreds inhabited the great canal at Versailles, where they breed familiarly with the swans; they were oftener on the grassy margins than in the water;" and adds, "there is at present a great number of them on the magnificent pools that decorate the charming gardens of Chantilly." Thus has America already added to the stock of domestic fowls two species, the turkey and the Canada goose, superior to most in size, and inferior to none in usefulness; for it is acknowledged by an English naturalist of good observation, that this last species "is as familiar, breeds as freely, and is in every respect as valuable as the common goose."*

The strong disposition of the wounded wild geese to migrate to the north in spring has been already taken notice of. Instances have occurred where, their wounds having healed, they have actually succeeded in mounting into the higher regions of the air, and joined a passing party to the north; and, extra-

* Bewick, vol. ii. p. 255.

ordinary as it may appear, I am well assured by the testimony of several respectable persons, who have been eye-witnesses to the fact, that they have been also known to return again in the succeeding autumn to their former habitation. These accounts are strongly corroborated by a letter which I some time ago received from an obliging correspondent at New York, which I shall here give at large, permitting him to tell his story in his own way, and conclude my history of this species :—

" Mr Platt, a respectable farmer on Long Island, being out shooting in one of the bays which, in that part of the country, abound with waterfowl, wounded a wild goose. Being wing-tipped, and unable to fly, he caught it, and brought it home alive. It proved to be a female; and turning it into his yard with a flock of tame geese, it soon became quite tame and familiar, and in a little time its wounded wing entirely healed. In the following spring, when the wild geese migrate to the northward, a flock passed over Mr Platt's barnyard; and just at that moment their leader happening to sound his bugle-note, our goose, in whom its new habits and enjoyments had not quite extinguished the love of liberty, and remembering the well-known sound, spread its wings, mounted into the air, joined the travellers, and soon disappeared. In the succeeding autumn, the wild geese, as was usual, returned from the northward in great numbers, to pass the winter in our bays and rivers. Mr Platt happened to be standing in his yard when a flock passed directly over his barn. At that instant, he observed three geese detach themselves from the rest, and, after wheeling round several times, alight in the middle of the yard. Imagine his surprise and pleasure when, by certain well-remembered signs, he recognised in one of the three his long-lost fugitive. It was she indeed! She had travelled many hundred miles to the lakes; had there hatched and reared her offspring; and had now returned with her little family to share with them the sweets of civilised life.

" The truth of the foregoing relation can be attested by many respectable people, to whom Mr Platt has related the circumstances as above detailed. The birds were all living, and in his possession, about a year ago, and had shown no disposition whatever to leave him."

The length of this species is three feet; extent, five feet two inches; the bill is black; irides, dark hazel; upper half of the neck, black, marked on the chin and lower part of the head with a large patch of white, its distinguishing character; lower part of the neck before, white; back and wing-coverts, brown, each feather tipt with whitish; rump and tail, black; tail-coverts and vent, white; primaries, black, reaching to the extremity of the tail; sides, pale ashy brown; legs and feet, blackish ash.

The male and female are exactly alike in plumage.

TUFTED DUCK. (*Anas fuligula.*)

PLATE LXVII.—Fig. 5, Male.

FULIGULA RUFITORQUES.—Bonaparte.

Fuligula rufitorques, *Bonap. Journ. Acad. Nat. Sc. Phil.—Synop.* p. 393.—
North. Zool. ii. p. 453.

This is an inhabitant of both continents; it frequents freshwater rivers, and seldom visits the sea-shore. It is a plump, short-bodied duck; its flesh generally tender and well tasted. They are much rarer than most of our other species, and are seldom seen in market. They are most common about the beginning of winter and early in the spring. Being birds of passage, they leave us entirely during the summer.

The tufted duck is seventeen inches long, and two feet two inches in extent; the bill is broad, and of a dusky colour, sometimes marked round the nostrils and sides with light blue; head, crested, or tufted, as its name expresses, and of a black colour, with reflections of purple; neck marked near its middle by a band of deep chestnut; lower part of the neck, black, which spreads quite round to the back; back and scapulars, black, minutely powdered with particles of white, not to be observed but on a near inspection ; rump and vent, also black ; wings, ashy brown ; secondaries, pale ash or bluish white ; tertials, black, reflecting green ; lower part of the breast and whole belly, white ; flanks crossed with fine zigzag lines of dusky ; tail, short, rounded, and of a dull brownish black ; legs and feet, greenish ash ; webs, black ; irides, rich orange ; stomach filled with gravel and some vegetable food.

In young birds, the head and upper part of the neck are purplish brown ; in some, the chestnut ring on the fore part of the middle of the neck is obscure, in others very rich and glossy, and, in one or two specimens which I have seen, it is altogether wanting.   The back is in some instances destitute of the fine powdered particles of white, while in others these markings are large and thickly interspersed.

The specimen from which the drawing was taken was shot on the Delaware on the 10th of March, and presented to me by Dr S. B. Smith of this city.   On dissection, it proved to be a male, and was exceeding fat and tender.   Almost every specimen I have since met with has been in nearly the same state ; so that I cannot avoid thinking this species equal to most others for the table, and greatly superior to many.

GOLDEN-EYE.   (*Anas clangula.*)

PLATE LXVII.—Fig. 6, Male.

Le Garrot, *Briss.* vi. p. 416, pl. 37, fig. 2.—*Buff.* ix. p. 222.—*Arct. Zool.* No. 486.—
Lath. Syn. iii. p. 535.

CLANGULA VULGARIS.—Fleming.*

Clangula vulgaris, *Flem. Br. Anim.* p. 120.—*North. Zool.* ii. p. 454.—Fuligula
clangula, *Bonap. Synop.* p. 393.—Subgen. Clangula.

This duck is well known in Europe, and in various regions of the United States, both along the sea-coast and about the

---

* The golden-eye is found on both continents, and in the northern parts of Europe during winter is one of the most common migratory ducks. The garrots are distinguished by a short, stout, and compact body ; the neck short, the head large, and apparently more so from its

lakes and rivers of the interior. It associates in small parties, and may easily be known by the vigorous whistling of its

thick plumage ; the bill short, but thick and raised at the base ; the feet placed far behind, and formed for swimming. The flight is short and rapid. In habit, they delight more in lakes and rivers than the sea ; are generally found in small flocks ; are very clamorous during the breeding season, and feed on fish, aquatic insects, moluscæ, &c. Richardson says, " *Clangula vulgaris* and *albeola* frequent the rivers and freshwater lakes throughout the Fur Countries in great numbers. They are by no means shy, allowing the sportsman to approach sufficiently near ; but dive so dexterously at the flash of a gun or the twang of a bow, and are consequently so difficult to kill, that the natives say they are endowed with some supernatural power. Hence their appellation of " conjuring," or " spirit-ducks."

In Britain, they are winter visitants, assembling in small parties on the lakes and rivers. On the latter, they may be generally found near the head or foot of the stream, diving incessantly for the spawn of salmon, with which I have often found their stomach filled. The party generally consists of from four to ten, and they dive together. At this time, it is not very difficult to approach them, by running forward while they are under water, and squatting when they rise. I have often, in this way, come to the very edge of the river, and awaited the arising of the flock. When taken by surprise, they dive on the instant of the first shot, but rise and fly immediately after.

The young of the first year has been made a nominal species, and is somewhat like the adult females, but always distinguished by larger size, darker colour of the plumage of the head, and the greater proportion of white on the wings. The males have the white spot on the cheek perceptible about the first spring, and the other parts of the plumage proportionally distinct. Among most of the flocks which visit our rivers in winter, it is rare to find more than one full plumaged male in each, sometimes not more than two or three are seen during the winter among fifty or sixty immature birds.

The American ducks belonging to this group are *C. vulgaris, albeola,* and *C. Barrovii,* or Rocky Mountain garrot, a new species, discovered by the Overland Arctic Expedition, and described and figured in the " Northern Zoology." The following is the description ; it has only yet been found in the valleys of the Rocky Mountains.

" Notwithstanding the general similarity in the form and markings of this bird and the common golden-eye, the difference in their bills evidently points them out to be a distinct species. The Rocky Mountain garrot is distinguished by the pure colour of its dorsal plumage, and the smaller portion of white on its wings and scapulars ; its long flank feathers are also much more broadly bordered all round with black.

wings as it passes through the air.   It swims and dives well,
but seldom walks on shore, and then in a waddling, awkward
manner.   Feeding chiefly on shellfish, small fry, &c., their
flesh is less esteemed than that of the preceding.   In the
United States, they are only winter visitors, leaving us again
in the month of April, being then on their passage to the north
to breed.   They are said to build, like the wood-duck, in hollow
trees.

The golden-eye is nineteen inches long, and twenty-nine
in extent, and weighs on an average about two pounds ; the
bill is black, short, rising considerably up in the forehead ; the
plumage of the head and part of the neck is somewhat tumid,
and of a dark green, with violet reflections, marked near the
corner of the mouth with an oval spot of white ; the irides are
golden yellow ; rest of the neck, breast, and whole lower parts,
white, except the flanks, which are dusky ; back and wings,
black ; over the latter a broad bed of white extends from the
middle of the lesser coverts to the extremity of the secondaries ;
the exterior scapulars are also white ; tail, hoary brown ; rump
and tail-coverts, black ; legs and toes, reddish orange ; webs
very large, and of a dark purplish brown ; hind toe and exterior edge of the inner one, broadly finned ; sides of the bill,
obliquely dentated ; tongue, covered above with a fine thick
velvety down, of a whitish colour.

The full plumaged female is seventeen inches in length, and
twenty-seven inches in extent ; bill, brown, orange near the tip ;
head and part of the neck, brown, or very dark drab, bounded
below by a ring of white ; below that, the neck is ash, tipt
with white ; rest of the lower parts, white ; wings, dusky, six
of the secondaries and their greater coverts, pure white, except
the tips of the last, which are touched with dusky spots ; rest
of the wing-coverts, cinereous mixed with whitish ; back and

The bases of the greater coverts in the golden-eye are black ; but they
are concealed, and do not form the black band so conspicuous in this
species." The total length of a male brought home by the Expedition
was twenty-two inches in length.—ED.

scapulars, dusky, tipt with brown ; feet, dull orange ; across the vent, a band of cinereous ; tongue, covered with the same velvety down as the male.

The young birds of the first season very much resemble the females, but may generally be distinguished by the white spot, or at least its rudiments, which mark the corner of the mouth ; yet, in some cases, even this is variable, both old and young male birds occasionally wanting the spot.

From an examination of many individuals of this species of both sexes, I have very little doubt that the morillon of English writers (*Anas glaucion*) is nothing more than the young male of the golden-eye.

The conformation of the trachea or windpipe of the male of this species is singular. Nearly about its middle it swells out to at least five times its common diameter, the concentric hoops or rings of which this part is formed falling obliquely into one another when the windpipe is relaxed; but when stretched, this part swells out to its full size, the rings being then drawn apart; this expansion extends for about three inches ; three more below this, it again forms itself into a hard cartilaginous shell of an irregular figure, and nearly as large as a walnut ; from the bottom of this labyrinth, as it has been called, the trachea branches off to the two lobes of the lungs ; that branch which goes to the left lobe being three times the diameter of the right. The female has nothing of all this. The intestines measure five feet in length, and are large and thick.

I have examined many individuals of this species, of both sexes and in various stages of colour, and can therefore affirm with certainty that the foregoing descriptions are correct. Europeans have differed greatly in their accounts of this bird, from finding males in the same garb as the females, and other full plumaged males destitute of the spot of white on the cheek; but all these individuals bear such evident marks of belonging to one peculiar species, that no judicious naturalist, with all these varieties before him, can long hesitate to pronounce them the same.

SHOVELLER. (*Anas clypeata.*)

PLATE LXVII.—Fig. 7, Male.

Le Souchet, *Briss.* vi. p. 329, 6, pl. 32, fig. 1.—*Buff.* ix. 191, *Pl. enl.* 971.—
    Arct. Zool. No. 485.—*Catesby*, i. pl. 96, female.—*Lath. Syn.* iii. p. 509.—
    Peale's Museum, No. 2734.

ANAS CLYPEATA.—Linnæus.*

Anas platyrhynchas, *Raii Synop.* p. 144.—Rynchaspis clypeata, *Leach.*—*Shaw's*
    Zool. Steph. Cont. xii. 115, pl. 48.—Spathulea clypeata, *Flem. Brit. Anim.*
    i. 123.—Anas clypeata, *Lath. Ind. Ornith.* ii. p. 856.—Shoveller, *Mont.*
    Ornith. Dict. and Sup.—*Bew.* ii. 345.—*Selby, m. and f. Illust.* pl. 48.—
    Canard Souchet, *Tem. Man.* ii. p. 842.—Anas clypeata, *Bonap. Synop.* p. 382.
    —*North. Zool.* ii. p. 439.

If we except the singularly formed and disproportionate size of the bill, there are few ducks more beautiful or more elegantly

---

* Mr Swainson, according to his views that the typical group should hold the typical name of the family, has restricted *Anas* (in that sense) to the shovellers. In fixing upon the typical representation of any large family, that gentleman goes upon the principle of taking the organ most peculiarly important to the whole, and selects that subordinate, or rather primary group, wherein that organ is most fully developed. Thus, in the ducks, he remarks there is nothing peculiar in diving, or living both on land and water, or endowments for rapid flight, for many others possess like powers ; but when we examine the dilated and softly textured bill, and more particularly the fine laminæ on the edges, we are struck with a formation at variance with our accustomed ideas of that member, and at once think that it must be applied to something equally peculiar in their economy. We shall thus be warranted in taking the bill as our criterion, and those birds where we find its structure most fully developed for the type. These are most decidedly to be seen in the shovellers, a group containing, as yet, only three or four known species ; in them we have the utmost dilatation of the bill towards its apex, and the laminæ upon its edges, and long and remarkably delicate. The bird itself possesses a powerful flight, and is a most expert diver and swimmer, but seems to prefer inland lakes or fens to the more open seas and rivers.

To this group will belong the curious pink-eared shoveller from New Holland, remarkable from the toothlike membrane projecting from the angles of the bill, and differing somewhat from the others in its brown and dusky plumage. Mr Swainson has formed on account of

marked than this. The excellence of its flesh, which is uniformly juicy, tender, and well tasted, is another recommendation to which it is equally entitled. It occasionally visits the sea-coast, but is more commonly found on our lakes and rivers, particularly along their muddy shores, where it spends great part of its time in searching for small worms and the larvæ of insects, sifting the watery mud through the long and finely-set teeth of its curious bill, which is admirably constructed for the purpose, being large, to receive a considerable quantity of matter, each mandible bordered with close-set, pectinated rows, exactly resembling those of a weaver's reed, which, fitting into each other, form a kind of sieve, capable of retaining very minute worms, seeds, or insects, which constitute the principal food of the bird.

The shoveller visits us only in the winter, and is not known to breed in any part of the United States. It is a common bird of Europe, and, according to M. Baillon, the correspondent of Buffon, breeds yearly in the marshes in France. The female is said to make her nest on the ground with withered grass, in the midst of the largest tufts of rushes or coarse herbage, in the most inaccessible part of the slaky marsh, and lays ten or twelve pale rust-coloured eggs; the young, as soon as hatched, are conducted to the water by the parent birds. They are said to be at first very shapeless and ugly, for the bill is then as broad as the body, and seems too great a weight for the little bird to carry. Their plumage does not acquire its full colours until after the second moult.

The blue-winged shoveller is twenty inches long, and two feet six inches in extent; the bill is brownish black, three inches in length, greatly widened near the extremity, closely pectinated on the sides, and furnished with a nail on the tip

---

this membrane a subgenus, *malacorhynchus*, but in which I am hardly yet prepared to coincide.

It may be mentioned here, that the only birds which possess the lamellated structure of the upper mandible is *pachyptila*, a genus coming near to the petrels, and *phœnicopterus* of Flamingo.—ED.

of each mandible; irides, bright orange; tongue, large and fleshy; the inside of the upper and outside of the lower mandible are grooved, so as to receive distinctly the long, separated reedlike teeth; there is also a gibbosity in the two mandibles, which do not meet at the sides, and this vacuity is occupied by the sifters just mentioned; head and upper half of the neck, glossy changeable green; rest of the neck and breast, white, passing round and nearly meeting above; whole belly, dark reddish chestnut; flanks, a brownish yellow, pencilled transversely with black, between which and the vent, which is black, is a band of white; back, blackish brown; exterior edges of the scapulars, white; lesser wing-coverts and some of the tertials, a fine light sky-blue; beauty spot on the wing, a changeable resplendent bronze green, bordered above by a band of white, and below with another of velvety black; rest of the wing, dusky, some of the tertials streaked down their middles with white; tail, dusky, pointed, broadly edged with white; legs and feet, reddish orange, hind toe not finned.

With the above another was shot, which differed in having the breast spotted with dusky and the back with white; the green plumage of the head intermixed with gray, and the belly with circular touches of white, evidently a young male in its imperfect plumage.

The female has the crown of a dusky brown; rest of the head and neck, yellowish white, thickly spotted with dark brown; these spots on the breast become larger, and crescent-shaped; back and scapulars, dark brown, edged and centred with yellow ochre; belly, slightly rufous, mixed with white; wing, nearly as in the male.

On dissection, the labyrinth in the windpipe of the male was found to be small; the trachea itself seven inches long; the intestines nine feet nine inches in length, and about the thickness of a crow-quill.

Drawn from Nature by J. Wilson

Engraved by W.H.Lizars

1. Goosander. 2. Female. 3. Pin-tail Duck. 4. Blue wing Teal. 5. Snow Goose.

68.

GOOSANDER. *(Mergus merganser.)*

PLATE LXVIII.—FIG. 1, MALE.

L'Harle, *Briss.* vi. p. 231, 1, pl. 23.—*Buff.* viii. p. 267, pl. 23.—*Arct. Zool.* 465.— *Lath. Syn.* iii. p. 418.—*Peale's Museum*, No. 2932.

MERGUS MERGANSER.—LINNÆUS.*

Goosander, or Merganser, *Mont. Ornith. Dict. and Supp.*—*Bew. Br. Birds*, ii. p. 254.—*Selby's Illust.* pl. 57.—Mergus merganser, *Bonap. Synop.* p. 397.— *Flem. Br. Anim.* p. 128.—Grande Harle, *Temm. Man. d'Ornith.* ii. 881.

THIS large and handsomely-marked bird belongs to a genus different from that of the duck, on account of the particular

---

* The genus *Mergus* has been universally allowed. It contains nine or ten species, allied in their general form, but easily distinguished by their plumage. They are truly aquatic, and never quit the sea or lakes except for a partial repose or pluming, or during the time of incubation. Their food is entirely fish, and they are necessarily expert divers ; the bill is lengthened and narrow, its edges regularly serrated with recurved points. The breeding places of many of them are yet unknown, but I believe that the greater proportion at that season retire inland to the more sequestered lakes. I am also of opinion that the male forsakes his mate so soon as she begins to sit, about which time he also loses the beautiful crest and plumage in which he is clothed during winter and spring, and assumes a duller garb. The males are remarkable for their difference from the other sex, whence the long-disputed point, now satisfactorily proved, of this and the following bird being different. That of the male is generally black or glossy green, contrasted with the purest white or rich shades of tawny yellow ; that of the female, the chaster grays and browns. Both are furnished with crests, composed of loose hackled feathers.

The distribution of the group seems to be European and both continents of America. I have seen none from India or New Holland, though from the former country they might be expected.

The goosander is a native of both continents, and is said to breed in the northern part of Scotland. This I have had no opportunity of verifying. It is frequent during winter on the larger rivers, in flocks of seven or eight, in which there is generally only one, or at most two, adult males—the others being in immature dress, or females ; thus the latter is said to be the most common. They fish about the bottoms of the streams and pools, and, I believe, destroy many fish. I have taken

form and serratures of its bill. The genus is characterised as follows :—" Bill, toothed, slender, cylindrical, hooked at the point; nostrils, small, oval, placed in the middle of the bill; feet, four-toed, the outer toe longest." Naturalists have denominated it *merganser.* In this country, the birds composing this genus are generally known by the name of fisherman, or fisher-ducks. The whole number of known species amount to only nine or ten, dispersed through various quarters of the world; of these, four species, of which the present is the largest, are known to inhabit the United States.

From the common habit of these birds in feeding almost entirely on fin and shell-fish, their flesh is held in little estimation, being often lean and rancid, both smelling and tasting strongly of fish; but such are the various peculiarities of tastes, that persons are not wanting who pretend to consider them capital meat.

The goosander, called by some the water-pheasant, and by others the sheldrake, fisherman, diver, &c., is a winter inhabitant only of the sea-shores, fresh-water lakes, and rivers of the United States. They usually associate in small parties of six or eight, and are almost continually diving in search of food. In the month of April they disappear, and return again early in November. Of their particular place and manner of breeding we have no account. Mr Pennant observes, that they continue the whole year in the Orkneys; and have been shot in the Hebrides, or Western Islands of Scotland, in summer. They are also found in Iceland and Greenland, and are said to breed there; some asserting that they build on trees; others, that they make their nests among the rocks.

The male of this species is twenty-six inches in length, and three feet three inches in extent; the bill, three inches long,

seven trout, about four or five inches in length, from the stomach of a female.

In Hudson's Bay (according to Hearne) they are called sheldrakes; the name by which they are also distinguished by the common people in all the rivers in the south of Scotland.—Ed.

and nearly one inch thick at the base, serrated on both mandibles, the upper overhanging at the tip, where each is furnished with a large nail ; the ridge of the bill is black ; the sides, crimson red ; irides, red ; head, crested, tumid, and of a black colour, glossed with green, which extends nearly half way down the neck, the rest of which, with the breast and belly, are white, tinged with a delicate yellowish cream ; back, and adjoining scapulars, black ; primaries, and shoulder of the wing, brownish black ; exterior part of the scapulars, lesser coverts, and tertials, white ; secondaries, neatly edged with black ; greater coverts, white ; their upper halves, black, forming a bar on the wing ; rest of the upper parts and tail, brownish ash ; legs and feet, the colour of red sealing-wax ; flanks, marked with fine semicircular dotted lines of deep brown ; the tail extends about three inches beyond the wings.

This description was taken from a full-plumaged male. The young males, which are generally much more numerous than the old ones, so exactly resemble the females in their plumage for at least the first and part of the second year, as scarcely to be distinguished from them ; and, what is somewhat singular, the crests of these and of the females are actually longer than those of the full-grown male, though thinner towards its extremities. These circumstances have induced some late ornithologists to consider them as two different species, the young or female having been called the dun diver. By this arrangement they have entirely deprived the goosander of his female ; for, in the whole of my examinations and dissections of the present species, I have never yet found the female in his dress. What I consider as undoubtedly the true female of this species is figured beside him. They were both shot in the month of April, in the same creek, unaccompanied by any other ; and, on examination, the sexual parts of each were strongly and prominently marked. The windpipe of the female had nothing remarkable in it ; that of the male had two very large expansions, which have been briefly described by Willoughby, who says, " It hath a large bony labyrinth on the windpipe, just above the divarications; and the windpipe hath, besides, two swellings out, one above another, each resembling a powder puff." These labyrinths are the distinguishing characters of the males; and are always found, even in young males who have not yet thrown off the plumage of the female, as well as in the old ones. If we admit these dun divers to be a distinct species, we can find no difference between their pretended females and those of the goosander, only one kind of female of this sort being known; and this is contrary to the usual analogy of the other three species, viz., the red-breasted merganser, the hooded, and the smew, all of whose females are well known, and bear the same comparative resemblance in colour to their respective males, the length of crest excepted, as the female goosander here figured bears to him.

Having thought thus much necessary on this disputed point, I leave each to form his own opinion on the facts and reasoning produced.

[* The goosander is a broad, long-bodied, and flat-backed bird. It is a great diver, and remains under water for a considerable time. It is very shy, and hard to be obtained, unless there is ice in the river, at which time it may be approached by stratagem, the shooter and his boat being clothed in white, so as to resemble floating ice. It appears to live chiefly upon fish, which its sharp-toothed and hooked bill is admirably calculated for securing. It rises from the water with considerable fluttering, its wings being small and short; but when in the air, it flies with great swiftness. It is a singular circumstance that those goosanders which are seen in the Delaware and Schuylkill, in the vicinity of Philadelphia, are principally old males.

The male goosander is twenty-six inches in length, and thirty-seven inches in breadth; the bill, to the angles of the mouth, is three inches long, nearly an inch thick at the base,

* From this to the end of the article, marked off with brackets, is an addition to Wilson's description by Mr Ord.—ED.

strongly toothed on both mandibles, the upper mandible with two corresponding rows of fine teeth within, the lower divided to the nail, and connected by a thin elastic membrane, which admits of considerable expansion, to facilitate the passage of fish ; nostrils, sub-ovate, broader on the hind part ; the bill is black above and below, its sides crimson ; the tongue is long, pointed, furnished with a double row of papillæ running along the middle, and has a hairy border ; irides, golden ; the front-let, lores, area of the eyes, and throat, jet black ; head, crested, tumid, and of a beautiful glossy bottle-green colour, extending nearly half-way down the neck, the remainder of which, with the exterior part of the scapulars, the lesser coverts, the greater part of the secondaries, the tertials and lining of the wings, white, delicately tinged with cream colour ; the breast and whole lower parts are of a rich cream colour ; the upper part of the back and the interior scapulars, a fine glossy black ; the primaries and exterior part of the secondaries, with their coverts, are brownish black ; the lower part of nearly all the coverts of the secondaries, white, the upper part, black, form-ing a bar across the wing ; the shoulder of the wing is brownish ash, the feathers tipt with black ; the middle and lower parts of the back and tail-coverts, ash, the plumage centred with brown ; tail, brownish ash, rounded, composed of eighteen feathers, and extends about three inches beyond the wings ; the flanks are marked with waving, finely-dotted lines of ash on a white ground ; tertials on the outer vanes, edged with black ; the legs and feet are of a rich orange ; toes, long, middle one somewhat the longest ; claws, flesh-coloured. The whole plumage is of a silky softness, particularly that of the head and neck, which feels like the most delicate velvet.

Naturalists represent the feet and legs of this species as of the colour of red sealing-wax. This is an error which arose from the circumstance of their having seen their specimens some time after they had been killed. When the bird is alive, these parts are of a beautiful orange, which changes after death to the colour they mention.

The above description was taken from a fine full-plumaged male, which was shot in the vicinity of Philadelphia in the month of January. It was in good condition, and weighed three pounds thirteen ounces avoirdupois.]

FEMALE GOOSANDER.

PLATE LXVIII.—Fig. 2.

Peale's Museum, No. 2933.—Dun Diver, *Lath. Syn.* iii. p. 240.—*Arct. Zool.* No. 465.—*Bewick's Brit. Birds*, ii. p. 23.—*Turt. Syst.* p. 335.—L'Harle Femelle, *Briss.* vi. p. 236, *Buff.* viii. p. 272, *Pl. enl.* 953.

MERGUS MERGANSER.—Linnæus.

Syn. of Fem. or Young.—Mergus castor, *Linn. Syst.* i. 209.—Merganser cinereus, *Briss. Orn.* vi. 254.—Dun Diver, or Sparling Fowl, *Mont. Bew.* &c.—Goosander Female, *Selby's Illust.* pl. lvii.

THIS generally measures an inch or two shorter than the male; the length of the present specimen was twenty-five inches; extent, thirty-five inches; bill, crimson on the sides, black above; irides, reddish; crested head and part of the neck, dark brown, lightest on the sides of the neck, where it inclines to a sorrel colour; chin and throat, white; the crest shoots out in long radiating flexible stripes; upper part of the body, tail, and flanks, an ashy slate, tinged with brown; primaries, black; middle secondaries, white, forming a large speculum on the wing; greater coverts, black, tipt for half an inch with white; sides of the breast, from the sorrel-coloured part of the neck downwards, very pale ash, with broad semi-circular touches of white; belly and lower part of the breast a fine yellowish cream colour—a distinguishing trait also in the male; legs and feet, orange red.

[It is truly astonishing with what pertinacity Montagu adheres to the opinion that the dun diver is a species distinct from the goosander. Had this excellent ornithologist had the same opportunities for examining these birds that we have, he would never have published an opinion which, in this quarter of the globe, would subject one, even from the vulgar, to the imputation of ignorance.*]

PINTAIL DUCK. (*Anas acuta.*)

PLATE LXVIII.—Fig. 3.

Le Canard à longue queue, *Briss.* vi. p. 369, 16, pl. 34, fig. 1, 2.—*Buff.* ix. p. 199, pl. 13; *Pl. enl.* 954.—*Arct. Zool.* No. 500.—*Lath. Syn.* iii. p. 526.— *Peale's Museum*, No. 2806.

DAFILA ACUTA.—Leach.†

Dafila caudacuta, *Shaw's Zool. Steph. Cont.* xii. p. 127.—Canard à longue queue au pillet, *Temm. Man. d'Ornith.* ii. 838.—Pintail, *Mont. Bew.*— *Selby's Illust.* pl. 42, *m.*—Anas acuta, *Cracker.*—*Flem. Br. Anim.* p. 124.— *Bonap. Synop.* p. 383.—Anas (Dafila) caudacuta, *North. Zool.* ii. p. 441.

THE pintail, or, as it is sometimes called, the sprigtail, is a common and well-known duck in our markets, much esteemed for the excellence of its flesh, and is generally in good order. It is a shy and cautious bird, feeds in the mud flats and shallow fresh-water marshes, but rarely resides on the sea-coast. It seldom dives, is very noisy, and has a kind of chattering note. When wounded, they will sometimes dive, and coming up, conceal themselves under the bow of the boat, moving round as it moves. Are vigilant in giving the alarm on the approach of the gunner, who often curses the watchfulness of

---

* The concluding paragraph, marked off with brackets, is an addition by Mr Ord.

† In this beautiful species we have the type of the subgenus *Dafila.* In it the marginal laminæ begin to disappear, and the bill to assume what may be called a more regular outline, approaching to that of *A. boschas,* our wild and domestic breed. Another peculiarity is the development of the tail, which becomes much lengthened, whence the name of *sea-pheasant.* In this country they are not very common, which may arise from their being more difficult to procure, by their frequenting the sea rather than any inland water; they are frequently taken, however, in decoys, and I once shot two feeding in the evening on a wet stubble field in company with the common wild duck.—ED.

the sprigtail. Some ducks, when aroused, disperse in different directions; but the sprigtails, when alarmed, cluster confusedly together as they mount, and thereby afford the sportsman a fair opportunity of raking them with advantage. They generally leave the Delaware about the middle of March, on the way to their native regions, the north, where they are most numerous. They inhabit the whole northern parts of Europe and Asia, and doubtless the corresponding latitudes of America; are said likewise to be found in Italy. Great flocks of them are sometimes spread along the isles and shores of Scotland and Ireland, and on the interior lakes of both these countries. On the marshy shores of some of the bays of Lake Ontario, they are often plenty in the months of October and November. I have also met with them at Louisville, on the Ohio.

The pintail duck is twenty-six inches in length, and two feet ten inches in extent; the bill is a dusky lead colour; irides, dark hazel; head and half of the neck, pale brown, each side of the neck marked with a band of purple violet, bordering the white; hind part of the upper half of the neck, black, bordered on each side by a stripe of white, which spreads over the lower part of the neck before; sides of the breast and upper part of the back, white, thickly and elegantly marked with transverse undulating lines of black, here and there tinged with pale buff; throat and middle of the belly, white, tinged with cream; flanks, finely pencilled with waving lines; vent, white; under tail-coverts, black; lesser wing-coverts, brown ash; greater, the same, tipt with orange; below which is the speculum, or beauty spot, of rich golden green, bordered below with a band of black and another of white; primaries, dusky brown; tertials, long, black, edged with white, and tinged with rust; rump and tail-coverts, pale ash, centred with dark brown; tail greatly pointed, the two middle tapering feathers being full five inches longer than the others, and black, the rest brown ash, edged with white; legs, a pale lead colour.

The female has the crown of a dark brown colour ; neck, of a dull brownish white, thickly speckled with dark brown ; breast and belly, pale brownish white, interspersed with white ; back, and root of the neck above, black, each feather elegantly waved with broad lines of brownish white, these wavings become rufous on the scapulars ; vent, white, spotted with dark brown ; tail, dark brown, spotted with white ; the two middle tail-feathers half an inch longer than the others.

The sprigtail is an elegantly formed, long-bodied duck, the neck longer and more slender than most others.

BLUE-WINGED TEAL.   (*Anas discors.*)

PLATE LXVIII.—Fig. 4.

Le Sarcelle d'Amerique, *Briss.* vi. p. 452, 35.—*Buff.* ix. p. 279, *Pl. enl.* 966.— *Catesby*, i. pl. 100.—White-faced Duck, *Lath. Syn.* iii. p. 502.—*Arct. Zool.* No. 503.—*Peale's Museum*, No. 2846.

BOSCHAS ? DISCORS.—Swainson.

Anas discors, *Cuv. Regn. Anim.* i. p. 539.—*Bonap. Synop.* p. 385.—Anas (Boschas) discors. *Swain. Journ. Royal Instit.*, No. iv. p. 22.—*North. Zool.* ii. p. 444.

The blue-winged teal is the first of its tribe that returns to us in the autumn from its breeding place in the north. They are usually seen early in September, along the shores of the Delaware, where they sit on the mud close to the edge of the water, so crowded together that the gunners often kill great numbers at a single discharge. When a flock is discovered thus sitting and sunning themselves, the experienced gunner runs his batteau ashore at some distance below or above them, and getting out, pushes her before him over the slippery mud, concealing himself all the while behind her ; by this method he can sometimes approach within twenty yards of the flock, among which he generally makes great slaughter. They fly rapidly, and, when they alight, drop down suddenly, like the snipe or woodcock, among the reeds or on the mud. They feed chiefly on vegetable food, and are eagerly fond of the seeds of the reeds or wild oats. Their flesh is excellent, and after their residence for a short time among the reeds, become very fat. As the first frosts come on, they proceed to the south, being a delicate bird, very susceptible of cold. They abound in the inundated rice-fields in the southern States, where vast numbers are taken in traps placed on small dry eminences that here and there rise above the water. These places are strewed with rice, and by the common contrivance called a *figure four*, they are caught alive in hollow traps. In the month of April they pass through Pennsylvania for the north, but make little stay at that season. I have observed them numerous on the Hudson opposite to the Katskill mountains. They rarely visit the sea-shore.

This species measures about fourteen inches in length, and twenty-two inches in extent; the bill is long in proportion, and of a dark dusky slate; the front and upper part of the head are black; from the eye to the chin is a large crescent of white; the rest of the head and half of the neck are of a dark slate, richly glossed with green and violet; remainder of the neck and breast is black or dusky, thickly marked with semicircles of brownish white, elegantly intersected with each other; belly, pale brown, barred with dusky in narrow lines; sides and vent, the same tint, spotted with oval marks of dusky; flanks elegantly waved with large semicircles of pale brown; sides of the vent pure white; under tail-coverts, black; back, deep brownish black, each feather waved with large semi-ovals of brownish white; lesser wing-coverts, a bright light blue; primaries, dusky brown; secondaries, black; speculum, or beauty spot, rich green; tertials, edged with black or light blue, and streaked down their middle with white; the tail, which is pointed, extends two inches beyond the wings; legs and feet, yellow, the latter very small; the two crescents of white, before the eyes, meet on the throat.

The female differs in having the head and neck of a dull dusky slate, instead of the rich violet of the male, the hind head is also whitish. The wavings on the back and lower parts more indistinct; wing nearly the same in both.

SNOW GOOSE. (*Anas hyperborea.*)

PLATE LXVIII.—Fig. 5, Male.

L'Oye de Neige, *Briss.* vi. p. 288, 10.—White Brant, *Lawson's Carolina*, p. 157.
—*Arct. Zool.* No. 477.—*Phil. Trans.* 62, p. 413.—*Lath. Syn.* iii. p. 445.—
Peale's Museum, No. 2635.

ANSER HYPERBOREUS.—Bonaparte.

Anser hyperboreus, *Bonap. Synop.* p. 376.—*North. Zool.* ii. p. 467.

This bird is particularly deserving of the further investigation of naturalists; for, if I do not greatly mistake, English writers have, from the various appearances which this species assumes in its progress to perfect plumage, formed no less than four different kinds, which they describe as so many distinct species, viz., the *snow goose*, the *white-fronted* or *laughing goose*, the *bean goose*, and the *blue-winged goose*, all of which, I have little doubt, will hereafter be found to be nothing more than perfect and imperfect individuals, male and female, of the snow goose, now before us.*

This species, called on the sea-coast the red goose, arrives in the river Delaware from the north early in November, sometimes in considerable flocks, and is extremely noisy, their notes being shriller and more squeaking than those of the Canada or common wild goose. On their first arrival they make but a short stay, proceeding, as the depth of winter approaches, farther to the south; but from the middle of February until the breaking up of the ice in March, they are frequently numerous along both shores of the Delaware about and below Reedy Island, particularly near Old Duck Creek,

---

* Mr Ord, in his reprint, adds the following note:—"This conjecture of our author is partly erroneous. The snow goose and the blue-winged goose are synonymous; but the other two named are distinct species, the characters of which are well defined by late ornithologists."

The blue-winged goose is our present bird in immature plumage, which they are said to retain for three years. The two other birds have since been added to the American Fauna. The young bird is described page 12 of the third volume.—Ed.

in the State of Delaware. They feed on the roots of the reeds there, tearing them up from the marshes like hogs. Their flesh, like most others of their tribe that feed on vegetables, is excellent.

The snow goose is two feet eight inches in length, and five feet in extent; the bill is three inches in length, remarkably thick at the base, and rising high in the forehead, but becomes small and compressed at the extremity, where each mandible is furnished with a whitish rounding nail; the colour of the bill is a purplish carmine; the edges of the two mandibles separate from each other in a singular manner for their whole length, and this gibbosity is occupied by dentated rows, resembling teeth, these and the parts adjoining being of a blackish colour; the whole plumage is of a snowy whiteness, with the exception, first, of the fore part of the head all round as far as the eyes, which is of a yellowish rust colour, intermixed with white; and, second, the nine exterior quill-feathers, which are black, shafted with white, and white at the root; the coverts of these last, and also the bastard wing, are sometimes of a pale ash colour; the legs and feet of the same purplish carmine as the bill; iris, dark hazel; the tail is rounded, and consists of sixteen feathers; that, and the wings, when shut, nearly of a length.

The bill of this bird is singularly curious; the edges of the upper and lower gibbosities have each twenty-three indentations, or strong teeth, on each side; the inside or concavity of the upper mandible has also seven lateral rows of strong projecting teeth; and the tongue, which is horny at the extremity, is armed on each side with thirteen long and sharp bony teeth, placed like those of a saw, with their points directed backwards; the tongue turned up, and viewed on its lower side, looks very much like a human finger with its nail. This conformation of the mandibles, exposing two rows of strong teeth, has probably given rise to the epithet *laughing*, bestowed on one of its varieties, though it might with as much propriety have been named the *grinning goose*.

The specimen from which the above figure and description was taken, was shot on the Delaware, below Philadelphia, on the 15th of February, and on dissection proved to be a male; the windpipe had no labyrinth, but, for an inch or two before its divarication into the lungs, was inflexible, not extensile, like the rest, and rather wider in diameter. The gullet had an expansion before entering the stomach, which last was remarkably strong, the two great grinding muscles being nearly five inches in diameter. The stomach was filled with fragments of the roots of reeds, and fine sand. The intestines measured eight feet in length, and were not remarkably thick. The liver was small. For the young and female of this species, see plate lxix., fig. 5.

Latham observes that this species is very numerous at Hudson's Bay, that they visit Severn river in May, and stay a fortnight, but go farther north to breed; they return to Severn Fort the beginning of September, and stay till the middle of October, when they depart for the south, and are observed to be attended by their young in flocks innumerable. They seem to occupy also the western side of America, as they were seen at Aoonalashka,* as well as Kamtschatka.† White brant, with black tips to their wings, were also shot by Captains Lewis and Clark's exploring party near the mouth of the Columbia river, which were probably the same as the present species. ‡ Mr Pennant says, " They are taken by the Siberians in nets, under which they are decoyed by a person covered with a white skin, and crawling on all-fours; when, others driving them, these stupid birds mistaking him for their leader, follow him, when they are entangled in the nets, or led into a kind of pond made for the purpose ! " We might here with propriety add—*this wants confirmation.*

* Ellis's Narrative.   † History of Kamtschatka.
‡ Gass's Journal, p. 161.